T0295174

Plant Biotechnology and Molecular Markers

Plant Biotechnology and Molecular Markers

Edited by Kohen Harris

www.statesacademicpress.com

States Academic Press,
109 South 5th Street,
Brooklyn, NY 11249, USA

Visit us on the World Wide Web at:
www.statesacademicpress.com

© States Academic Press, 2023

This book contains information obtained from authentic and highly regarded sources. Copyright for all individual chapters remain with the respective authors as indicated. All chapters are published with permission under the Creative Commons Attribution License or equivalent. A wide variety of references are listed. Permission and sources are indicated; for detailed attributions, please refer to the permissions page and list of contributors. Reasonable efforts have been made to publish reliable data and information, but the authors, editors and publisher cannot assume any responsibility for the validity of all materials or the consequences of their use.

ISBN: 978-1-63989-752-0

Trademark Notice: Registered trademark of products or corporate names are used only for explanation and identification without intent to infringe.

Cataloging-in-Publication Data

Plant biotechnology and molecular markers / edited by Kohen Harris.
p. cm.
Includes bibliographical references and index.
ISBN 978-1-63989-752-0
1. Plant biotechnology. 2. Genetic polymorphisms. 3. Genetic markers.
4. Plant genetic engineering. I. Harris, Kohen.
SB106.B56 P53 2023
631.523 3--dc23

Table of Contents

Preface

The world is advancing at a fast pace like never before. Therefore, the need is to keep up with the latest developments. This book was an idea that came to fruition when the specialists in the area realized the need to coordinate together and document essential themes in the subject. That's when I was requested to be the editor. Editing this book has been an honour as it brings together diverse authors researching on different streams of the field. The book collates essential materials contributed by veterans in the area which can be utilized by students and researchers alike.

A molecular marker is a fragment or sequence of DNA that is associated with a certain location within the genome. The development of molecular marker technology has contributed significantly in crop improvement. In plant biotechnology, markers are used to reflect random sequences, genomic variants, or genes. Molecular markers are used majorly in the construction of DNA and genome maps. They also play a potential role in detecting crop tolerance to abiotic stresses, pathogen resistance, genetic diversity and agronomical characters. Marker-assisted selection (MAS) is a novel breeding tool used for more accurate and useful selections in breeding populations. There are several advantages of MAS including selection of genotypes at seedling stage, selecting traits with low heritability, and the ability to discriminate between homozygotes and heterozygotes. This book provides comprehensive insights into molecular marker technology and its applications in plant biotechnology. It will serve as a valuable source of reference for graduate and postgraduate students.

Each chapter is a sole-standing publication that reflects each author's interpretation. Thus, the book displays a multi-facetted picture of our current understanding of application, resources and aspects of the field. I would like to thank the contributors of this book and my family for their endless support.

Editor

An Improved Oil Palm Genome Assembly as a Valuable Resource for Crop Improvement and Comparative Genomics in the *Arecoideae* Subfamily

Ai-Ling Ong [1,2,*], **Chee-Keng Teh** [1,2], **Sean Mayes** [2], **Festo Massawe** [3], **David Ross Appleton** [1] and **Harikrishna Kulaveerasingam** [1]

1   Biotechnology & Breeding Department, Sime Darby Plantation R&D Centre, Serdang 43400, Selangor Darul Ehsan, Malaysia; teh.chee.keng@simedarbyplantation.com (C.-K.T.); david.ross.appleton@simedarbyplantation.com (D.R.A.); harikrishna.k@simedarbyplantation.com (H.K.)
2   School of Biosciences, University of Nottingham, Sutton Bonington Campus, Leicestershire LE12 5RD, UK; sean.mayes@nottingham.ac.uk
3   School of Biosciences, University of Nottingham Malaysia, Semenyih 43500, Selangor Darul Ehsan, Malaysia; festo.massawe@nottingham.edu.my
*   Correspondence: ong.ailing.sdtc@simedarbyplantation.com

Abstract: Oil palm (*Elaeis guineensis* Jacq.) is the most traded crop among the economically important palm species. Here, we report an extended version genome of *E. guineensis* that is 1.2 Gb in length, an improvement of the physical genome coverage to 79% from the previous 43%. The improvement was made by assigning an additional 1968 originally unplaced scaffolds that were available publicly into the physical genome. By integrating three ultra-dense linkage maps and using them to place genomic scaffolds, the 16 pseudomolecules were extended. As we show, the improved genome has enhanced the mapping resolution for genome-wide association studies (GWAS) and permitted further identification of candidate genes/protein-coding regions (CDSs) and any non-coding RNA that may be associated with them for further studies. We then employed the new physical map in a comparative genomics study against two other agriculturally and economically important palm species—date palm (*Phoenix dactylifera* L.) and coconut palm (*Cocos nucifera* L.)—confirming the high level of conserved synteny among these palm species. We also used the improved oil palm genome assembly version as a palm genome reference to extend the date palm physical map. The improved genome of oil palm will enable molecular breeding approaches to expedite crop improvement, especially in the largest subfamily of *Arecoideae*, which consists of 107 species belonging to *Arecaceae*.

Keywords: *Elaeis guineensis*; high-density linkage maps; physical maps; comparative genomics; genome-enabled breeding; *Arecoideae*; pseudomolecules

1. Introduction

The *Arecaceae* (formerly known as *Palmeae*) family of the monocot order *Arecales* is a plant family hosting perennial palm species that mostly grow across the equatorial belt. The family is made of 181 genera with 2600 species [1]. Among them, the genus *Elaeis* (oil palm) and genus *Cocos* (coconut) together are the world's largest source of oils and fats (about 35%), while date palm plays a major role in food security and agricultural production in arid and fragile ecosystems worldwide.

Two species of *Elaeis* including *E. guineensis* Jacq. and *E. oleifera* (HBK) Cortes originating from West Africa and Latin America, respectively, produce edible oil from fruit mesocarp and kernel. However, only the African oil palm is typically commercially planted due to its superior oil yield with an average oil yield in Malaysia of 3.5 tons hectare^{-1} year^{-1}, compared to only 0.4 tons hectare^{-1}

year^{-1} produced by its South American counterpart [2]. The average oil yield of commercial oil palm is also roughly 8 to 10 times higher than other major temperate oil crops [3], making it the most efficient oil crop in the world. However, with increasing demand for fats and oils from a growing population, yield improvement is essential to meet food requirements at the same time as halting deforestation. Crop improvement through breeding and genomics is the basis for increasing yield and therefore productivity without the use of more land.

Although coconut (*C. nucifera* L.) is not as productive as oil palm, it generates more products such as copra, coir, and timber to serve traditional markets, particularly in Asia. Nut yield in Asia is approximately 5.5 tons hectare^{-1} year^{-1}, while the copra yield ranges from 10 to 20 percent of the nut weight, equivalent to around 1 ton hectare^{-1} year^{-1} [4]. The recent rise in demand for coconut products has resulted from coconut water and virgin coconut oil being promoted as an energy drink and health supplement. The market size of coconut products was reported to be USD 11.5 billion in 2018 and is anticipated to reach USD 31.1 billion by 2026 [5]. Another related palm species, the date palm (*Phoenix dactylifera* L.) does not produce oil but stores mostly carbohydrate, which is converted to sugar in its fruit mesocarp [6]. Hence, the crop has been domesticated in the Middle East and Indus Valley as a major staple food, and the earliest cultivation can be traced back to 3700 BC [7]. Date palm thrives in some of the most arid environments on earth and is able to create a local microclimate, which also allows other crops to be grown [8]. Similarly, higher global demand for inherent medicinal, nutritional, and health advantages of date palm products has boosted the market value of the date palm industry from USD 9.2 billion in 2014 to USD 13 billion in 2018 [9].

Within the *Arecaceae*, there are many other palms that could make a significant contribution to economic development, food, and nutritional security and provide ecosystem services for sustainable cultivation, including sago palm, nipa palm, peach palm, etc. However, it is unlikely that companies will be willing to make major investment in developing genetic improvement programs without being sure that there will be sizable returns. To date, oil palm receives the most research attention due to its significant role in the world economy, although coconut and date palm have seen steep increases in research focus with their recent emergence. In general, genomic research is centered on the development of marker-assisted selection (MAS) for expediting breeding selection for these outcrossing perennial palms that typically require more than a decade to complete a breeding cycle [10–14]. In the 90s, the first restriction fragment length polymorphism (RFLP)-based linkage map of oil palm [10] was reported and subsequently deployed for mapping quantitative trait loci (QTL) responsible for fruit characteristics and vegetative traits [15]. The effort was continued and expanded using various DNA marker systems: random amplified polymorphic DNA (RAPD), amplified fragment length polymorphism (AFLP), diversity arrays technology (DArT), simple sequence repeat (SSR), and single nucleotide polymorphism (SNP) [10,16–21]. The reported QTL, however, were only applicable to a few populations closely related to the mapping populations. In oil palm, the Deli *dura* that produces a thick kernel shell, high bunch number, and high mesocarp oil content and is widely used in different programs and subsequently has been developed in breeding populations to different sub-populations, such as *Johore Labis, Ulu Remis*, and *Gunung Melayu*.

Mapping resolution, especially for complex traits (e.g., oil yield and disease tolerance), is limited by insufficient recombination and a lack of statistical power within a small bi-parental family, which usually consists of only 16 to 96 field-planted palms in which traits are recorded in breeding programs. To improve mapping resolution, research efforts gradually shifted towards genome-wide association studies (GWAS) to access total meiotic recombination accumulated in a large admixed population. Nevertheless, the mapping method was of limited power until the reference *E. guineensis* genome (AVROS *pisifera*), namely P5-build, was sequenced and made publicly available in 2013 [22]. We then reported the first GWAS for mesocarp oil content based on 2045 *tenera* oil palms [23] using a high-density OP200K SNP array [24]. However, only 60% of the linked SNPs were located across 16 chromosomes (or pseudomolecules) of the reference oil palm genome, with the P5-build (2n = 32), leaving room for improvement. The same limitations on GWAS were also observed in the date palm [25].

A good genome assembly does not only rely on assembly of contigs into scaffolds but also the placement and contiguity of scaffolds on chromosomes to form the physical map. To summarize the latest status of assembled genomes of oil palm, coconut, and date palm (as available in National Centre for Biotechnology Information; NCBI): 1535.18 Mb of P5-build oil palm genome first achieved a chromosome coverage level with 43% of scaffolds anchored [22], although no improvement has been published since 2013. On the other hand, the 2202.46 Mb of assembled coconut genome and the 772.32 Mb of assembled date palm genome have recently surpassed the oil palm genome, with 46% [26] and 50% [25] of scaffolds anchored to the chromosomes. The genome size of oil palm is estimated at about 1.8 Gb [27,28]. We previously reported on the possibility of further extending the physical map of the current P5-build reference genome using one ultra-dense linkage map of a commercial Deli *dura* × AVROS *pisifera* population, and the total length of additional scaffolds anchored in our previous study has improved by 311 Mb [29]. Hence, our aim in this study was to maximize the improvement of the oil palm genome assembly by integrating multiple linkage maps from various origins. Here, we also report how the improved oil palm assembly, namely using Physical Map version 6 (PMv6), can serve as an important reference for comparative genomic studies among coconut and date palms, with expansion to other *Arecaceae* members in the future.

2. Results

2.1. Construction of Multiple High-Density Oil Palm Linkage Maps

Three bi-parental populations, including the reported commercial Deli *dura* × AVROS *pisifera* (295 *tenera* palms) [29], semi-wild Deli *dura* × Nigerian *dura* (112 *dura* palms), and *Johore Labis dura* × *Johore Labis dura* (214 *dura* palms) were used to construct individual linkage maps. We obtained 17 to 24 linkage groups (LGs; Table 1) with one to six chromosomes being represented by two or three linkage groups representing the expected 16 major LGs for total oil palm chromosomes number. The *Johore Labis dura* population showed the lowest number of linked SNPs (6920 loci; due to the use of refined SNP chip of 20K loci), while the highest number of linked SNPs (32,650 loci) was observed for the Deli *dura* × Nigerian *dura* population. The Deli *dura* × AVROS *pisifera* map remained the densest with a mean inter-marker distance of 0.04 cM, as reported [29], compared to the sparsest *Johore Labis* map (0.18 cM). In this study, a total of 11,421 common SNPs was identified and adopted for integration of the multiple linkage maps.

Table 1. Statistics of linkage maps developed for each oil palm population.

Population	Number of Linked Markers	Number of Linkage Groups (LGs)	Length of Linkage Map (cM)	Marker Interval (cM)	Genome-Wide Recombination Rate of P5-Build (cM/Mb)	Genome-Wide Recombination Rate of PMv6 (cM/Mb)
Deli *dura* × AVROS *pisifera* *	27,890	19	1151.70	0.04	1.75	0.98
Deli *dura* × Nigerian *dura*	32,650	17	1646.95	0.05	2.50	1.40
Johore Labis dura	6920	24	1268.26	0.18	1.93	1.08

* previously reported [29].

2.2. Improvement of the Oil Palm Physical Genome Assembly by Integration of Multiple Oil Palm Linkage Maps

Linkage-directed genome assembly improvement not only anchored previously unplaced scaffolds to the pseudomolecules but also split and reassigned a range of the scaffolds flanked by N bases in the published physical genome, P5-build [22]. The newly assembled genome, termed PMv6 with an N50 of 83.1 Mb, improved the scaffold assignment to oil palm pseudomolecules from 43% to 77% (1.2 Gb, Figure 1). On average, each pseudomolecule in the PMv6 genome was assembled using 142 scaffolds with an average length of 73.7 Mb, whereas only 19 scaffolds giving an average length of 41.1 Mb were reported in Singh et al., 2013 [22]. Chromosome 2 was also extended by 1.2-fold and now is the largest pseudomolecule, accounting for 12% of the total PMv6 assembly length (Figure 1a,b).

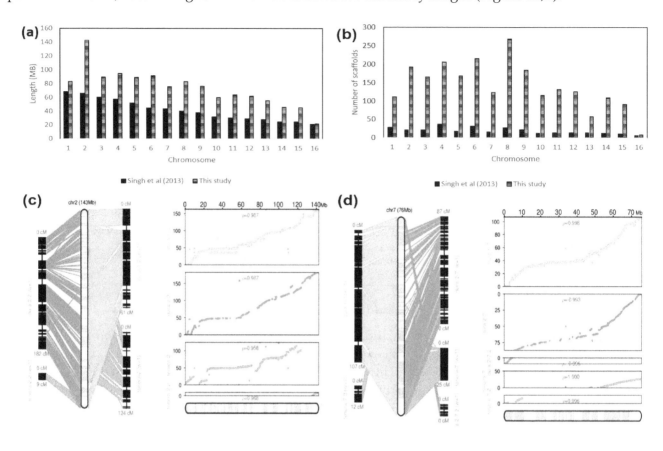

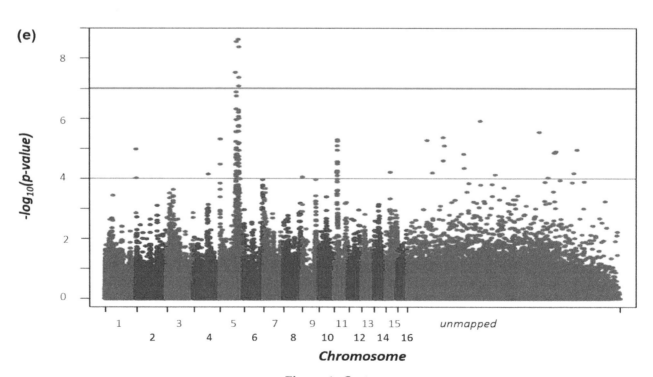

Figure 1. *Cont.*

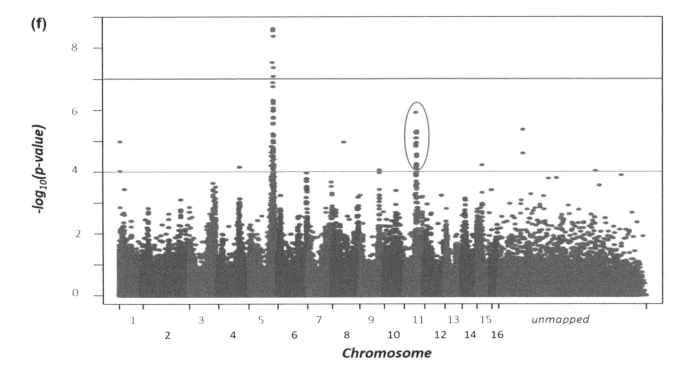

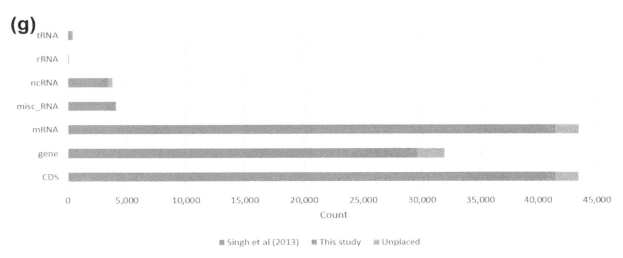

Figure 1. Refined assembly of oil palm genome (PMv6). (**a**) Chromosome length (Mb) comparison between P5-build, Singh et al., (2013), and PMv6 (this study). (**b**) Number of scaffolds from previous physical map [22]) and the improved physical map developed in this study. (**c**) Buildup of PMv6 by scaffold anchoring using multiple linkage maps; left panel represents the central combined linkage map from three mapping populations, and the right panel shows the correlation plots of each populations' genetic versus physical order; presented in the order of Deli *dura* × Nigerian *dura* (*dura* 1—green), *Johore Labis dura* (*dura* 2—orange), and Deli *dura* × AVROS *pisifera* (*tenera*—blue) for Chromosome 2 and (**d**) Chromosome 7 as examples. (**e**) Genome-wide association studies (GWAS) Manhattan plot for a commercial oil palm population for one of the oil yield trait [23] for the genome published by P5-build. (**f**) GWAS Manhattan plot for the same commercial oil palm population for one of oil yield trait after assembly improvement in this study, circle in red represent newly assigned and significant SNPs. (**g**) Statistics of genomic features improvement after placement of scaffolds into chromosomes for P5-build and this study.

In Figure 1c,d, we illustrate how the scaffolds were anchored to Chromosome 2 and 7 based on the integrated linkage maps, and how they were validated by correlation between linkage and physical positions (Figure 1c,d). The results show that assembly improvement using the commercial Deli *dura*

× AVROS *pisifera* map [29] (blue) was only limited to one end of the pseudomolecule, especially on Chromosome 7. Nevertheless, the shortcomings were solved when integrating the semi-wild Deli *dura* × Nigerian *dura* map (green) and the *Johore Labis dura* map (orange). The sigmoidal curves of Chromosome 2 and 7 clearly distinguishing both telomere regions from centromeric regions where limited recombination events were observed. The correlation between linkage and physical positions reached more than 0.90 in each mapping family, and the collinearity between each linkage maps can be observed with consistence marker orderings within each LG. Similar results were observed for the remaining chromosomes and are shown in Supplementary Figure S1. The details of total SNPs residing within genomic scaffolds used to build PMv6 genome assembly can be found in Supplementary Table S1. Using the correlation outputs, the genome-wide recombination rate was estimated to be 1.15 cM/Mb, which is about half of the previously reported estimation based on the P5-build genome [29] (Table 1). This investigation may infer overestimation of the recombination rate in our previous estimation, mainly due to the length extension of the PMv6 genome assembly reported here. Understanding recombination differences has distinct implications for population structure, gene evolution, and genetic improvement [30].

The improved PMv6 genome was used to evaluate the improvement of mapping resolution of the previously reported GWAS for mesocarp oil content [23], reducing the unmapped SNPs from 40% (37,003) (Figure 1e) to only 13% (12,175) (Figure 1f). The major QTLs still remain at Chromosome 5 and Chromosome 11, but the significance level on Chromosome 11 was increased to a −log *p*-value = 5.9 from 5.2 and more SNPs surpassed genome-wide -log *p*-value cutoff at 4.0, due to SNPs newly assigned to Chromosome 11, improving trait location. From the perspective of genomic features and gene annotations, the PMv6 genome assembly now includes 5769 genes, 7705 protein-coding regions (CDS), and 1563 non-coding RNA, which could not be placed into the P5-build genome (Figure 1g). Interestingly, among the non-coding RNA, 48% of total predicted ribosomal RNA (rRNA) were mapped for the first time in PMv6.

2.3. Comparative Genomic Analysis of Three Palm Species

The improved oil palm genome, PMv6 was subsequently used for genome comparison with coconut [26] and date palm [25]. The results confirmed a high level of syntenic block conservation between the oil palm and the other two palm species (Figure 2). For coconut, the 16 homologous chromosome groups (2n = 32) appeared to be highly coherent with the oil palm genome, but we observed possible fusion and fission event differences on oil palm-based Chromosome 2 (eg-2) and 10 (eg-10), respectively. Compared to coconut, a 30 Mb fragment of eg-2 fused with eg-7 accounts for coconut-based Chromosome 2 (cn-2), suggesting a fission/fusion event since the separate evolution of the two species. In a similar comparative way, a 20 Mb fragment of eg-10 has fused with eg-16 to form cn-8.

In general, the genome conserved synteny between date palm and oil palm was more complicated. By referring to Ziwen et al. (2015), oil palm and coconut were estimated to have diverged from date palm approximately 62.5 million years ago [31]. Chromosome fusion/fission events between date palm-based Chromosome 4 (pd-4) and pd-16 and between pd-1 and pd-10 may have led to formation eg-1 and eg-2 in the oil palm genome, respectively. This possibly explains why date palm (2n = 36) has two additional homologous chromosome groups compared to oil palm. In addition, pd-11 compared to oil palm has two chromosome fragments with one of them fused with pd-12 to form eg-10 and the remaining pd-11 fragment being maintained as eg-6 in the oil palm genome.

We then further compared the *SHELL* gene, which is a homologue of *SEEDSTICK* (NCBI Gene ID: LOC105034563)—a type II MADS-box transcription factor reported in oil palm (at 3.05 Mb of eg-2) with the orthologs of coconut (at 116 Mb of cn-1) and date palm (at 38 Mb of pd-1). In the oil palm genome, the *pisifera* allele (sh^{AVROS}) results in an amino acid change from lysine to asparagine, causing the absence of shell surrounding the fruit kernel observed *pisifera* fruit [32] (Figure 3a). The analysis was also extended to other candidate genes including Cytochrome P450 (*CYP703*) (LOC105059962), glycerol-3-phosphate acyltransferase 3-like (*GPAT3*) (LOC105059961), Cytidine deaminase-like (*CYT DA*) (LOC105059743), and Lonely-Guy (*LOG*) (LOC105055182) reported to be involved in sex determination of date palm [33]. All the candidate genes (except *LOG*, located at 10.86 Mb of eg-12) were located at 44–45 Mb of eg-10 in the oil palm genome but have yet to have the scaffolds containing them placed in the date palm genome.

In Figure 3b, a phenogram based on the sex-determination-related *LOG* genes revealed that coconut has the shortest evolutionary distance to oil palm followed by date palm, among other monocots outgroup taxa (Figure 3b)

2.4. Oil-Palm-Guided Improvement of the Date Palm Genome

Due to the high levels of collinearity of gene order and genomic sequences between the palm species, the PMv6 genome was adopted as a reference to improve the published genome assemblies of date palm. We started with the earlier version, namely PDK_30 (GenBank assembly accession: GCA 000181215.2) [34]. Genomic scaffolds from date palm were tiled following oil palm ordering for each pair of syntenic chromosomes, then the joined scaffolds were split if required following the published date palm genetic map [35]. This resulted in anchoring 78% of scaffolds to the physical map from the total length of the full assembly (381.5 Mb). The next test was performed on the more recent assembly of date palm genome (GCA_009389715.1) [25] using the same approach. The second improvement was not as apparent as the first version but did result in a total length of 458 Mb, equivalent to an extra 10% of total scaffold length anchored to chromosomes, compared to the published assembly. The placement of scaffolds in each of the date palm chromosomes extended their total length except for pd-10 and pd-16 (Figure 4). The assembly improvement for both versions of the date palm genome were further compared using date palm genetic markers [35] located within each assembly. An average correlation of 0.80 was observed for the 18 chromosomes when comparing physical location of the genetic markers between both improved assemblies, ranging from the lowest on pd-16 ($r = 0.38$) to the highest on pd-4 and pd-15 (equally $r = 0.98$) (Figure S2).

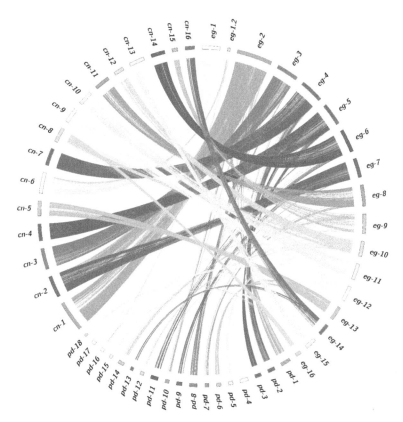

Figure 2. Synteny analysis including three palm species: oil palm (eg), coconut (cn), and date palm (pd). Chromosome pairs with strong syntenic relations were linked by the same ribbon color.

(a)

```
CLUSTAL 2.1 multiple sequence alignment


Elaeis_guineensis          MGRGKIEIKRIENTTSRQVTFCKRRNGLLKNAYELSVLCDAEVALIVFSSRGRLYEYANN
Cocos_nucifera             MGRGKIEIKRIENTTSRQVTFCKRRNGLLKKAYELSVLCDAEVALIVFSSRGRLYEYANN
Phoenix_dactylifera        MGRGKIEIKRIENTTSRQVTFCKRRNGLLKKAYELSVLCDAEVALIVFSSRGRLYEYANN
                           ****************************  :*****************************

Elaeis_guineensis          SIRSTIDRYKKACANSSNSGATIEINSQQYYQQESAKLRHQIQILQNANRHLMGEALSTL
Cocos_nucifera             SIRSTIDRYKKACANSSSSGATIEINSQQYYQQESAKLRHQIQILQNANRHLMGEALSSL
Phoenix_dactylifera        SIRSTIDRYKKACANSSNSSAAIEINSQQYYQQESAKLRHQIQILQNANRHLMGESLSSL
                           *****************.*.*:**********************************:**:*

Elaeis_guineensis          TVKELKQLENRLERGITRIRSKKHELLFAEIEYMQKREVELQNDNMYLRAKIAENERAQQ
Cocos_nucifera             TVKELKQLENRLERGITRIRSKKHELLFAEIEYMQKREVDLQNDNMYLRAKIAENERAQQ
Phoenix_dactylifera        TVKELKQLENRLERGITRIRSKKHELLFAEIEYMQKKEVELQNDNMYLRAKIAENERAQQ
                           ************************************:**:*******************

Elaeis_guineensis          AGIVPAGPDFDALPTFDTRNYYHVNMLEAAQHYSHHQDQTTLHLGYEMKADPAAKNLL
Cocos_nucifera             AGIVQAGPEFDTLPTFDSRNYYHVNMLEAAQHYSHHQDQTTLHLGYEMKADPAAKTLL
Phoenix_dactylifera        AGIVPAGPEFDTLPTFDSRNYYHVNLLEAAQHYSQHQDQTTLHLGYEMKADPAAKNLL
                           **** ***:**:*****:******:*******:*****************.**
```

(b)

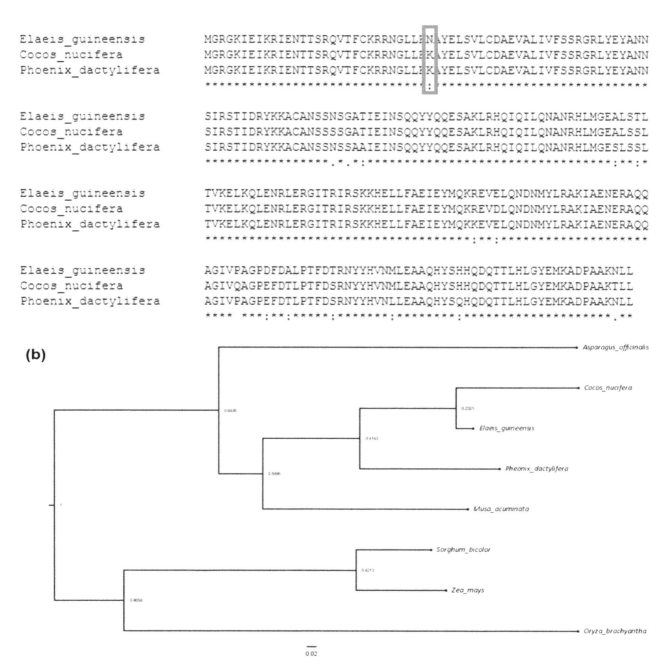

Figure 3. High conservation of genes among palm species with minimal amino acid changes. (**a**) Alignments of *SHELL* gene encoding for the layer of shell surrounding the kernel for oil palm and the orthologues from coconut and date palm. (**b**) Phenogram tree for *LOG* gene orthologues, which play a role in sex determination for date palms; orthologues of the gene among palms and other monocots outgroup taxa were included.

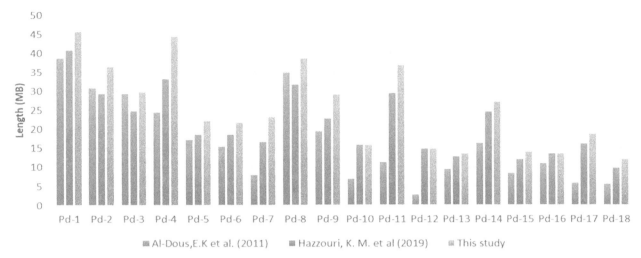

Figure 4. Oil-palm-directed date palm genome assembly improvement by length (Mb) across 18 pseudomolecules. Anchoring of scaffolds was performed on the two published assemblies of date palm genome by Al-Dous et al. (2011) [34] and later by Hazzouri et al. (2019) [25].

3. Discussion

In our recent publication, we demonstrated an extension of 311 Mb of the published oil palm pseudomolecules (P5-Build [22]) using a linkage map of a commercial Deli *dura* × AVROS *pisifera* family (295 palms) [29]. In this study, we constructed two denser linkage maps from a semi-wild Deli *dura* × Nigerian *dura* population (112 palms) and *Johore Labis dura* population (214 palms). The Deli *dura* origin is derived from four palms planted at Bogor Botanical Garden in 1848 [36] and exhibits a thick kernel shell, high bunch number, and high mesocarp oil content. It has become the most important genetic resource for major breeding programs in Southeast Asia. Separate breeding programs using Deli *dura* have led to sub-populations, such as *Ulu Remis* and *Johore Labis* [37]. Breeding history and selection have likely resulted in the lowest number of linked SNPs and the lowest mapping density being observed in the pure *Johore Labis dura* population (Table 1). To avoid inbreeding depression, Sime Darby Plantation R&D also introgressed the Deli *dura* with a Nigerian *dura* to widen the genetic base, reflected by the higher mapping density and number of linked SNPs observed.

By integrating the three linkage maps, a total of 11,421 family-transferable/common SNPs were used to anchor 77% of the assembled 1.5 Gb genome, in order to generate the new oil palm physical map, PMv6 (Figure 1), compared to 43% in the previously published P5-build physical map. Using Chromosome 2 and 7 as examples, the quality of scaffold anchoring based on the integrated linkage maps can be visualized and was validated through a correlation between genetic and physical positions, allowing the recombination landscape throughout each chromosome to be determined (Figure 1c,d). Species with larger genomes often have higher levels of recombination at the telomeres and lower recombination in the heterochromatic and centromeric regions despite genes being present within transposable element (TE) [38,39] clusters, as reported in wheat [40,41]. In such a scenario, a sigmoidal curve of the recombination rate should be expected, but this was not clearly observed in both chromosomes when assembled from the commercial Deli *dura* × AVROS *pisifera* map (blue), suggesting only a partial improvement. However, the integration of the semi-wild Deli *dura* × Nigerian *dura* map (green) and the *Johore Labis dura* map (orange) enabled extension of both chromosomes with a sigmoidal curve of recombination from telomere-to-telomere, although whether this represents the full chromosome needs further validation, through techniques such as Fluorescence in situ hybridization (FISH) technique.

The PMv6 genome enabled an increase in the mapping resolution of the reported GWAS for mesocarp oil content [23] by locating an additional 65% (24,828) previously unplaced SNPs (Figure 1f). Among the total SNP markers mapped in PMv6 assembly, there are 32,552 SNPs located within genic/non-coding RNA regions, equivalent to 25% of the total SNP assigned to scaffolds/genome sequence. Increased marker density allows better capture of linkage disequilibrium between origins and render GWAS more powerful. Hence, PMv6 enables a more complete set of annotated genes and non-coding RNA to be located that are potentially responsible for phenotypic changes residing within QTL to be identified after conducting GWAS.

In general, most of the chromosomes between the oil palm, coconut, and date palm genomes are highly syntenic (Figure 2). They also share agriculturally desirable traits, such as tall or short varieties in coconut palm, which would be useful in oil palm and date palm. It would be possible to use the causative gene from coconut to screen for allelic differences in the gene orthologues in other species. For instance, the *SHELL* gene in the oil palm genome (eg-2) was found to be highly conserved with the genomic sequence orthologs in coconut (cn-1) and date palm (pd-1). Only the oil palm genome sequence carries sh^{AVROS} variants causing the lysine-to-asparagine amino acid change, which results in disruption of DNA binding within the MADS-box domain [32], thus no shell is found in this *pisifera* oil palm. Indeed, two more fruit forms including thick-shelled dura (sh^{Deli}/sh^{Deli} as wild-type) and thin-shelled *tenera* (sh^{Deli}/sh^{AVROS}) also naturally exist because of haploinsufficiency effect on the *SHELL* gene [32,42]. The *SHELL* variants are extremely important for oil palm breeding, because crossing between dura and *pisifera* (usually female sterile) produce *tenera* with fertility restored and 30% more oil than the dura [43,44]. Unlike oil palm, shell-less fruits in coconut and date palm have not yet been reported. The shell layer may be essential to coconut and date palm for maintaining their endosperm moisture content to allow seed germination, especially in arid habitats. Consequently, the mutation probably has been naturally selected out. Nevertheless, we still identified a strong QTL for shell thickness of *tenera* oil palms located within 2.5 Mb to 3.9 Mb sharing the same region with the *SHELL* gene on eg-2 [24], even when the later was fixed. Hence, the results obtained from oil palm research could lead to the same region in coconut palm being screened for the orthologs responsible for shell variation in coconut within cn-1 and eventually deployed in MAS for selecting thicker- or thinner-shelled coconuts. Coconut shell normally occupies 15% of the total weight of a fruit and approximately 9 million tons of coconut shell is globally produced as an agricultural waste every year [45]. Nevertheless, the waste can be turned into activated charcoal and commands a premium [46,47].

Oil palm and coconut are estimated to have diverged from date palm around 62.5 million year ago [31]. Fission or fusion of two chromosomes into one or vice versa is known in many lineages [48]; this event is observed in this study too, whereby date palm has 18 pairs of chromosomes, while oil palm and coconut have only 16 pairs. Torres et al. found that the sex-determination-related genes are now located at 44–45 MB of eg-10 [33], where fusion between pd-11 and pd-12 probably occurred, comparative to oil palm. This also suggests that eg-10 contains this cluster of key genes for inflorescence gender and that the split into two chromosomes in the date palm may be an important event leading eventually to a diecious palm. The finding is supported by a shorter evolutionary distance of LOG gene (suppression on female inflorescence) between oil palm and coconut, compared to date palm (Figure 3b). Of the three palms, date palm is the only diecious palm. The cycle of male and female inflorescences of monoecious oil palm is mainly affected by environmental and genetic factors. Water stress and excessive frond pruning lead to high induction of male inflorescences and abortion of both sexes when the stress becomes extreme [49]. Understanding the genetic control underlying sex determination in oil palm can provide a new avenue to address abiotic stress tolerance in oil palm. Understanding the basis of inflorescence sex determination may allow breeding for specific environments, to optimize sex ratio to the target environment, especially in the context of global climate change and of plantations in drought-prone areas. In terms of aptitude for branching, date palm and sago palm can also propagate vegetatively by producing suckers, which would be an extremely valuable trait in oil palm, where this currently does not happen.

Only 19% of the scaffolds were anchored based on a linkage map of the Khalas cultivar to the first date palm assembled genome, PDK_30 [34,35]. Subsequently, the genome has recently been greatly improved to 50% anchored scaffolds of the expected 772.32 Mb genome size using third generation sequencing PacBio long reads by Hazzouri et al. [25]. In the present study, we have shown that the new oil-palm-guided date palm assembly is still able to anchor a further 10% of scaffolds, and the oil-palm-directed assembly is highly consistent with the latest date palm assembly. Hence, a full utilization of available linkage maps, especially with high mapping density should be incorporated in genome improvement. The oil palm PMv6 genome assembly can serve as a reference to date palm and coconut, and to other palm members in the future. Moreover, based on the taxonomic distances between oil palm and date palm, we would expect similar results from any palm species with a divergence time less than 62.5 million years [31] ago within *Arecoideae*, a subfamily of *Arecaceae* that consists of 107 species [50,51], using the oil palm PMv6 genome as a reference. Improved genome sequences can in turn lead to crop improvement strategies for these important food, feed, and fuel sources that will ultimately lead to the higher productivity required for agriculture to meet growing demands in the years to come without the use of more forest land.

4. Materials and Methods

4.1. Mapping Populations and DNA Preparations

Three mapping populations from oil palm breeding populations of restricted origins (BPROs) including the 295 commercial *tenera* palms derived from a Deli *dura* × AVROS *pisifera* family (278 × TT41/4), previously described in Ong et al. [29], were used. The other two populations were maintained as female parental breeding sources, namely the Deli *dura* × Nigerian *dura* and *Johore Labis dura* × *Johore Labis dura* with population sizes of 112 and 214 individuals, respectively. Leaf samples were collected from each palm planted in different breeding trials maintained by Sime Darby Plantation Research and Development Centre, Sime Darby Plantation Sdn Bhd, Malaysia. Fresh leaf tissue was sampled from the third frond of each palm. Samples of genomic DNA were isolated from 0.1 g of the leaf tissue using the DNAeasy Plant Mini Kit (Qiagen, Germany). The DNA concentration and purity were quantified on a 0.8% agarose gel using known standards.

4.2. SNPs Identification and Genotyping

The extracted genomic DNA (25 ng/µL) samples from Deli *dura* × Nigerian *dura* and Deli *dura* × AVROS *pisifera* were genotyped using the 200K SNPs included on the published OP200K Infinium array (Illumina, USA) [24], and the population *Johore Labis dura* × *Johore Labis dura* was genotyped using another 20K SNP panel, selected as a subset of the 200K panel on the Infinium iScan platform according to the manufacturer's recommendations. Raw genotyping data were analyzed with GenomeStudio version 20011.1 with genotyping module version 1.8.4. Using a GenCall score cutoff of 0.15, auto-clustering of the SNPs was done. The SNP clustering was confirmed manually by visual inspection. The SNP calls were exported into the PLINK program for minor allelic frequency (MAF) cutoff at 0.01 [52]. Polymorphic SNPs were identified where at least one of the parents was heterozygous. Subsequently, the polymorphic SNPs were dropped where the call rate <95% or where there was significant segregation distortion, using a false discovery rate (FDR)-corrected method with an effective cutoff at p-value < 0.05.

4.3. Construction of Linkage Maps

All linkage maps were constructed using Lep-MAP3 [53,54] for large SNP datasets, especially developed for outbred families such as oil palm with an underlying maximum likelihood (ML) method. Filtering module was used for marker quality checking, followed by running the SeparateChromosomes module for binning the assayed markers into linkage groups with optimized LOD values ranging from 5 to 15 with intervals of 5. The JoinSingles module assigned singular markers to existing linkage groups to maximize the map ability of the total input marker set. Lastly, the OrderMarkers module ordered the binned markers according to the Kosambi mapping function for conversion of recombination frequencies into map distances (in unit centiMorgan, cM). The option "sexAveraged = 1" was selected along with OrderMarkers to join the maps of both parents.

4.4. Anchoring of Scaffolds to Linkage Groups

Linked SNP markers (from OP200K) with flanking sequences of 60mers located in the oil palm scaffolds (placed and unplaced in pseudomolecules) were identified using BLASTN [55] with e-value cutoff at 1×10^{-150} and sequence identity at least of 95%. High-density linkage maps built from the three mapping populations were merged using ALLMAPS assembly module, together with equal weighting among the linkage maps [56]. Scaffolds were then ordered and orientated according to the merged and weighted linkage maps, thus resulting in the new genome content. The same parameters have been applied to date palm SNP markers with flanking sequences of 100mers to identify the locations of the markers in two of the date palm genome versions.

4.5. Genome and Gene Comparisons

Palm genomic sequences were downloaded from NCBI. Alignments of the coconut [26] and date palm [57] genome against the current oil palm genome (PMv6) was done using Nucmer, which is a module developed in Mummer [58,59], an ultrafast sequences aligner, made for genome alignments. The syntenic relationship between both oil palm and coconut chromosomal alignments were visualized using the CIRCOS software [60]. Chromosome pairs with strong syntenic relations between oil palm, coconut, and date palm were linked by same ribbon colors. Candidate genes of interest were also aligned using multiple sequence alignment tool, clustalw [61,62] via translated protein and orthologs of other closely related monocots as outgroup: asparagus (*Asparagus officinalis*), banana (*Musa acuminata*), sorghum (*Sorghum bicolor*), corn (*Zea mays*), and turf (*Oryza brachyantha*) protein sequences were identified from NCBI. The phenogram was build using neighbor-joining (NJ) with 1000 bootstrap iterations and visualized using a phylogenetic tree viewer, namely Figtree (v1.4.4) [63].

4.6. Reconstruction of Date Palm Pseudomolecules

Mummer (Nucmer) output from the step above tiled the date palm's query scaffolds according to the sequence order in the newly extended oil palm pseudochromosomes. Those scaffolds were then ordered and orientated into GenBank agp formats, and the pseudomolecules were built using the agp module in ALLMAPS. SNP markers obtained from the published date palm genetic maps [35] were aligned to the new pseudomolecules using the same BLASTN criteria, as above. The integration of SNP marker locations in the genetic map against physical map location was plotted using ALLMAPs path module. Chimeric contigs, which contained ambiguous markers that mapped to more than two linkage groups, were then processed using the split module. Ambiguous short contigs were excluded during the second round of rebuild of pseudomolecules by running the ALLMAPs path module again.

Supplementary Materials:
Figure S1: Integration of genetic with physical maps for oil palm, Chromosomes 1 to 16; Figure S2: Correlation plots for locations of DNA marker located within date palm chromosome (pd-15) for genome assembly improvement versions by Al-Dous et al. (2011) and Hazzouri et al. (2019); Table S1: Summary for consensus map generated from multiple linkage maps and statistics for SNP markers and scaffolds used to build genome assembly for PMv6. The assembled genome has been uploaded in NCBI with submission number: SUB7516020/ BioProject: PRJNA636092.

Author Contributions: Conceptualization, A.-L.O., C.-K.T., and S.M.; investigation, A.-L.O., C.-K.T., and S.M.; manuscript writing, C.-K.T. and A.-L.O.; writing—review and editing, S.M., F.M., D.R.A., and H.K.; supervision, F.M., S.M., D.R.A., and H.K.; funding acquisition, D.R.A. and H.K. All authors read and agreed to the published version of the manuscript.

Acknowledgments: We would like to acknowledge the contribution of Oil Palm Breeding Section, Sime Darby Plantation R&D Centre for conducting the breeding trials and providing the oil palm materials. We also would like to thank the Molecular Breeding and Bioinformatics team for enhancement of figures in the manuscript. We would like to thank the University of Nottingham Malaysia and Yayasan Sime Darby for the PhD fee scholarship to A.-L.O. and C.-K.T.

References

1. Christenhusz, M.J.M.; Byng, J.W. The number of known plants species in the world and its annual increase. *Phytotaxa* **2016**, *261*, 201–217. [CrossRef]

2. Hardon, J.J. Interspecific hybrids in the genus Elaeis II. vegetative growth and yield of F1 hybrids E. guineensis x E. oleifera. *Euphytica* **1969**, *18*, 380–388. [CrossRef]

3. Basiron, Y.; Balu, N.; Chandramohan, D. Palm oil: The driving force of world oils and fats economy. *Oil Palm Ind. Econ. J.* **2004**, *4*, 1–10.

4. Prades, A.; Salum, U.N.; Pioch, D. New era for the coconut sector. What prospects for research? *OCL Oilseeds Fats Crop. Lipids* **2016**, *23*. [CrossRef]

5. LLP, A.A. Coconut Products Market by Type Application and Form: Global Opportunity Analysis and Industry Forecast, 2019–2026. 2019. Available online: https://www.globenewswire.com/news-release/2020/02/21/1988574/0/en/Coconut-Products-Market-Study-2019-2026-World-Market-Projected-to-Cross-31-Billion-by-2026.html (accessed on 22 January 2020).

6. Bourgis, F.; Kilaru, A.; Cao, X.; Ngando-Ebongue, G.F.; Drira, N.; Ohlrogge, J.B.; Arondel, V. Comparative transcriptome and metabolite analysis of oil palm and date palm mesocarp that differ dramatically in carbon partitioning. *Proc. Natl. Acad. Sci. USA* **2011**, *108*, 12527–12532. [CrossRef] [PubMed]

7. Munier, P. Le palmier-dattier. In *Munier 1973 Palmier*; Maisonneuve&Larose: Paris, France, 1973; Volume 24.

8. Al-Yahyai, R.; Manickavasagan, A. *Dates: Production, Processing, Food, and Medicinal Values*; CRC Press: Boca Raton, FL, USA, 2012. [CrossRef]

9. Shahbandeh, M. Global Date Palm Industry Value 2014–2023. Available online: https://www.statista.com/statistics/960213/date-palm-market-value-worldwide/ (accessed on 22 January 2020).

10. Mayes, S.; Jack, P.L.; Corley, R.H.; Marshall, D.F. Construction of a RFLP genetic linkage map for oil palm (*Elaeis guineensis* Jacq.). *Genome* **1997**, *40*, 116–122. [CrossRef]

11. Batugal, P.; Bourdeix, R.; Baudouin, L. Coconut breeding. In *Breeding Plantation Tree Crops: Tropical Species*; Springer: Berlin, Germany, 2009; pp. 327–375. ISBN 9780387712031.

12. Batugal, P.; Bourdeix, R. Conventional coconut breeding. *Coconut Genet. Resour.* **2005**, 251.

13. Hadrami, I.E.; Hadrami, A. El; Hadrami, A. El Breeding date palm. In *Breeding Plantation Tree Crops: Tropical Species*; Springer: New York, NY, USA, 2009; pp. 191–216. ISBN 9780387712031.

14. Hardon, J.J. Oil palm breeding, Introduction. *Oil Palm Breed. Introd.* **1976**, 98–107.

15. Rance, K.A.; Mayes, S.; Price, Z.; Jack, P.L.; Corley, R.H.V. Quantitative trait loci for yield components in oil palm (*Elaeis guineensis* Jacq.). *Theor. Appl. Genet.* **2001**, *103*, 1302–1310. [CrossRef]

16. Moretzsohn, M.C.; Nunes, C.D.M.; Ferreira, M.E.; Grattapaglia, D. RAPD linkage mapping of the shell thickness locus in oil palm (*Elaeis guineensis* jacq.). *Theor. Appl. Genet.* **2000**, *100*, 63–70. [CrossRef]

17. Ting, N.C.; Jansen, J.; Mayes, S.; Massawe, F.; Sambanthamurthi, R.; Ooi, L.C.L.; Chin, C.W.; Arulandoo, X.; Seng, T.Y.; Alwee, S.S.R.S.; et al. High density SNP and SSR-based genetic maps of two independent oil palm hybrids. *BMC Genom.* **2014**, *15*. [CrossRef] [PubMed]

18. Gan, S.T.; Wong, W.C.; Wong, C.K.; Soh, A.C.; Kilian, A.; Low, E.T.L.; Massawe, F.; Mayes, S. High density SNP and DArT-based genetic linkage maps of two closely related oil palm populations. *J. Appl. Genet.* **2018**, *59*, 23–34. [CrossRef] [PubMed]

19. Billotte, N.; Marseillac, N.; Risterucci, A.M.; Adon, B.; Brottier, P.; Baurens, F.C.; Singh, R.; Herrán, A.; Asmady, H.; Billot, C.; et al. Microsatellite-based high density linkage map in oil palm (*Elaeis guineensis* Jacq.). *Theor. Appl. Genet.* **2005**, *110*, 754–765. [CrossRef]

20. Seng, T.Y.; Saad, S.H.; Chin, C.W.; Ting, N.C.; Singh, R.S.H.; Zaman, F.; Tan, S.G.; Alwee, S.S.R. Genetic linkage map of a high yielding FELDA Deli×Yangambi oil palm cross. *PLoS ONE* **2011**, *6*. [CrossRef] [PubMed]

21. Lee, M.; Xia, J.H.; Zou, Z.; Ye, J.; Rahmadsyah; Alfiko, Y.; Jin, J.; Lieando, J.V.; Purnamasari, M.I.; Lim, C.H.; et al. A consensus linkage map of oil palm and a major QTL for stem height. *Sci. Rep.* **2015**, *5*, 8232. [CrossRef] [PubMed]

22. Singh, R.; Ong-Abdullah, M.; Low, E.-T.L.; Manaf, M.A.A.; Rosli, R.; Nookiah, R.; Ooi, L.C.-L.; Ooi, S.; Chan, K.-L.; Halim, M.A.; et al. Oil palm genome sequence reveals divergence of interfertile species in old and new worlds. *Nature* **2013**, *500*, 335–339. [CrossRef]

23. Teh, C.K.; Ong, A.L.; Kwong, Q.B.; Apparow, S.; Chew, F.T.; Mayes, S.; Mohamed, M.; Appleton, D.; Kulaveerasingam, H. Genome-wide association study identifies three key loci for high mesocarp oil content in perennial crop oil palm. *Sci. Rep.* **2016**, *6*. [CrossRef]

24. Kwong, Q.B.; Teh, C.K.; Ong, A.L.; Heng, H.Y.; Lee, H.L.; Mohamed, M.; Low, J.Z.B.; Apparow, S.; Chew, F.T.; Mayes, S.; et al. Development and Validation of a High-Density SNP Genotyping Array for African Oil Palm. *Mol. Plant* **2016**, *9*, 1132–1141. [CrossRef]

25. Hazzouri, K.M.; Gros-Balthazard, M.; Flowers, J.M.; Copetti, D.; Lemansour, A.; Lebrun, M.; Masmoudi, K.; Ferrand, S.; Dhar, M.I.; Fresquez, Z.A.; et al. Genome-wide association mapping of date palm fruit traits. *Nat. Commun.* **2019**, *10*, 4680. [CrossRef]

26. Xiao, Y.; Xu, P.; Fan, H.; Baudouin, L.; Xia, W.; Bocs, S.; Xu, J.; Li, Q.; Guo, A.; Zhou, L.; et al. The genome draft of coconut (*Cocos nucifera*). *Gigascience* **2017**, *6*. [CrossRef]

27. Rival, A.; Beule, T.; Barre, P.; Hamon, S.; Duval, Y.; Noirot, M. Comparative flow cytometric estimation of nuclear DNA content in oil palm (*Elaeis guineensis* jacq) tissue cultures and seed-derived plants. *Plant Cell Rep.* **1997**, *16*, 884–887. [CrossRef] [PubMed]

28. Pellicer, J.; Leitch, I.J. The Plant DNA C-values database (release 7.1): An updated online repository of plant genome size data for comparative studies. *New Phytol.* **2020**, *226*, 301–305. [CrossRef]

29. Ong, A.L.; Teh, C.K.; Kwong, Q.B.; Tangaya, P.; Appleton, D.R.; Massawe, F.; Mayes, S. Linkage-based genome assembly improvement of oil palm (*Elaeis guineensis*). *Sci. Rep.* **2019**, *9*, 6619. [CrossRef]

30. Kianian, P.M.A.; Wang, M.; Simons, K.; Ghavami, F.; He, Y.; Dukowic-Schulze, S.; Sundararajan, A.; Sun, Q.; Pillardy, J.; Mudge, J.; et al. High-resolution crossover mapping reveals similarities and differences of male and female recombination in maize. *Nat. Commun.* **2018**, *9*, 2370. [CrossRef]

31. He, Z.; Zhang, Z.; Guo, W.; Zhang, Y.; Zhou, R.; Shi, S. De Novo Assembly of Coding Sequences of the Mangrove Palm (*Nypa fruticans*) Using RNA-Seq and Discovery of Whole-Genome Duplications in the Ancestor of Palms. *PLoS ONE* **2015**, *10*, e0145385. [CrossRef] [PubMed]

32. Singh, R.; Low, E.T.L.; Ooi, L.C.L.; Ong-Abdullah, M.; Ting, N.C.; Nagappan, J.; Nookiah, R.; Amiruddin, M.D.; Rosli, R.; Manaf, M.A.A.; et al. The oil palm SHELL gene controls oil yield and encodes a homologue of SEEDSTICK. *Nature* **2013**, *500*, 340–344. [CrossRef] [PubMed]

33. Torres, M.F.; Mathew, L.S.; Ahmed, I.; Al-Azwani, I.K.; Krueger, R.; Rivera-Nuñez, D.; Mohamoud, Y.A.; Clark, A.G.; Suhre, K.; Malek, J.A. Genus-wide sequencing supports a two-locus model for sex-determination in Phoenix. *Nat. Commun.* **2018**, *9*, 3969. [CrossRef] [PubMed]

34. Al-Dous, E.K.; George, B.; Al-Mahmoud, M.E.; Al-Jaber, M.Y.; Wang, H.; Salameh, Y.M.; Al-Azwani, E.K.; Chaluvadi, S.; Pontaroli, A.C.; Debarry, J.; et al. De novo genome sequencing and comparative genomics of date palm (*Phoenix dactylifera*). *Nat. Biotechnol.* **2011**, *29*, 521. [CrossRef] [PubMed]

35. Mathew, L.S.; Spannagl, M.; Al-Malki, A.; George, B.; Torres, M.F.; Al-Dous, E.K.; Al-Azwani, E.K.; Hussein, E.; Mathew, S.; Mayer, K.F.X.; et al. A first genetic map of date palm (*Phoenix dactylifera*) reveals long-range genome structure conservation in the palms. *BMC Genom.* **2014**, *15*, 285. [CrossRef]

36. Pamin, K. *A Hundred and Fifty Years of Oil Palm Development in Indonesia from the Bogor Botanical Garden to the Industry*; Medan IOPRI: Medan Maimun, Kota Medan, Indonesia, 1998.

37. Rajanaidu, N.; Kushairi, A.; Rafii, M.; Mohd, D.A.; Maizura, I.; Jalani, B.S. Oil palm breeding and genetic resources. *Adv. Oil Palm Res.* **2000**, *1*, 171–237.

38. Shen, C.; Li, X.; Zhang, R.; Lin, Z. Genome-wide recombination rate variation in a recombination map of cotton. *PLoS ONE* **2017**, *12*. [CrossRef] [PubMed]

39. Milo, R.; Phillips, R. *Cell Biology by the Numbers*; Garland Science: New York, NY, USA, 2016; pp. 334–338. [CrossRef]

40. Sidhu, D.; Gill, K.S. Distribution of genes and recombination in wheat and other eukaryotes. *Plant Cell. Tissue Organ Cult.* **2005**, *79*, 257–270. [CrossRef]

41. Jordan, K.W.; Wang, S.; He, F.; Chao, S.; Lun, Y.; Paux, E.; Sourdille, P.; Sherman, J.; Akhunova, A.; Blake, N.K.; et al. The genetic architecture of genome-wide recombination rate variation in allopolyploid wheat revealed by nested association mapping. *Plant J.* **2018**, *95*, 1039–1054. [CrossRef]

42. Teh, C.K.; Muaz, S.D.; Tangaya, P.; Fong, P.Y.; Ong, A.L.; Mayes, S.; Chew, F.T.; Kulaveerasingam, H.; Appleton, D. Characterizing haploinsufficiency of SHELL gene to improve fruit form prediction in introgressive hybrids of oil palm. *Sci. Rep.* **2017**, *7*, 3118. [CrossRef] [PubMed]

43. Corley, R.H.V.; Tinker, P.B. *The Oil Palm: Fifth Edition*; John Wiley&Sons: Hoboken, NJ, USA, 2015.

44. Hardon, J.J.; Corley, R.H.V.; Lee, C.H. Breeding and Selecting the Oil Palm. In *Improving Vegetatively Propagated Crops*; Academic Press: London, UK, 1987.

45. Bellow, S.A.; Agunsoye, J.O.; Adebisi, J.A.; Kolawole, F.O.; Hassan, S.B. Physical properties of coconut shell nanoparticles. *Kathmandu Univ. J. Sci. Eng. Technol.* **2018**, *12*, 63–79. [CrossRef]

46. Budi, E.; Umiatin, U.; Nasbey, H.; Bintoro, R.A.; Wulandari, F.; Erlina, E. Adsorption and Pore of Physical-Chemical Activated Coconut Shell Charcoal Carbon. In *IOP Conference Series: Materials Science and Engineering*; Management Science and Engineering: Stanford, CA, USA, 2018.

47. Mohd Iqbaldin, M.N.; Khudzir, I.; Mohd Azlan, M.I.; Zaidi, A.G.; Surani, B.; Zubri, Z. Properties of coconut shell activated carbon. *J. Trop. For. Sci.* **2013**, *25*, 497–503.

48. Tang, H.; Bowers, J.E.; Wang, X.; Ming, R.; Alam, M.; Paterson, A.H. Synteny and collinearity in plant genomes. *Science* **2008**, *320*, 486–488. [CrossRef]

49. Corley, R.H.V. Oil palm physiology-a review. In Proceedings of the International Oil Palm Conference, Kuala Lumpur, Malaysia, 16–18 November 1973.

50. Couvreur, T.L.P.; Forest, F.; Baker, W.J. Origin and global diversification patterns of tropical rain forests: Inferences from a complete genus-level phylogeny of palms. *BMC Biol.* **2011**, *9*, 44. [CrossRef]

51. Barrett, C.F.; Bacon, C.D.; Antonelli, A.; Cano, Á.; Hofmann, T. An introduction to plant phylogenomics with a focus on palms. *Bot. J. Linn. Soc.* **2016**, *182*, 234–255. [CrossRef]

52. Purcell, S.; Neale, B.; Todd-Brown, K.; Thomas, L.; Ferreira, M.A.R.; Bender, D.; Maller, J.; Sklar, P.; de Bakker, P.I.W.; Daly, M.J.; et al. PLINK: A Tool Set for Whole-Genome Association and Population-Based Linkage Analyses. *Am. J. Hum. Genet.* **2007**, *81*, 559–575. [CrossRef]

53. Rastas, P.; Paulin, L.; Hanski, I.; Lehtonen, R.; Auvinen, P.; Brudno, M. Lep-MAP: Fast and accurate linkage map construction for large SNP datasets. *Bioinformatics* **2013**, *29*, 3128–3134. [CrossRef] [PubMed]

54. Rastas, P. Lep-MAP3: Robust linkage mapping even for low-coverage whole genome sequencing data. *Bioinformatics* **2017**, *33*, 3726–3732. [CrossRef] [PubMed]

55. Altschul, S.F.; Gish, W.; Miller, W.; Myers, E.W.; Lipman, D.J. Basic Local Alignment Search Tool. *J. Mol. Biol.* **1990**, *215*, 403–410. [CrossRef]

56. Tang, H.; Zhang, X.; Miao, C.; Zhang, J.; Ming, R.; Schnable, J.C.; Schnable, P.S.; Lyons, E.; Lu, J. ALLMAPS: Robust scaffold ordering based on multiple maps. *Genome Biol.* **2015**, *16*, 3. [CrossRef]

57. Al-Mssallem, I.S.; Hu, S.; Zhang, X.; Lin, Q.; Liu, W.; Tan, J.; Yu, X.; Liu, J.; Pan, L.; Zhang, T.; et al. Genome sequence of the date palm Phoenix dactylifera L. *Nat. Commun.* **2013**, *4*, 2274. [CrossRef]

58. Kurtz, S.; Phillippy, A.; Delcher, A.L.; Smoot, M.; Shumway, M.; Antonescu, C.; Salzberg, S.L. Versatile and open software for comparing large genomes. *Genome Biol.* **2004**, *5*, R12. [CrossRef] [PubMed]

59. Marçais, G.; Delcher, A.L.; Phillippy, A.M.; Coston, R.; Salzberg, S.L.; Zimin, A. MUMmer4: A fast and versatile genome alignment system. *PLoS Comput. Biol.* **2018**, *14*, e1005944. [CrossRef]

60. Krzywinski, M.; Schein, J.; Birol, I.; Connors, J.; Gascoyne, R.; Horsman, D.; Jones, S.J.; Marra, M.A. Circos: An information aesthetic for comparative genomics. *Genome Res.* **2009**, *19*, 1639–1645. [CrossRef]

61. Clustalw, U.; To, C.; Multiple, D.O. ClustalW and ClustalX. *Options* **2003**, 1–22. [CrossRef]

62. Larkin, M.A.; Blackshields, G.; Brown, N.P.; Chenna, R.; Mcgettigan, P.A.; McWilliam, H.; Valentin, F.; Wallace, I.M.; Wilm, A.; Lopez, R.; et al. Clustal W and Clustal X version 2.0. *Bioinformatics* **2007**, *23*, 2947–2948. [CrossRef]

63. Rambaut, A. Fig Tree v. 1.4.4. 2018. Available online: http://tree.bio.ed.ac.uk/software/figtree/ (accessed on 12 June 2020).

Rice Breeding in Russia using Genetic Markers

Elena Dubina [1], Pavel Kostylev [2,*], Margarita Ruban [1], Sergey Lesnyak [1], Elena Krasnova [2] and Kirill Azarin [3]

[1] Federal Scientific Rice Centre, Belozerny, 3, 350921 Krasnodar, Russia; lenakrug1@rambler.ru (E.D.); arrri_kub@mail.ru (M.R.); Sergei_l.a@mail.ru (S.L.)

[2] Agrarian Research Center "Donskoy", Nauchny Gorodok, 3, 347740 Zernograd, Russia; krasnovaelena67@mail.ru

[3] Department of Genetics, Southern Federal University, 344006 Rostov-on-Don, Russia; azarinkv@sfedu.ru

* Correspondence: p-kostylev@mail.ru

Abstract: The article concentrates on studying tolerance to soil salinization, water flooding, and blast in Russian and Asian rice varieties, as well as hybrids of the second and third generations from their crossing in order to obtain sustainable paddy crops based on domestic varieties using DNA markers. Samples IR 52713-2B-8-2B-1-2, IR 74099-3R-3-3, and NSIC Rc 106 were used as donors of the *SalTol* tolerance gene. Varieties with the *Sub1A* locus were used as donors of the flood resistance gene: Br-11, CR-1009, Inbara-3, TDK-1, and Khan Dan. The lines C101-A-51 (Pi-2), C101-Lac (Pi-1, Pi-33), IR-58 (Pi-ta), and Moroberekan (Pi-b) were used to transfer blast resistance genes. Hybridization of the stress-sensitive domestic varieties Novator, Flagman, Virazh, and Boyarin with donor lines of the genes of interest was carried out. As a result of the studies carried out using molecular marking based on PCR in combination with traditional breeding, early-maturing rice lines with genes for resistance to salinity (*SalTol*) and flooding (*Sub1A*), suitable for cultivation in southern Russia, were obtained. Introgression and pyramiding of the blast resistance genes *Pi-1, Pi-2, Pi-33, Pi-ta*, and *Pi-b* into the genotypes of domestic rice varieties were carried out. DNA marker analysis revealed disease-resistant rice samples carrying 5 target genes in a homozygous state. The created rice varieties that carry the genes for blast resistance (Pentagen, Magnate, Pirouette, Argamac, Kapitan, and Lenaris) were submitted for state variety testing. The introduction of such varieties into production will allow us to avoid epiphytotic development of the disease, preserving the biological productivity of rice and obtaining environmentally friendly agricultural products.

Keywords: rice; salinity; submergence tolerance; blast; SSR markers; PCR analysis

1. Introduction

Rice (*Oryza sativa* L.) is the most important food crop for more than half of the world's population (China, Japan, India, Bangladesh, etc.). Biotic and abiotic stressors are the main obstacles to increasing global crop production and expanding rice production. It was found that only about 10% of the world's agricultural land is located in areas that do not suffer from stress factors [1].

Decreased rice yields in adverse climatic conditions threaten global food security. Genetic loci that ensure productivity in difficult conditions exist in the germplasm of cultivated plants, their wild relatives, and species adapted to extreme conditions [2].

One-fifth of the world's irrigated land (North Africa, Central and South-East Asia, etc.) is adversely affected by high soil salinity [3]. About 45 million hectares in the world are subject to soil salinization [4]. In the Russian Federation, rice is grown on an area of about 200 thousand hectares, of which about 80 thousand hectares are saline [5]. The decline in productivity on saline soils can be overcome by developing rice varieties tolerant to salinity and introducing them into agricultural production. Several non-allelic genes provide tolerance to salinity during ontogenesis [6]. The main locus of salt tolerance is *SalTol*, which was first identified in some rice varieties [7,8]. This locus is mapped on chromosome 1 and its main function is to control the balance of Na+/K+ ions in rice plants [9].

One of the serious abiotic stress factors for rice, which inhibits plant growth and affects crop yield, is prolonged submersion of plants under water, which often happens to large areas of land in the rice-growing regions of South-East Asia [5]. Rice dies if total flooding lasts more than two weeks. A negative effect on the growth and development of rice plants at this time is exerted by a lack of oxygen (O_2) and limited diffusion of carbon dioxide (CO_2). Lack of light due to turbid flood water in the rice paddies during this period limits the ability of plants to photosynthesize and can even lead to their death [4–6].

Scientists in Asia have found the *Sub1A* gene, which regulates the response of plant cells to ethylene and gibberellin, leading to restriction of carbohydrate intake and dormancy of shoots under water, which contributes to tolerance to immersion [10,11]. In Russia, this gene can be used to develop varieties resistant to a large layer of water during the germination phase, which will become an effective way to protect rice from weeds without herbicides. Most weeds die under water without oxygen, and rice can survive. To develop such varieties, it is necessary to combine in one genotype genes with increased energy of initial growth, the ability to anaerobic germination, resistance to prolonged flooding and lodging.

In all countries of the world, including Russia, blast is among the most dangerous fungal diseases of rice and causes large yield losses in the years of epiphytoty. The most effective way to protect rice without fungicides is to grow blast-resistant varieties. More than 50 genes of resistance to this pathogen are known: *Pi-1*, *Pi-2*, *Pi-33*, *Pi-b*, *Pi-ta*, *Pi-z*, etc. [12]. Combining several effective resistance genes with their contribution on the genetic basis of the best varieties is an effective breeding strategy for resistance to variable fungal pathogens. Lines with a combination of 3–5 resistance genes show an increase and broadening of the spectrum of blast resistance in comparison with lines with separate genes. A number of successful breeding programs have already been carried out abroad to develop blast-resistant rice varieties by the gene pyramiding method using marker breeding [13].

Resistance to various biotic and abiotic factors is one of the traits that are difficult to assess when the assessment of the breeding material is possible only in the presence of an appropriate stress factor. At present, during the breeding of agricultural plants for resistance, the splitting population obtained from the crossing of resistance sources with genotypes that have productivity is tested against a natural background, or artificial infection is carried out under controlled conditions. This procedure, although it gives excellent results, is quite lengthy and costly. In addition, there are always susceptible plants that have escaped damage [14].

The use of DNA markers allows us to speed up the assessment and conduct selection without phenotypic assessment, at an early stage, regardless of the external conditions. In recent years, great progress has been made in the development of molecular marking technologies and their application to control complex agronomic traits using marker breeding [15]. The technology of molecular marking of resistance loci makes it possible to quickly select plant forms with target genes without using provocative backgrounds [16]. The identification of molecular markers linked to genes of resistance to these factors facilitates breeding work. The use of DNA markers brings the breeding of agricultural plants to a qualitatively new level, making it possible to evaluate genotypes directly and not through phenotypic manifestations, which, ultimately, is realized in the accelerated development of varieties with a complex of valuable traits [14]. Therefore, it is relevant to develop new rice varieties by marking [17].

The purpose of the study was the development of initial rice material using DNA markers for breeding highly productive varieties resistant to biotic and abiotic environmental stress factors: soil salinity, prolonged flooding, and blast.

2. Materials and Methods

We used samples from the collection of the Institute of Agricultural Genetics (Vietnam) as donors of the transferred salt tolerance gene: IR 52713-2B-8-2B-1-2, IR 74099-3R-3-3, and NSIC Rc 106, which were crossed with the early–maturing Krasnodar variety Novator. These varieties carry the *SalTol* locus, which has been mapped near the centromeric region of the first chromosome. RM493 and RM7075 [18] were used as flanking SSR-markers of this locus, with the greatest difference in the length of microsatellite repeats between the parental forms.

Varieties with the *Sub1A* locus were used as donors of the flooding resistance gene: BR-11, CR-1009, Inbara-3, TDK-1, and Khan Dan. The early-ripening variety Novator and rice lines with the introgressed genes for blast resistance *Pi-2* and *Pi-33* were also taken as recipients. The *Sub1A* locus is mapped to an interval of 0.06 morganides in chromosome 9 [11]. We used microsatellite markers for the *Sub1A* gene, CR25K and SSR1A. The *Sub1A* gene was identified by molecular marking based on PCR using specific primers.

When transferring blast resistance genes, lines C101-A-51 (*Pi-2*), C101-Lac (*Pi-1, Pi-33*), IR-58 (*Pi-ta*), and Moroberekan (*Pi-b*) were used. To identify the Pi-1 gene, we used primer pairs of the flanking microsatellite SSR markers RM224 and RM144; for the Pi-2 gene, we used RM527 and SSR140; for the Pi-33 gene, RM310 and RM72; for the *Pi-b* and *Pi-ta* genes, intragenic markers developed in the laboratory of biotechnology, Federal Scientific Rice Centre. They are localized on chromosomes 11, 6, 8, 2, and 12, respectively (Table 1) [19,20].

Table 1. Nucleotide sequences of codominant markers for identification of the allelic status of genes Pi-1, Pi-2, Pi-a, and Pi-b.

Resistance Gene	Chromosomal Localization of Gene	Marker		Sequence
Pi-2	6	Rm 527	F	GGC TCG ATC TAG AAA ATC CG
			R	TTG CAC AGG TTG CGA TAG AG
		SSR140	F	AAG GTG TGA AAC AAG CTA GCA
			R	TTC TAG GGG AGG GGT GTA GAA
Pi-33	8	Rm 72	F	CCG GCG ATA AAA CAA TGA G
			R	GCA TCG GTC CTA ACT AAG GG
		Rm310	F	CCA AAA CAT TTA AAA TAT CATG
			R	GCT TGT TGG TCA TTA CCA TTC
Sub1A	9	Sub1A203	F	GAT GT GT GGAGGAGAAGT GA
			R	GGTAGAT GCCGAGAAGT GTA
		Rm 7481	F	CGACCCAATATCTTTCTGCC
			R	ATTGGTCGTGCTCAACAAG
SalTol	1	Rm 493	F	GTACGTAAACGCGGAAGGTGACG
			R	CGACGTACGAGATGCCGATCC
		Rm 7075	F	TATGGACTGGAGCAAACCTC
			R	GGCACAGCACCAATGTCTC

The early-ripening released rice varieties Boyarin, Flagman, and Virage served as the paternal form. During plant hybridization, pneumocastration of maternal forms and pollination by the Twell method were used [21]. Hybrid plants were grown on checks of Federal State Unitary Enterprise "Proletarskoe" (Rostov region) and the Federal State Unitary Elite Seed-growing Enterprise "Krasnoe" of the Federal Scientific Rice Centre, Krasnodar region. From the selected rice leaves, genomic DNA was isolated under laboratory conditions at the Federal Scientific Rice Centre, the Academy of Biology and Biotechnology of the Southern Federal University, and the All-Russian Research Institute of Agricultural Biotechnology. PCR products were separated by electrophoresis in 2.5% agarose and 8% acrylamide gels. The experimental data were statistically processed using Microsoft Excel and Statistica 6 software.

The account of the degree of damage to plants (in percentages) was carried out on the 14th day after inoculation, in accordance with the express method for assessing rice varietal resistance to blast. The assessment was carried out by taking two indicators into account: the type of reaction (in points) and the degree of damage (in percentages), using the ten-point scale of the International Rice Research Institute [12]:

- resistant: 0–1 point—no damage, small brown spots, covering less than 25% of the total leaf surface;
- medium resistant: 2–5 points—typical elliptical blast spots, 1–2 cm long, covering 26–50% of the total leaf surface;
- susceptible: 6–10 points—typical blast spots of elliptical shape, 1–2 cm long, covering 51% or more of the total leaf surface.

The intensity of disease development (IDD, %) was calculated by the formula (Equation (1)):

$$IDD = \sum (a \times b)/n \times 9 \qquad (1)$$

where IDD is the intensity of disease development (%), $\sum (a \times b)$ is the sum of the products of the number of infected plants multiplied by the corresponding damage point, and n is the number of recorded plants (pcs).

Depending on the damage points, all varieties wee conventionally divided into 4 groups: resistant, intermediate, susceptible, and strongly susceptible.

3. Results and Discussion

The development of blast-resistant varieties and their rapid introduction into production is the most promising solution in the fight against this disease. However, the development of resistant varieties is one of the most difficult areas of breeding. The causative agents of the disease have a great potential for variability, which, combined with its colossal reproduction capabilities, provides the pathogen with the highest adaptive capabilities. Combining several effective genes of resistance on a genetic basis of the best varieties widely used in production is an effective breeding strategy for long-term resistance to variable fungal pathogens.

Based on the use of DNA marker breeding (marker-assisted selection (MAS)—breeding with use of DNA markers towards genes of interest), we introduced 5 blast resistance genes into domestic rice varieties adapted to the agro-climatic conditions of rice cultivation in southern Russia.

A series of crosses made it possible to obtain rice lines based on the varieties Boyarin, Flagman and Virage with the introgressed and pyramided blast resistance genes *Pi-1*, *Pi-2*, *Pi-33*, *Pi-ta*, and *Pi-b* in a homozygous state. During all cycles of backcrossing, the transfer of the dominant alleles of each such gene in the offspring was controlled by closely linked molecular markers. Plants with no resistance alleles in the genotype were discarded.

At the first stage of work in 2005 at Agrarian research center "Donskoy", 6 hybrids were obtained from crossing the varieties Boyarin and Virage with three donors of blast resistance carrying the *Pi-l*, *Pi-2*, and *Pi-33* genes. After analysis at the Federal Scientific Rice Centre, homozygous forms were identified for the dominant alleles.

At the second stage of work (2008), after crossing the *Pi-1 + 33* × Boyarin and *Pi-2* × Boyarin hybrids between themselves, it was possible to obtain forms with three pyramided genes simultaneously: *Pi-1*, *Pi-2*, and *Pi-33* in a homozygous state.

At the third stage of work (2010), they were hybridized with varieties—donors of the *Pi-ta* and *Pi-b* genes—for combining 5 genes. There were two types of crosses: ((*Pi-1 + 2 + 33*) × *Pi-ta*) × *Pi-b* and *Pi-1 + 2 + 33* × (*Pi-ta* × *Pi-b*).

Leaves were selected from the best F_2 hybrid plants for DNA analysis at All-Russian Research Institute of Agricultural Biotechnology and the Federal Scientific Rice Centre using one marker for each gene. Based on the analysis results, it was possible to isolate two rice samples that were homozygous for all five dominant alleles. Reanalysis of the leaves of these samples confirmed last year's results, i.e., homozygosity for the dominant alleles of all five loci.

Figures 1 and 2 show the panicles of two lines, 1225/13 and 1396/13, which show the presence of dominant alleles at five loci in the homozygous state: *Pi-1*, *Pi-2*, *Pi-33*, *Pi-b*, and *Pi-ta*.

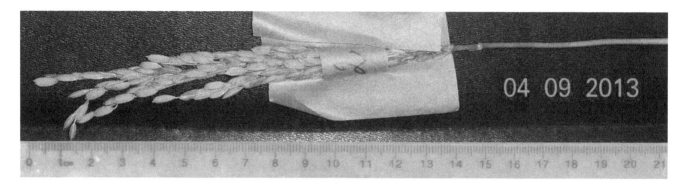

Figure 1. Panicle of the early-ripening line 1225/13.

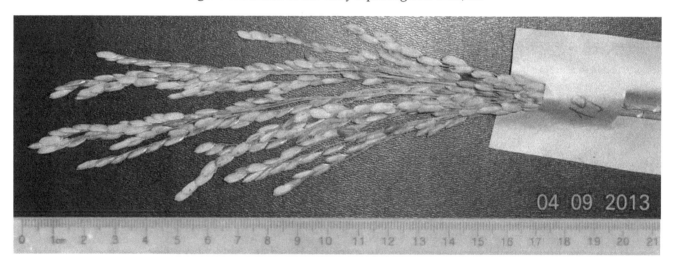

Figure 2. Panicle of the mid-ripening line 1396/13.

Line 1225/13 is early maturing, matures in 110 days, and dwarfish (80 cm), with a small panicle (15 cm) (Figure 1).

The second line 1396/13 is mid-ripening, the period to maturity is 120 days, and it is taller (100 cm), with a large long panicle (22 cm) (Figure 2).

Against the infectious background in the Federal Scientific Rice Centre, the index of disease development (IDD) in this line was only 1.4%, while the variety Novator was damaged by 67.7%. The results of the analysis made it possible to send these lines to the breeding nursery in 2014–2015 for testing for yield and blast resistance. The variety **Pentagen** (1396/13), carrying 5 genes for blast resistance, is widely used in hybridization with high-yielding Russian varieties.

In the process of work at the Federal Scientific Rice Centre in 2007–2008, crosses were carried out and F_1 hybrids were obtained from the combination (Flagman × C101-Lac) × (Flagman × C-101-A-51), which have the blast resistance genes *Pi-33* and *Pi-2* in their genotypes, respectively. The resulting F_1 generation was used in backcrosses with the recipient parental forms. It should be noted that the F_1 plants had a high degree of sterility (up to 95%). After the first series of backcrosses in 2008, the BC_1 and BC_2 generations were obtained in artificial climate chambers. In BC_1 populations, fertility increased and averaged about 50%. Starting from the first backcrossing, marker control was carried out for the presence of transferred donor alleles in the hybrid offspring. In 2009, plants of the BC_3 and BC_4 generation were obtained. Among these plants, we selected the forms with the shortest growing season and the highest panicle fertility. From the BC_4F_1 stage (the first self-pollination of rice plants, which makes it possible to transfer the donor allele to a homozygous state), individual selection was carried out. Segregation for Pi-2, Pi-33, and Sub1A genes fit into the Mendelian framework: in the second generation as a result of DNA analysis of the obtained plants, the ratio was 1:2:1 by genotype and 3:1 by phenotype.

Plants were selected that were closest in morphotype to the recipient parental form and had donor genes for resistance to the pathogen *Pyricularia oryzae* Cav. in their genotype in a homozygous state [22].

Figure 3 shows the results of PCR analysis for identification of the *Pi-33* blast resistance gene in the BC_4F_3 hybrid material.

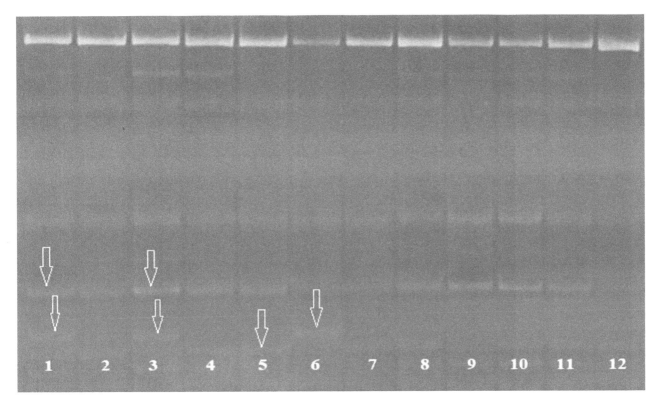

Figure 3. Electrophoregram of genomic DNA amplification products at the loci RM310 and RM72: 1–4, 7–12, analyzed hybrid BC_4F_3 plants; 5, donor line of the Pi-33 gene C101-Lac; Flagman, maternal form.

The figure shows that plants Number 2, 4, and 7–12 are homozygotes for the dominant allele; plants Number 1 and 3 are heterozygous. The size of the PCR product in varieties with the dominant allele of the Pi-33 gene, which determines the resistance, is 198 bp; in varieties with a recessive allele, it is 152 bp.

In 2015–2016, the resulting rice lines with introgressed blast resistance genes *Pi-2* and *Pi-33* were crossed with the variety Khan Dan (Vietnamese breeding): the donor of the *Sub1A* gene. This work was performed for obtaining breeding material with combined genes for disease resistance and tolerance to prolonged immersion of plants under water. In 2017–2020, F_4 and BC_2F_3 generations were obtained using climate chambers at the Federal Scientific Rice Centre (All-Russian Rice Research Institute, Krasnodar, Russia).

To increase economic efficiency and reduce labor costs, multi-primer systems have been developed to identify two genes (*Pi* and *Sub1A*) in a hybrid material simultaneously.

At the first stage, when we selected DNA markers for reliable interpretation of PCR products and identification of non-specifically amplified fragments, the following parameters were taken into account: the annealing temperature of specific pairs introduced into the reaction mixture, the difference in the size of PCR products synthesized during amplification with specific primer pairs (at least 100 base pairs), and the self-complementarity of the primer sequences.

The results of testing the combination of primer pairs flanking the marker regions of the *Pi-2 + Sub1A* genes are shown in Figure 4.

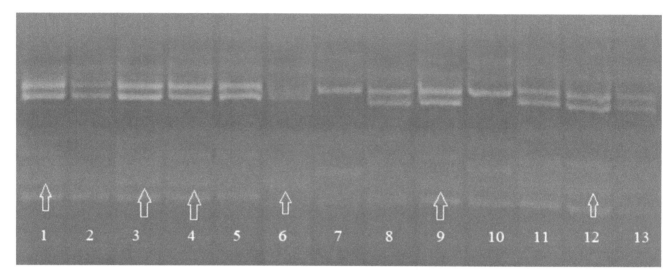

Figure 4. Electrophoregram of multiplex PCR of genomic DNA amplification products at the loci RM527 and SSR140 for the Pi-2 gene and at the Sub1A203 locus for the *Sub1A* gene: 1–5, 9–13, analyzed hybrid plants of the BC_2F_3 generation; 6, Khan Dan, donor of the *Sub1A* gene; 7, Flagman, maternal form; 8, C101Lac-A-51, donor line of the *Pi-2* gene.

The electrophoregram shows that when PCR with such a combination of molecular markers is carried out, the target products specific for DNA markers of the desired genes are reliably amplified. Samples Number 3 and 12 have dominant alleles of the genes *Pi-2* and *Sub1A* in a homozygous state in their genotype; Samples 1, 4, 5, and 9 are homozygous for the *Sub1A* gene and have the *Pi-2* gene in the genotype in a heterozygous state; Sample 10 is a recessive homozygote for two target genes and was rejected. The size of the PCR product in varieties with the dominant allele of the *Pi-2* gene, which determines the resistance, is 233 bp. The size of the PCR product in varieties with the dominant allele of the *Sub1A* gene, which determines the resistance, is 118 bp. Clear identification on the electrophoregram makes it possible to reliably identify the presence of dominant alleles of the target genes.

The introduction of such varieties into production will allow us to avoid epiphytic development of the disease, preserving the biological rice productivity, and obtaining environmentally friendly agricultural products.

Magnat is the first cultivar in Russia created at the Agrarian Research Center Donskoy together with the Federal Scientific Rice Centre by the method of marker selection from a hybrid population (C101A-51 × Boyarin) × (C101 LAC × Boyarin) with transfer of blast resistance genes. Sample C101 LAC is a donor of the genes *Pi-1* and *Pi-33*, and C101A-51 is a *Pi-2*. The growing season is 125 ± 1 days and the plant height is 96 ± 2 cm. The panicle is erect and compact, 17.5 ± 0.5 cm long, and bears 185 ± 5 spikelets. The grain is oval, 8.3 ± 0.2 mm long, 3.1 ± 0.1 mm wide, and 2.2 ± 0.1 mm thick and weighs 24.0 ± 2.0 mg. The yield of the Magnat variety was 8.25 t/ha, which is 1.1 t/ha higher than that of the Boyarin standard.

The rice variety **Pirouette** was bred at the Agrarian Research Center Donskoy, together with the Federal Scientific Rice Centre, by the method of stepwise hybridization and marker breeding from a hybrid population (C101-A-51 (*Pi-2*) × Boyarin) × (C101-Lac (*Pi-1 + 33*) × Virazh). It contains three blast resistance genes: *Pi-1*, *Pi-2*, and *Pi-33*. The variety is mid-ripening, the growing season from flooding to full ripeness is 124 ± 1 days. The average yield of the variety Pirouette was 9.57 t/ha, which is

1.13 t/ha higher than that of the standard variety Yuzhanin. Plant height is 88 ± 2 cm; the panicle is erect, compact, and 17.5 ± 0.5 cm long and carries 165 ⊥ 5 spikelets. The spikelets are oval, 8.9 ± 0.2 mm long, and 3.7 ± 0.1 mm wide. The weight of 1000 grains is 31.6 ± 2.0 g. The variety is resistant to lodging and shedding, is cold-tolerant, and germinates well from under a layer of water. It has been ncluded in the Register of Breeding Achievements of the Russian Federation for the North Caucasus region since 2020.

The rice variety **Kapitan** was bred at the Agrarian Research Center Donskoy in cooperation with the Federal Scientific Rice Centre by the method of triple backcrossing and marker breeding from the Flagman × IR-36 hybrid population. The variety is mid-ripening and the growing season from the flooding to full ripeness is 120 ± 1 days. On average, over the years of competitive testing, the yield of the variety Kapitan was 8.13 t/ha, which is 0.64 t/ha higher than that of the variety Yuzhanin. A higher yield of this variety is formed due to more grain in the panicle and an increased weight of the caryopsis. The average height of plants is 90 ± 2 cm; the panicle is semi-inclined, compact, and 18.5 ± 0.5 cm long; and the average number of spikelets is 140 ± 10 pieces (Figure 5). The grains are oval, 9.5 ± 0.2 mm long, and 3.6 ± 0.1 mm wide. The average weight of 1000 grains is 35.0 ± 2.1 g. The variety carries the *Pi-ta* gene and is resistant to blast, lodging, and shedding. The variety has been under state testing since 2019.

Figure 5. Rice panicles of the variety Kapitan.

The rice variety **Argamak** was bred at the Agrarian Research Center Donskoy by individual multiple selection of plants with the largest panicles from a hybrid population Il. 14 (*Pi-1*, *Pi-2*, *Pi-33*) × Kuboyar. The variety belongs to the mid-ripening group, and the growing season from flooding to full ripeness is 119 days. On average, over the years of competitive testing (2017–2019), the yield of the variety was 8.79 t/ha, which is 1.59 t/ha higher than that of the variety Yuzhanin. The maximum yield was formed in 2019: 10.1 t/ha, 2.55 more than the standard. The high yield of this variety was formed due to the greater grain content of the panicle than that of the standard and the increased density of the stem. Plant height is 93 ± 2 cm on average; the panicle is erect, compact, and 16 ± 0.5 cm long; the average number of spikelets is 142 ± 6 pieces. The grains are oval,

8.4 ± 0.2 mm long, 3.3 ± 0.1 mm wide. Weight of 1000 grains—31.1 ± 1.9 g. The variety is resistant to blast, lodging, and shedding. It has been tested at state varietal testing since 2020.

The rice variety **Lenaris** (Federal Scientific Rice Centre) had shown high adaptability, non-lodging, and the possibility for straight combine harvesting. Its yield was 10.6 t/ha. Plants had high spikelet fertility and short stems (77 ± 5 cm) and were resistant to the Krasnodar population of *P. oryzae* as well. Their panicle is slightly drooping and compact; its length is 18 ± 1.0 cm. The mass of 1000 grains is about 30.4 ± 1.8 g.

In 2013–2014, the Agrarian Research Center Donskoy conducted crosses and obtained F_1–F_2 hybrids of the variety Novator with Asian donor rice varieties carrying the *SalTol* and *Sub1A* genes. The hybrids of the second generation varied significantly in terms of quantitative traits: growing season (from early ripening to non-flowering), plant height (75–122 cm), panicle length (14–25 cm), number of filled grains (80–206 pcs), number of spikelets (99–300 pcs), panicle density (4.4–16.6 pcs/cm), 1000-grain weight (26.3–34.9 g), grain weight per panicle (2.1–5.5 g), etc.

Hybridization of the salt-sensitive domestic variety Novator with the lines IR52713-2B-8-2B-1-2, IR74099-3R-3-3, and IR61920-3B-22-2-1 (NSIC Rc 106)—*SalTol* locus donors—was carried out. The first generation of hybrids was used to generate an F_2 hybrid population. From the populations of plants of the second generation, 90 early-ripening samples with well-ripened grains (30 in each combination of crossing) were selected, which were analyzed by PCR for the presence of introduced *SalTol* alleles. As an example, Figure 2 shows the data of electrophoretic analysis of PCR products with the Rm493 marker. The donor allele of the parental line NSIC Rc 106, designated as 2.2, was found in a homozygous state in Sample 282. The rest of the plants, whose amplification spectra are presented in this electrophoregram, carried the alleles of the donor and the variety Novator; that is, they were heterozygous for the *SalTol* locus (Figure 2). Similar results were obtained during DNA analysis of the studied rice samples with the RM7075 marker (Figures 6 and 7).

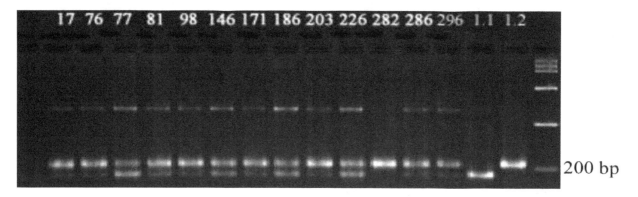

Figure 6. Electrophoregram of genomic DNA amplification products with RM 493: 1.1, Novator; 1.2, NSIC Rc 106; 17–296, hybrid plants NSIC Rc 106 × Novator; DNA marker (100–1500 bp).

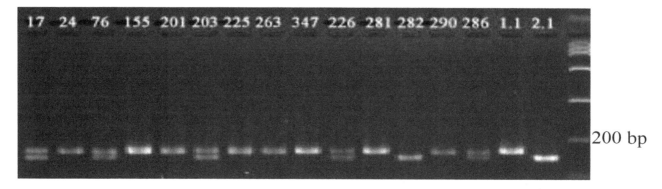

Figure 7. Electrophoregram of genomic DNA amplification products with RM 7075: 1.1, Novator; 2.1, NSIC Rc 106; 17–286, hybrid plants NSIC Rc 106 × Novator; DNA marker (100–1500 bp).

In general, according to the results of DNA analysis of F_2 hybrids, 17 plants homozygous for the dominant allele of the *SalTol* locus were identified, 29 samples carried *SalTol* in a heterozygous state, and 44 plants showed only recessive alleles inherited from the variety Novator.

Segregation for *SalTol* genes did not fit into the Mendelian framework, since the sample was unrepresentative due to selection. Plants with recessive alleles of the gene prevailed, and the number of salt-tolerant dominant homozygotes was less than the expected number. This is due to the linkage of *SalTol* genes with genes unfavorable for plants in our conditions: photosensitivity, late maturity, spikelet shedding, and spinosity.

Testing plants under salinity in the early stages of development is a quick, common method based on simple criteria. It was shown that at the initial vegetation stage, the length of the root and shoots and seed germination are potential indicators of resistance to the effects of increased salt concentrations [18,19]. Evaluation of the potential salt tolerance of the studied rice hybrids and their parental forms revealed significant variations in salinity tolerance depending on the genotype. The greatest decrease in seed germination—52%—was found in the salt-sensitive variety Novator. The line NSIC Rc 106 and second-generation plants, which were homo- and heterozygous for the *SalTol* locus, showed the highest resistance by seed germination (germination decrease of less than 5%). The donor lines IR52713-2B-8-2B-1-2, and IR74099-3R-3-3 and hybrid combinations obtained on their basis with the *SalTol* gene in a homozygous state also showed high resistance for this trait.

The least suppression of growth indices, as well as in the case of seed germination, was recorded in the lines NSIC Rc 106, IR52713-2B-8-2B-1-2, IR74099-3R-3-3, and *SalTol* homozygous plants from the F_2 generation; the greatest decrease in the length of shoots and roots under salt stress was shown in the variety Novator and in hybrid plants that did not inherit the *SalTol* locus according to molecular analysis data. Thus, DNA analysis made it possible to simplify the breeding scheme and obtain salt-tolerant F_2 hybrids carrying the *SalTol* locus in a homozygous state. These results indicate that the developed codominant markers of the *SalTol* locus RM 493 and RM 7075 are an effective tool for marker-assisted selection of salt-tolerant forms based on domestic rice genotypes.

Rice samples with *SalTol* genes in 2018–2020 were studied in a control nursery and in competitive variety testing; productive forms were identified.

At the same time, in 2013, hybrids were obtained by crossing the variety Novator with donors of the *Sub1A* gene. The Asian varieties turned out to be late-ripening and photosensitive and did not flower under our conditions. Hybridization was carried out only with the help of artificial climate chambers. The first generation in 2013 was characterized by a high degree of sterility (90–95%) and brown color of the flowering scales during maturation, which indicates significant genetic differences between the parental forms. In the second generation in 2014, a very large spectrum of splitting was observed in terms of the growing season, plant height, panicle length and shape, number of spikelets, and spinosity (Table 2).

Table 2. Variations in the quantitative traits in F_2 hybrids from crossing submergence-resistant samples with the variety Novator, 2014.

Trait	Crossing Combination			
	BR 11 × Novator	CR-1009 × Novator	Inbara 3 × Novator	TDK-1 × Novator
Plant height, cm	71–129 (97.5)	57–131 (89.4)	60–149 (100.2)	45–138 (99.9)
Panicle length, cm	11.5–27 (18.4)	10–26 (17.7)	9.5–32 (19.1)	9–27 (18.9)
Number of grains, pcs	10–220 (77.1)	2–201 (50.3)	4–343 (60.2)	4–180 (55.3)
Number of spikelets, pcs	57–322 (174.6)	38–273 (133.8)	18–411 (137.0)	17–261 (122.2)
Spikelet length, mm	6.1–10.1 (8.1)	6.8–9.8 (8.0)	6.1–11.9 (9.1)	7.2–11.3 (9.3)
Spikelet width, mm	2.3–3.8 (3.1)	2.5–4.0 (3.1)	2.1–3.8 (2.9)	2.3–3.9 (3.0)
Mass of 1000 grains, g	11–38 (25.4)	10–35 (23.2)	12–37 (25.9)	13–39 (25.8)
Mass of grain from the panicle, g	0.72–5.54 (1.98)	0.03–5.90 (1.22)	0.06–5.42 (1.54)	0.07–6.09 (1.79)

Note: The average value is indicated in brackets.

This wide range of variability is not observed in other crops. This is due to the genetic and ecological-geographical remoteness of the crossed forms. In each combination, about 400 plants were selected for morphometric and genetic analysis. Among the F_2 hybrids, we managed to select the best plants according to many traits, combining early maturity, optimal plant height, good grain size in panicles, non-shattering, and fertility of spikelets (Table 3).

Table 3. Selected F_2 hybrid plants from crossing submergence-resistant samples with Novator, 2014.

Hybrid	Duration, Days	Plant Height, cm	Panicle Length, cm	Number of Grains in Panicle, pcs	Mass of 1000 Grains, g
Novator, st	112	97.5	16.5	110	31.8
176(BxN) *	120	108.0	21.5	146	27.0
334 (BxN)	118	96.7	17.3	145	21.6
34 (CxN)	121	81.2	16.5	109	25.3
390 (CxN)	122	82.0	17.5	122	27.7
273 (IxN)	120	95.4	15.1	151	24.4
507 (IxN)	119	97.2	19.0	138	29.3
81 (TxN)	123	96.5	17.2	159	32.0
393 (TxN)	121	97.3	14.9	152	25.2

Note *: (BxN), BR 11 × Novator; (CxN), CR-1009 × Novator; (IxN, Inbara 3 × Novator; (TxN), TDK-1 × Novator.

PCR analysis of leaves was carried out in 20 plants of each of the four hybrids, as a result of which, forms with the *Sub1A* flood resistance gene were isolated. The electrophoretic analysis of PCR products with the RM 7481 marker is shown in Figure 8. The donor allele of the parental line CR-1009 was homozygous in Samples 2, 3, 5, 9, 13 and 17. Plants 2, 4, 6–8, 10, 11, 16, 18, and 19 were heterozygous at the *Sub1A* locus; that is, they carried both the alleles of the donor and the recessive alleles inherited from the variety Novator. Thus, according to the results of PCR analysis with the RM 7481 marker, 14 homozygotes of F_2 plants at the *Sub1A* locus were identified, 40 samples carried *Sub1A* in a heterozygous state, and 22 plants inherited only the recessive allele from the variety Novator.

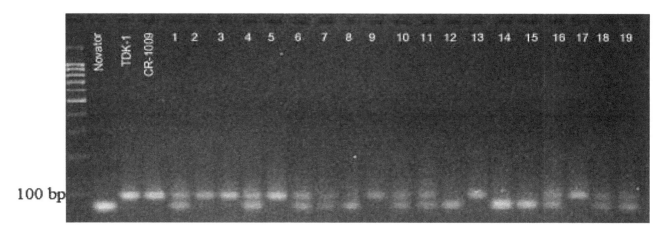

Figure 8. Electrophoregram of the amplification products of rice genomic DNA with the primer RM 4781. 1–19, F_2 (Novator × CR-1009); TDK-1 and CR-1009, donor of the *Sub1A* gene. Molecular weight marker, 1 kb.

Of the analyzed BR-11 × Novator hybrid plants, the Sub1A gene (in homo- and heterozygous state) was present in nine, i.e., in a ratio of 9:11, although with monohybrid segregation, it should have been 15:5. In the hybrid combination CR-1009 × Novator, F_2 segregated in a ratio of 18:2, i.e., almost all of the selected plants had the *Sub1A* gene. In the hybrids Inbara-3 × Novator and TDK-1 × Novator, segregation took place in a ratio of 14:6 or approximately 3:1, i.e., close to Mendelian.

The deviations in segregation of the two combinations can be explained by the influence of selection and gene linkage. A total of 55 plants with the target gene in the homo- and heterozygous state were isolated from 80 plants of four hybrids. The selected samples with the $Sub1A$ gene in 2015 were reproduced in the Federal State Unitary Enterprise "Proletarskoye"of the Rostov Region, where the best F_3 plants were selected from them for DNA analysis.

In F_3 plants, significant morphological and biological segregation continued. Significant variation was noted for the growing season, plant height, size of panicles and caryopses, fertility, grain shedding, etc. The best forms were selected from them and leaves were taken for DNA analysis. At the next stage of work, in 2016–2020, constant lines carrying the $Sub1A$ gene in a homozygous state were selected and tested for yield and resistance to prolonged water flooding. As a result, rice varieties for herbicide-free technologies will be developed, vigorously overcoming a deep layer of water in the germination phase with minimal seed loss.

4. Conclusions

1. As a result of the studies carried out using molecular marking based on PCR in combination with traditional breeding, early-maturing rice lines with genes for resistance to salinization ($SalTol$) and to flooding ($Sub1A$), which are suitable for cultivation in the south of Russia, were isolated.

2. Rice lines have been developed, the genotype of which contains five effective blast resistance genes ($Pi-1$, $Pi-2$, $Pi-33$, $Pi-ta$, and $Pi-b$). The introduction of such varieties into production will allow us to avoid epiphytotic development of the disease, preserving the biological productivity of rice and obtaining environmentally friendly agricultural products.

3. Samples of the F_4 and BC_2F_3 generations were obtained with combined blast resistance (Pi) and prolonged flooding tolerance ($Sub1A$) genes as a factor in the control of weeds in the homo- and heterozygous state, which was confirmed by the data of their DNA analysis. The testing of the obtained rice breeding resources for submergence tolerance under laboratory conditions made it possible to select tolerant rice forms that will be studied in the breeding process for a complex of agronomically valuable traits. Their use will reduce the use of chemical plant protection products against diseases and weeds, thereby increasing the ecological status of the rice-growing industry.

The research was carried out with the financial support of the Kuban Science Foundation in the framework of the scientific project № 20.1/1.

Author Contributions: Conceptualization, P.K.; methodology, E.D. and K.A.; validation, P.K.; formal analysis, E.D.; investigation, E.D., P.K., M.R., S.L. and E.K.; resources, E.D.; writing—original draft preparation, E.D., P.K. and K.A.; project administration, E.D.; funding acquisition, E.D. All authors have read and agreed to the published version of the manuscript.

References

1. Kumar, V.; Bhagwat, S.G. Microsatellite (SSR) Based Assessment of Genetic Diversity among the Semi-dwarf Mutants of Elite Rice Variety WL112. *Int. J. Plant Breed. Genet.* **2012**, *6*, 195–205. [CrossRef]
2. Singh, S.; Mackill, D.; Ismail, A. Responses of Sub1 rice introgression lines to submergence in the field: Yield and grain quality. *Field Crops Res.* **2009**, *113*, 23. [CrossRef]
3. Negrao, S.; Courtois, B.; Ahmadi, N.; Abreu, I.; Saibo, N.; Oliveira, M.M. Recent updates on salinity stress in rice: From physiological to molecular responses. *Crit. Rev. Plant Sci.* **2011**, *30*, 329–377. [CrossRef]
4. Food and Agriculture Organization. Report of Salt Affected Agriculture. Available online: http://www.fao.org/ag/agl/agll/spush (accessed on 16 October 2010).
5. Ladatko, N.A. The growth of rice plants under conditions of chloride salinity. *Rice Breed.* **2006**, *9*, 37–41.
6. Mekawy, A.M.; Assaha, D.V.; Yahagi, H.; Tada, Y.; Ueda, A.; Saneoka, H. Growth, physiological adaptation, and gene expression analysis of two Egyptian rice cultivars under salt stress. *Plant Physiol. Biochem.* **2011**, *187*, 17–25. [CrossRef] [PubMed]

7.   Ren, Z.H.; Gao, J.P.; Li, L.G.; Cai, X.L.; Huang, W.; Chao, D.Y.; Zhu, M.Z.; Wang, Z.Y.; Luan, S.; Lin, H.X. A rice quantitative trait locus for salt tolerance encodes a so-dium transporter. *Nat. Genet.* **2005**, *37*, 1141–1146. [CrossRef] [PubMed]

8.   Takehisa, H.; Shimodate, T.; Fukuta, Y.; Ueda, T.; Yano, M.; Yamaya, T.; Kameya, T.; Sato, T. Identification of quantitative trait loci for plant growth of rice in paddy field flooded with salt water. *Field Crops Res.* **2004**, *89*, 85–95. [CrossRef]

9.   Platten, J.D.; Egdane, J.A.; Ismail, A.M. Salinity tolerance, Na$^+$ exclusion and allele mining of HKT1; 5 in *Oryza sativa* and *O. glaberrima*: Many sources, many genes, one mechanism? *BMC Plant Biol.* **2013**, *13*, 2–16. [CrossRef] [PubMed]

10.   Xu, K.; Mackill, D.J. A major locus for submergence tolerance mapped on rice chromosome 9. *Mol. Breed.* **1996**, *2*, 219–224. [CrossRef]

11.   Xu, K.; Xu, X.; Fukao, T.; Canlas, P.; Maghirang-Rodriguez, R.; Heuer, S.; Ismail, A.M.; Bailey-Serres, J.; Ronald, P.C.; Mackill, D.J. Sub1A is an ethylene-response-factor like gene that confers submergence tolerance to rice. *Nature* **2006**, *442*, 705–708. [CrossRef] [PubMed]

12.   Kovalenko, E.D.; Gorbunov, Y.B.; Kovalyova, A.A.; Dudenko, V.P.; Kolomiec, T.M. *Guidelines for Assessing the Resistance of Rice Varieties to the Blast Causative Agent*; VASKHNIL: Moscow, Russia, 1988; 30p.

13.   Choi, H.C.; Kim, Y.G.; Hong, H.C.; Hwang, H.G.; Ahn, S.N.; Han, S.S.; Ryu, J.D.; Moon, H.P. Development of blast-resistant rice multi-line cultivars and their stability to blast resistance and yield performance. *Korean J. Breed.* **2006**, *38*, 83–89.

14.   Mukhina, Z.M. *Using DNA Markers to Study the Genetic Diversity of Plant Resources: Monographic*; Education-South: Krasnodar, Russian, 2008; p. 84.

15.   Singh, D.; Kumar, A.; Kumar, A.S.; Chauhan, P.; Kumar, V.; Kumar, N.; Singh, A.; Mahajan, N.; Sirohi, P.; Chand, S.; et al. Marker assisted selection and crop management for salt tolerance: A review. *Afr. J. Biotechnol.* **2011**, *10*, 14694–14698. [CrossRef]

16.   Sabouri, H.; Rezai, A.M.; Moumeni, A.; Kavousi, A.; Katouzi, M.; Sabouri, A. QTLs mapping of physiological traits related to salt tolerance in young rice seedlings. *Biol. Plant* **2009**, *53*, 657–662. [CrossRef]

17.   Kostylev, P.I.; Redkin, A.A.; Krasnova, E.V.; Dubina, E.V.; Ilnitskaya, E.T.; Yesaulova, L.V.; Mukhina, Z.M.; Kharitonov, E.M. Creating rice varieties, resistant to pirikulariosis by means of DNA markers. *Vestn. Russ. Acad. Agric. Sci.* **2014**, *1*, 26–28.

18.   Linh, L.H.; Xuan, T.D.; Ham, L.H.; Ismail, A.M.; Khanh, T.D. Molecular Breeding to Improve Salt Tolerance of Rice (*Oryza sativa* L.) in the Red River Delta of Vietnam. *Int. J. Plant Genom.* **2012**, *2012*. [CrossRef] [PubMed]

19.   Fukuta, Y.; Xu, D.; Kobayashi, N. Genetic characterization of universal differential varieties for blast resistance developed under the IRRI-Japan Collaborative Reseaarch Project using DNA markers in rice (*Oryza sativa* L.). *JIRCAS Work. Rep.* **2009**, *63*, 35–68.

20.   Kostylev, P.I. *Methods of Breeding, Seed Farming Techniques and High-Quality Rice: Monographic*; Rostov-on-Don: Zernograd, Russia, 2011; 267p.

21.   Dubina, E.; Kostylev, P.; Garkusha, S.; Ruban, M.; Pischenko, D. Marker assisted rice breeding for resistance to biotic and abiotic stressors. In Proceedings of the BIO Web of Conference, XI International Scientific and Practical Conference "Biological Plant Protection is the Basis of Agroecosystems Stabilization", Krasnodar, Russia, 21–24 September 2020. [CrossRef]

22.   Dubina, E.; Mukhina, Z.; Kharitonov, E.; Shilovskiy, V.; Kharchenko, E.; Esaulova, L.; Korkina, N.; Maximenko, E.; Nikitina, I. Creation of blast disease-resistant rice varieties with modern DNA-markers. *Russ. J. Gen.* **2015**, *8*, 752–756. [CrossRef]

3

SNP- and Haplotype-Based GWAS of Flowering-Related Traits in Maize with Network-Assisted Gene Prioritization

Carlos Maldonado [1], **Freddy Mora** [1,*], **Filipe Augusto Bengosi Bertagna** [2], **Maurício Carlos Kuki** [2] and **Carlos Alberto Scapim** [3]

[1]   Institute of Biological Sciences, University of Talca, 2 Norte 685, Talca 3460000, Chile; camaldonado@utalca.cl
[2]   Genetic and Plant Breeding Post-Graduate Program, Universidade Estadual de Maringá, Maringá PR 87020-900, Brazil; filipeabbertagna@gmail.com (F.A.B.B.); mcarloskuki@gmail.com (M.C.K.)
[3]   Departamento de Agronomia, Universidade Estadual de Maringá, Maringá PR 87020-900, Brazil; cascapim@uem.br
*   Correspondence: fmora@utalca.cl

Abstract: Maize (*Zea mays* L.) is one of the most crucial crops for global food security worldwide. For this reason, many efforts have been undertaken to address the efficient utilization of germplasm collections. In this study, 322 inbred lines were used to link genotypic variations (53,403 haplotype blocks (HBs) and 290,973 single nucleotide polymorphisms (SNPs)) to corresponding differences in flowering-related traits in two locations in Southern Brazil. Additionally, network-assisted gene prioritization (NAGP) was applied in order to better understand the genetic basis of flowering-related traits in tropical maize. According to the results, the linkage disequilibrium (LD) decayed rapidly within 3 kb, with a cut-off value of $r^2 = 0.11$. Total values of 45 and 44 marker-trait associations (SNPs and HBs, respectively) were identified. Another important finding was the identification of HBs, explaining more than 10% of the total variation. NAGP identified 44, 22, and 34 genes that are related to female/male flowering time and anthesis-silking interval, respectively. The co-functional network approach identified four genes directly related to female flowering time ($p < 0.0001$): *GRMZM2G013398*, *GRMZM2G021614*, *GRMZM2G152689*, and *GRMZM2G117057*. NAGP provided new insights into the genetic architecture and mechanisms underlying flowering-related traits in tropical maize.

Keywords: gene prioritization; linkage disequilibrium; marker-trait association; tropical maize

1. Introduction

Maize (*Zea mays* L.) plays an important role in the human diet and accounts for a large proportion of the global cereal demand. Together with rice and wheat, these three cereals account for more than 40% and 35% of the world's calorie and protein supply, respectively [1,2]. Maize is among the few crops grown on almost every continent and has diverse uses, including food, animal feed, and ethanol production [3]. The United States, China, and Brazil are the top three largest maize-producing countries in the world, representing more than 70% of total maize production [4].

Since maize is one of the most important crops for global food security, several efforts have been undertaken addressing the efficient utilization of germplasm materials. In fact, the development of maize germplasm collections has been beneficial to capture and maintain the high levels of genetic diversity that exist locally and globally [5–9]. These efforts have allowed the methodical exploration of the genetic architecture of complex traits in maize, which benefit from the high diversity [8]. Liu [10], for instance, performed a genome-wide association study (GWAS; a standard forward genetic technique) using a population comprised of a global core collection of maize inbred lines, and found several candidate genes associated with starch synthesis, of which one gene (*Glucose-1-phosphate adenylyltransferase*) is known as an important regulator of kernel starch content. Li et al. [11] identified several genetic variants associated with maize flowering time using an extremely large multigenetic background population (>8000 maize lines). The associated single nucleotide polymorphisms (SNPs) detected in this large panel exhibited high accuracy for predicting flowering time.

In an effort to overcome certain limitations present in forward and reverse genetic techniques, for example lacking in functional clues of trait-associated candidate genes derived from forward genetics studies and in silico strategies for candidate gene selection in targeted mutagenesis in reverse genetics approaches, Lee et al. [12] recently presented a network-assisted gene prioritization system (MaizeNet), which facilitates genetic analysis through supporting candidate genes based on network neighbors with known traits or functions, and aids in identifying potential candidate genes that are highly likely to be causal to the phenotype of interest. This network-based resource provides new insights into the genetic architecture and mechanisms underlying complex traits in maize and promises to accelerate the discovery of trait-associated genes for crop improvement. In this study, an integrated approach using GWAS (based on 53,403 haplotype blocks (HBs) and 290,973 SNPs) and network-assisted gene prioritization was applied in order to better understand the genetic basis of flowering-related traits in tropical maize. To this end, marker-trait association analyses were performed using a multigenetic background population comprising 322 inbred lines of field corn, popcorn, and sweet corn.

2. Materials and Methods

2.1. Trial Conditions and Phenotyping

A total of 322 inbred lines of tropical maize were used in this genome-wide association study, which were derived from three genetic backgrounds collected in Brazil: Field corn (178), popcorn (128), and sweet corn (16). This maize panel was evaluated during the growing season of 2017–2018 in two locations (Cambira and Sabaudia) situated in Southern Brazil, Parana State. The experimental design was an alpha-lattice with 24 incomplete blocks and 3 replications per line. Female and male flowering time (FF and MF, respectively) were measured in each line as the number of days from sowing to anther extrusion from the tassel glumes (MF) or to visible silks (FF). Additionally, the anthesis-silking interval (ASI) was calculated as the difference between MF and FF.

2.2. Population Structure, Linkage Disequilibrium (LD), and Haplotype Blocks

Genomic DNA was isolated from young leaves of five plants from each inbred line of tropical maize (319 in Cambira and 293 in Sabaudia), approximately 30 days after germination. The DNA extraction was carried out by Cetyl trimethyl ammonium bromide (CTAB) according to the protocol established by Chen and Ronald [13]. The quality of DNA was evaluated and quantified using 1% agarose gel and Nanodrop, respectively. The DNA samples were sent to the University of Wisconsin-Madison—Biotechnology Center for SNP discovery via genotyping by sequencing (GBS), which is described in Elshire et al. [14] and Glaubitz et al. [15]. The raw database was filtered considering a minor allele frequency (MAF) > 0.05, resulting in a genotype file of 291,633 high-quality SNPs. The LD kNNi imputation (linkage disequilibrium k-nearest neighbor imputation) was performed to impute missing data in the dataset [16]. Finally, SNPs with a MAF < 0.01 and a proportion of missing data per location >90% were eliminated from the imputed dataset [17]. Subsequently, 290,973 SNPs were retained after filtering for MAF and missing data.

The population structure was inferred using the model-based Bayesian clustering approach implemented in the program InStruct [18]. For each K value (where K is the number of genetically differentiated groups, K = 1–6), 10 runs were performed separately, each with 100,000 Monte Carlo Markov Chain replicates and a burn-in period of 10,000 iterations. The optimal K value was determined with the highest ΔK method [19] and the lowest deviance information criterion (DIC).

The extent of LD was estimated using the correlation coefficients of the allelic frequencies (r^2) considering all the possible combinations of the alleles. The critical r^2 value was calculated according to the method used by Breseghello and Sorells [20].

The HBs were constructed for each chromosome according to the confidence interval algorithm developed by Gabriel [21], implemented in the software Haploview v.4.2 [22]. This method considers the 95% confidence intervals of the disequilibrium coefficient (D′) values and builds a haplotype block if the LD is classified as a "strong LD" type (D′ higher than 0.98 and lower interval limit of >0.7). Finally, HBs were later transformed into multiallelic markers, considering the allelic combinations within each block to be independent alleles [5,23].

2.3. SNP- and Haplotype-Based GWAS

The HB- and SNP-based association analyses were performed using a mixed linear model (MLM) in TASSEL 3.0 and TASSEL 5.2, respectively [24], which considers the effects of the population structure (Q) and genetic relationships or matrix kinship (K) among inbred lines. The kinship matrix was calculated based on identity by state (IBS) [25] in TASSEL. The Adjusted-Entry Means of the general linear model (experimental design) were used as the adjusted phenotypes according to Contreras-Soto et al. [26] and Arriagada et al. [27]. Correlations between each pair of traits were calculated using a Bayesian bi-trait model [28–30]. The statistical analysis was performed using the R package MCMCglmm.

2.4. Prioritization of GWAS Candidate Genes and Inference of Co-Functional Networks for Flowering Traits in Maize

The candidate genes were chosen from the genes around the significant loci (SNP or haplotype blocks) identified by GWAS. To this end, a window (or threshold) of twice the distance indicated by the LD analysis was established, placing the marker in the center of the window. The gene prioritization was performed using MaizeNet [12] based on the connections of the candidate genes to the genes in one estimated network with previously associated genes with flowering-time in *Zea mays*. New candidate genes were then ranked by closeness to the "guide genes" (derived from estimated network in MaizeNet) measured for each candidate gene (derived from GWAS) as the sum of network edge scores from that gene to the guide genes [12]. The estimated co-functional network was carried out through the association of genes (candidate genes and genes identified in prioritization of MaizeNet) with subnetworks enriched by gene ontology annotations related to the biological processes (GOBP) of flowering in MaizeNet. Finally, the given genes are related to the flowering-time if the subnetworks of MaizeNet significantly associated with these genes, and if are also enriched for on the relevant GOBP term for flowering.

3. Results and Discussion

3.1. Genetic Structure

The Bayesian clustering analysis (InStruct) of the population structure indicated that the 322 inbred lines from the Brazilian germplasm represent two main genetic clusters (k = 2; Figure S1A), inferred from both the lowest DIC value and the second-order change rate of the probability function with respect to Q (ΔQ) [19]. Cluster I contained 221 lines (68.6%), over 80% (177/221) of which were genotypes of field corn, while all sweet corn lines (16) were within this cluster. On the other hand, cluster II consisted of 101 lines, over 99% (100/101) of which were genotypes of popcorn (Figure S1). Similar results were obtained by the PCA method for this association mapping panel, as shown in Figure S1B. The first component explained 12.1% of the total variation and most of the inbred lines were separated in the same genetically differentiated groups (Figure S1B). These results are in accordance with the previous findings of Maldonado et al. [5] and Coan et al. [6], in which tropical maize inbred lines were grouped into two genetically differentiated clusters, separating field corn and popcorn lines.

3.2. Linkage Disequilibrium

The genome-wide LD decay pattern is shown in Table 1 and Figure S2. The LD statistic r^2 showed a clear nonlinear trend with physical distance. According to these results, the LD decayed rapidly within 3 kb, with a cut-off value of $r^2 = 0.11$. The average LD on all chromosomes (Chr) was $r^2 = 0.09$. On the other hand, 0.63% of the total pairs of linked SNPs were in a complete LD ($r^2 = 1$), and 4.4% had an r^2 value >0.5 (strong LD). The LD of Chr 3 and 7 decayed faster than the other chromosomes, with ~2.2 kb for a cut-off value of $r^2 = 0.11$. Past studies have found that this LD pattern (i.e., rapid decay with increasing physical distance) is typical in tropical maize germplasms [6,9,32]. The LD decay pattern in this study was similar to the findings of Yan et al. [33] and Coan et al. [6], who reported that the LD pattern (in tropical maize germplasms) decreases rapidly in the range of 0.1–10 kb.

Table 1. Summary of information on linkage disequilibrium (LD) and haplotype blocks (HBs) determined in inbred lines of tropical maize. Chr corresponds to the chromosome number; $N°_{SNP}$ indicates the number of single nucleotide polymorphisms (SNPs) detected on each chromosome; $N°_{HB}$ is the number of haplotype blocks; $Size_{HB}$ and Max(kb) correspond to the maximum number of SNPs forming a haplotype block and the maximum size (in kb) for a haplotype block, respectively.

Chr	LD	$N°_{SNP}$	Position (pb) First-Last SNP	Cambira			Sabaudia		
				$N°_{HB}$	$Size_{HB}$	Max(kb)	$N°_{HB}$	$Size_{HB}$	Max(kb)
1	2.87	39,148	38,222–275,861,066	7180	32	498	7087	32	497
2	2.68	37,341	40,724–244,417,305	6874	35	500	6730	36	500
3	2.21	34,889	191,169–235,520,333	6370	35	487	6256	33	491
4	6.55	26,908	217,040–246,840,261	4790	42	500	4693	42	499
5	2.61	35,691	12,711–223,658,670	6584	33	493	6472	32	500
6	3.75	23,441	169,964–173,881,702	4368	53	466	4317	53	466
7	2.23	24,958	180,204–182,128,999	4683	28	500	4546	28	500
8	4.1	25,537	204,228–181,043,617	4703	50	498	4577	50	498
9	2.94	22,404	61,292–159,668,042	4091	35	500	3991	36	500
10	2.85	20,656	128,669–150,847,940	3760	52	498	3698	51	498
Mean	2.94	29,097	-	5340	40	494	5238	39	495

3.3. Haplotype Blocks

Total values of 53,403 and 52,377 HBs were identified in all chromosomes for Cambira and Sabaudia, respectively (Table 1), over 47%, 20%, and 33% of which contained two, three, and four (or more) SNPs, respectively (Figure S3). These HBs were constructed considering 319 and 293 inbred lines in Cambira and Sabaudia (respectively), and 290,973 SNPs distributed on all chromosomes (Table 1). An average of ~20,000 SNPs per chromosome satisfied the criteria of the 95% confidence interval proposed by Gabriel et al. [21]. Particularly, the largest number of HBs in both locations was determined by combinations of SNPs located on Chr 1, while the smaller amount was constructed by SNPs located on Chr 10 (Table 1). In this study, several genomic regions were detected in strong disequilibrium, up to ~0.1 Mb. Therefore, as these regions have a strong LD, it is possible to suggest that they will be inherited together across generations. Moreover, about 2.3% of the HBs formed in both locations had an extension over 0.1 Mb, with a D' value between 0.7 and 0.98 [21]. Analysis of the LD pattern enabled the identification and characterization of several HBs (or strongly linked genomic regions), because there is a strong LD among the SNPs that compose it. This indicates that recombination events within these HBs are unlikely, thus, these HBs should inherit together across generations.

3.4. Genome-Wide Association Study and Network-Assisted Gene Prioritization

Total values of 45 and 44 associations (SNPs and HBs, respectively) were identified for the studied traits, which are distributed in all chromosomes of maize (Table 2 and Table S1). Four SNPs were jointly associated with the FF and MF traits. In Cambira, four haplotype blocks—two loci on Chr 8 (bin

8.03) and two on Chr 9 (bin 9.06)—were jointly associated with FF and MF. In turn, Chr 3 presented two genomic regions associated with FF in Cambira (one SNP and one HB) and three in Sabaudia (one SNP and two HBs), while various SNPs (five) and HBs (three) were associated with some the three traits on Chr 9 (bin 9.06). Interestingly, all associations were environment-specific, confirming the existence of a significant and complex genetic-by-environment interaction. The results from Bayesian bi-trait analyses showed a high correlation between FF and MF, which was significantly different from zero in both locations (r = 0.94 and 0.92), justifying the fact that FF and MF share significant loci. In accordance with our findings, Xu et al. [34] found a very high amount of quantitative trait loci (QTL) significant on bins 1.03, 8.05, and 9.06 for photoperiod sensitivity and flowering time (traits highly correlated in maize; [35]), while Chardon et al. [36], through a meta-analysis, detected hot-spot QTL regions for flowering time on bins 8.03 and 8.05. On the other hand, 64 QTLs related to maize flowering time were identified by Liu et al. [37], which were distributed on chromosomal bins 1.01, 1.03, 1.1, 2.02, 3.02, 3.04, 4.05, 6.06, 7.02, 7.03, 7.04, 8.03, 8.05, 9.01, and 9.07. Like these previous studies, this study also identified significant marker-trait associations on bins 1.01, 1.03, 1.1, 2.02, 3.02, 3.04, 4.05, 6.06, 7.02, 7.03, 7.04, 8.03, 8.05, 9.01, and 9.07. This result suggests that these regions should contain important genes controlling the flowering time in maize. In addition, chromosomes 8 and 9 had the main associations for all three traits, which is consistent with studies that considered other environments and genetic materials [34,36–38].

Table 2. Summary of the associations detected by a genome-wide association study (GWAS), based on in haplotype blocks and SNP for the traits of female/male flowering time (FF and MF, respectively) and anthesis–silking interval (ASI) measured in inbred lines of tropical maize.

Marker	Trait	Cambira			Sabaudia		
		N_M	Chr(N_M)	PV%	N_M	Chr(N_M)	PV%
SNP	FF	10	2(2), 3(1), 6(1), 7(4), 8(1), and 9(1)	5.6–6.3	7	2(1), 3(1), 5(3), and 6(2)	6.5–10.1
	MF	5	2(2), 8(1), and 9(2)	5.7–6.4	6	3(3), 5(1), and 6(2)	6.5–8.5
	ASI	8	1(2), 3(1), 5(1), 6(1), 7(1), and 8(2)	5.6–6.0	9	1(2), 3(2), 8(3), 9(1), and 10(1)	6.5–9.9
Haplotype Blocks	FF	11	1(1), 3(1), 5(1), 6(2), 7(1), 8(3), and 9(2)	5.6–17.0	3	3(2) and 9(1)	6.3–8.6
	MF	7	2(1), 4(1), 7(1), 8(2), and 9(2)	4.6–13.0	12	1(2), 4(4), 5(1), 7(3), 8(1), and 9(1)	5.0–6.9
	ASI	4	1(1), 3(1), 8(1), and 10(1)	5.6–7.8	7	2(2), 5(1), 7(2), 8(1), and 9(1)	5.1–9.5

PV%: Percentage of the phenotype variation explained by the marker; N_M: Number of significant marker-trait associations.

In Cambira, the proportion of the phenotypic variance (PV%) explained by SNP markers was ~6%, while haplotype blocks explained 6–17%, 5–13%, and 6–8% of the phenotypic variation of FF, MF, and ASI, respectively (Table 2 and Table S1). On the other hand, in Sabaudia, the PV% explained by SNPs was similar to that detected by HBs. In Sabaudia, the PV% values were moderate (either SNPs or HBs), which varied between 5 and 10%, while in Cambira, HBs showed higher PV% values (>10%) in comparison with SNPs. Moreover, the HB HChr9B2943 (in Cambira) was jointly associated with FF and MF, accounting for 17% and 13% of the total variation of FF and MF, respectively (Table 2 and Table S1). Several studies reported PV% values of flowering time smaller than 10% [34,37,39,40]. In fact, numerous QTLs with small effects would be contributing to genetic variation in flowering time across diverse maize germplasms [34,37,41]. In accordance with this, 93% (41/44) of the significant HB and all SNP associations did not explain more than 10% of the total variation. Importantly, three HBs had PV% values higher than 10%, indicating the potential effectiveness of haplotypes over individual

SNP analysis, an aspect emphasized by Maldonado et al. [5] and Contreras-Soto et al. [26]. Twenty-five of the 45 SNPs detected by GWAS (i.e., 56%) were found to be part of a haplotype block, which in turn were significantly associated with a given trait. Moreover, 14 HBs contained at least 1 significant SNP, and 9 HBs contained 2 or 3 SNPs significantly associated with some trait. On the other hand, 68% (30/44) of the HBs detected did not contain any associated SNPs, which suggests that haplotype blocks are useful for discovering genomic regions that are not detected by SNP markers. On the other hand, the use of haplotype blocks in GWAS reduces the number of multiple tests, compared with SNP-based association analysis [5]. Moreover, the use of haplotype blocks as multiallelic markers might improve marker-trait association analyses, compensating the biallelic limitation of SNP markers [5,26].

Based on the physical position of the maize reference genome (http://www.maizegdb.org//), 51 candidate genes were identified neighboring the significant SNPs and HBs (Table S1), of which 11 were present in more than one trait (FF and MF) (Table S1). The network-assisted gene prioritization performed by MaizeNet [12] identified 100 additional genes based on biological processes of flowering and reproduction. Forty-four, 22, and 34 genes that were identified by MaizeNet are related to FF, MF, and ASI, respectively (Table S1). Co-functional networks determined by MaizeNet [12] are shown in Figure 1 and Figure S4. The co-functional networks identified the following genes directly related to FF—*GRMZM2G013398*, *GRMZM2G021614*, *GRMZM2G152689*, and *GRMZM2G117057*—with statistical significance of $p < 0.0001$ (Figure 1). On the other hand, the co-functional networks of MF presented significances of 2.2×10^{-11} and 2.3×10^{-5} using HBs and SNPs, respectively. The gene *GRMZM2G013398* has an ortholog in *Arabidopsis thaliana* that encodes CONSTANS-LIKE 9 (COL9), which has light-controlled functions and is crucial to inducing the day-length specific expression of the *FLOWERING LOCUS T* (*FT*) gene in leaves [42]. FT protein is the main component of florigen that strongly influences the timing of flowering [43]. Notably, the CONSTANS protein strongly influences the performance of maize flowering time in response to photoperiod, directly inducing the transcription of *FT* genes in *Arabidopsis* [42,43]. On the other hand, the genes *GRMZM2G021614*, *GRMZM2G152689*, and *GRMZM2G117057* encode phosphatidylethanolamine-binding proteins (pebp9, pebp10, and pebp11, respectively), which play important roles in floral transition in angiosperms [44]. Moreover, Kikuchi et al. [45] and Wickland and Hanzawa [46] showed that the presence and structure of these genes, together with their roles in the regulation of flowering, are well conserved among cereal plants.

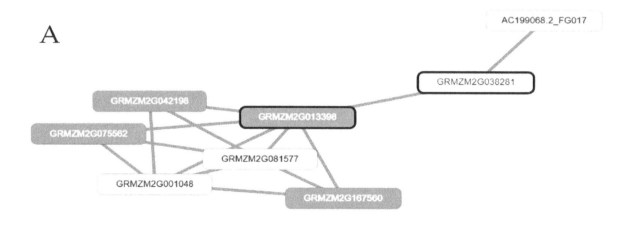

Figure 1. *Cont.*

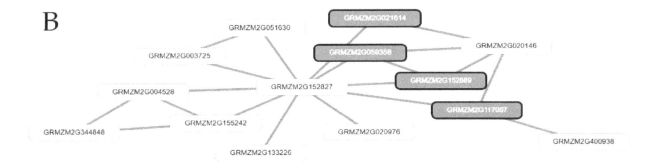

Figure 1. Co-functional networks estimated using genes identified by SNP- and haplotype-based GWAS (**A**,**B**, respectively), genes identified by network-assisted gene prioritization (in MaizeNet) for flowering time and subnets enriched by gene ontology annotations related to the biological processes of female flowering (FF) in MaizeNet. (**A**) Gene *GRMZM2G013398* identified by the prioritization analysis (nodes orange highlighted with bold borderline) connected with all the subnetwork genes of MaizeNet (white nodes) and genes associated with the ontology annotations related to flowering time (orange nodes). (**B**) Genes *GRMZM2G117057*, *GRMZM2G021614*, *GRMZM2G059358*, and *GRMZM2G152689* identified by prioritization analysis (nodes orange highlighted with bold borderline) connected with all the subnetwork genes of MaizeNet (white nodes).

4. Conclusions

In the present study, we identified several loci (SNPs and haplotype blocks) with variable contributions to phenotypic expression, which were located in regions that play important roles in the control of flowering time in maize. The GWAS based on haplotype blocks was beneficial to identify loci with major effects in comparison to SNP-based GWAS. The co-functional network approach identified four genes that strongly influence the timing of flowering in tropical maize. In general, network-assisted gene prioritization provides new insights into the genetic architecture and mechanisms underlying flowering-related traits in tropical maize.

Supplementary Materials
Figure S1. Inferred population structure in a collection of maize germplasm (322 inbred lines). (**A**) Genetic structure inferred by a Bayesian clustering model using InStruct and a dendrogram carried out using the neighbor-joining (Nei's genetic distances). The light gray and dark gray indicate the proportion of the genome extracted from the two main genetic clusters estimated by InStruct. (**B**) Principal components analysis (PCA) with two major groups identified, which correspond closely to InStruct results. Values in parentheses indicate the percentage of variation explained by each main component; Figure S2. Linkage disequilibrium (LD) decay pattern in all chromosomes of maize. Chromosomes 3 and 7 decayed faster than the other chromosomes, while chromosome 4 presented the slowest decay (lower and upper margins, respectively); Figure S3. Frequency distribution of the size of haplotype blocks consisting of two or more SNPs, in the locations Cambira and Sabaudia; Figure S4. Co-functional networks estimated using genes identified by SNP- and Haplotype-based GWAS (**A** and **B**, respectively), genes identified in prioritization of MaizeNet for flowering time and subnets enriched by gene ontology annotations related to the biological processes of male flowering (MF) in MaizeNet. White nodes represent all the subnetwork genes of MaizeNet, orange nodes are genes associated with the ontology annotations related to flowering time, and nodes highlighted with bold borderline correspond to genes identified by GWAS or the prioritization analysis; Table S1. Details of the associations and candidate genes detected in SNP- and Haplotype-based GWAS for the traits of Female/Male Flowering time (FF and MF, respectively) and Anthesis–Silking Interval (ASI) measured in inbred lines of tropical maize in two locations (Cambira and Sabaudia).

Author Contributions: Conceptualization, F.M., C.M., M.C.K., and F.A.B.B.; methodology, F.M., C.A.S., and C.M.; software, C.M., M.C.K., and F.A.B.B.; validation, F.M., C.M., M.C.K., F.A.B.B., and C.A.S.; formal analysis, F.M., C.A.S., and C.M.; investigation, M.C.K., F.A.B.B., and C.A.S.; resources, F.M. and C.M.; writing—original draft preparation, F.M. and C.M.; writing—review and editing, F.M. and C.M.; visualization, M.C.K., F.A.B.B., and C.A.S.; supervision, F.M.; project administration, C.A.S.; funding acquisition, C.A.S.

Acknowledgments: Freddy Mora thanks FONDECYT (grant number 1170695). Carlos Maldonado thanks CONICYT-PCHA/Doctorado Nacional/2017-21171466.

References

1. FAO. *Save and Grow in Practice: Maize, Rice, Wheat—A Guide to Sustainable Production*; Food and Agriculture Organization of the United Nations: Rome, Italy, 2016.
2. Begcy, K.; Sandhu, J.; Walia, H. Transient heat stress during early seed development primes germination and seedling establishment in rice. *Front. Plant. Sci.* **2019**, *9*, 1768. [CrossRef] [PubMed]
3. Rashid, S.; Rosentrater, K.A.; Bern, C. Effects of deterioration parameters on storage of maize: A review. *J. Nat. Sci. Res.* **2013**, *3*, 147.
4. Ranum, P.; Pena-Rosas, J.P.; Garcia-Casal, M.N. Global maize production, utilization, and consumption. *Ann. N. Y. Acad. Sci.* **2014**, *1312*, 105–112. [CrossRef] [PubMed]
5. Maldonado, C.; Mora, F.; Scapim, C.A.; Coan, M. Genome-wide haplotype-based association analysis of key traits of plant lodging and architecture of maize identifies major determinants for leaf angle: *hapLA4*. *PLoS ONE* **2019**, *14*, e0212925. [CrossRef]
6. Coan, M.; Senhorinho, H.J.; Pinto, R.J.; Scapim, C.A.; Tessmann, D.J.; Williams, W.P.; Warburton, M.L. Genome-Wide Association Study of Resistance to Ear Rot by *Fusarium verticillioides* in a Tropical Field Maize and Popcorn Core Collection. *Crop. Sci.* **2018**, *58*, 564–578. [CrossRef]
7. Xiao, Y.; Liu, H.; Wu, L.; Warburton, M.; Yan, J. Genome-Wide Association Studies in Maize: Praise and Stargaze. *Mol. Plant.* **2017**, *10*, 359–374. [CrossRef]
8. Zhu, X.M.; Shao, X.Y.; Pei, Y.H.; Guo, X.M.; Li, J.; Song, X.Y.; Zhao, M.A. Genetic Diversity and Genome-Wide Association Study of Major Ear Quantitative Traits Using High-Density SNPs in Maize. *Front. Plant. Sci.* **2018**, *9*, 966. [CrossRef]
9. Romay, M.C.; Millard, M.J.; Glaubitz, J.C.; Peiffer, J.A.; Swarts, K.L.; Casstevens, T.M.; Elshire, R.J.; Acharya, C.B.; Mitchell, S.E.; Flint-Garcia, S.A.; et al. Comprehensive genotyping of the USA national maize inbred seed bank. *Genome Biol.* **2013**, *14*, R55. [CrossRef]
10. Liu, N.; Xue, Y.; Guo, Z.; Li, W.; Tang, J. Genome-Wide Association Study Identifies Candidate Genes for Starch Content Regulation in Maize Kernels. *Front. Plant. Sci.* **2016**, *7*, 1046. [CrossRef]
11. Li, Y.; Li, C.; Bradbury, P.; Liu, X.; Lu, F.; Romay, C.; Glaubitz, J.; Wu, X.; Peng, B.; Shi, Y.; et al. Identification of genetic variants associated with maize flowering time using an extremely large multi-genetic background population. *Plant. J.* **2016**, *86*, 391–402. [CrossRef]
12. Lee, T.; Lee, S.; Yang, S.; Lee, I. MaizeNet: A co-functional network for network-assisted systems genetics in *Zea mays. Plant. J.* **2019**, *99*, 571–582. [CrossRef] [PubMed]
13. Chen, D.H.; Ronald, P.C. A rapid DNA minipreparation method suitable for AFLP and other PCR applications. *Plant. Mol. Biol. Rep.* **1999**, *17*, 53–57. [CrossRef]
14. Elshire, R.J.; Glaubitz, J.C.; Sun, Q.; Poland, J.A.; Kawamoto, K.; Buckler, E.S.; Mitchell, S.E. A robust, simple genotyping-by-sequencing (GBS) approach for high diversity species. *PLoS ONE* **2011**, *6*, e19379. [CrossRef] [PubMed]
15. Glaubitz, J.C.; Casstevens, T.M.; Lu, F.; Harriman, J.; Elshire, R.J.; Sun, Q.; Buckler, E.S. TASSEL-GBS: A high capacity genotyping by sequencing analysis pipeline. *PLoS ONE* **2014**, *9*, e90346. [CrossRef]
16. Money, D.; Gardner, K.; Migicovsky, Z.; Schwaninger, H.; Zhong, G.Y.; Myles, S. LinkImpute: Fast and accurate genotype imputation for nonmodel organisms. *G3 Genes Genom. Genet.* **2015**, *5*, 2383–2390. [CrossRef]
17. Yu, L.X.; Zheng, P.; Bhamidimarri, S.; Liu, X.P.; Main, D. The Impact of Genotyping-by-Sequencing Pipelines on SNP Discovery and Identification of Markers Associated with Verticillium Wilt Resistance in Autotetraploid Alfalfa (*Medicago sativa* L.). *Front. Plant. Sci.* **2017**, *8*, 89. [CrossRef]
18. Gao, H.; Williamson, S.; Bustamante, C.D. A Markov chain Monte Carlo approach for joint inference of population structure and inbreeding rates from multilocus genotype data. *Genetics* **2007**, *176*, 1635–1651. [CrossRef]
19. Evanno, G.; Regnaut, S.; Goudet, J. Detecting the number of clusters of individuals using the software STRUCTURE: A simulation study. *Mol. Ecol.* **2005**, *14*, 2611–2620. [CrossRef]
20. Breseghello, F.; Sorrells, M.E. Association Mapping of Kernel Size and Milling Quality in Wheat (*Triticum aestivum* L.) Cultivars. *Genetics* **2006**, *172*, 1165–1177. [CrossRef]
21. Gabriel, S.B.; Schaffner, S.F.; Nguyen, H.; Moore, J.M.; Roy, J.; Blumenstiel, B.; Higgins, J.; DeFelice, M.; Lochner, A.; Faggart, M.; et al. The structure of haplotype blocks in the human genome. *Science* **2002**, *296*, 2225–2229. [CrossRef]

22. Barrett, J.C.; Fry, B.; Maller, J.D.M.J.; Daly, M.J. Haploview: Analysis and visualization of LD and haplotype maps. *Bioinformatics* **2004**, *21*, 263–265. [CrossRef] [PubMed]
23. Ballesta, P.; Maldonado, C.; Pérez-Rodríguez, P.; Mora, F. SNP and Haplotype-Based Genomic Selection of Quantitative Traits in *Eucalyptus globulus*. *Plants* **2019**, *8*, 331. [CrossRef] [PubMed]
24. Bradbury, P.J.; Zhang, Z.; Kroon, D.E.; Casstevens, T.M.; Ramdoss, Y.; Buckler, E.S. TASSEL: Software for association mapping of complex traits in diverse samples. *Bioinformatics* **2007**, *23*, 2633–2635. [CrossRef] [PubMed]
25. Endelman, J.B.; Jannink, J.L. Shrinkage estimation of the realized relationship matrix. *G3 Genes Genom. Genet.* **2012**, *2*, 1405–1413. [CrossRef]
26. Contreras-Soto, R.I.; Mora, F.; de Oliveira, M.A.R.; Higashi, W.; Scapim, C.A.; Schuster, I. A genome-wide association study for agronomic traits in soybean using SNP markers and SNP-based haplotype analysis. *PLoS ONE* **2017**, *12*, e0171105. [CrossRef]
27. Arriagada, O.; Amaral-Júnior, A.T.; Mora, F. Thirteen years under arid conditions: Exploring marker-trait associations in *Eucalyptus cladocalyx* for complex traits related to flowering, stem form and growth. *Breed. Sci.* **2018**, *68*, 367–374. [CrossRef]
28. Hadfield, J. MCMCglmm Course Notes. 2012. Available online: http://cran.r-project.org/web/packages/MCMCglmm/vignettes/CourseNotes.pdf (accessed on 7 July 2019).
29. Mora, F.; Zúñiga, P.E.; Figueroa, C.R. Genetic variation and trait correlations for fruit weight, firmness and color parameters in wild accessions of *Fragaria chiloensis*. *Agronomy* **2019**, *9*, 506. [CrossRef]
30. Mora, F.; Ballesta, P.; Serra, N. Bayesian analysis of growth, stem straightness and branching quality in full-sib families of *Eucalyptus globulus*. *Bragantia* **2019**, *78*. [CrossRef]
31. Hadfield, J.D. MCMC methods for multi-response generalized linear mixed models: The MCMCglmm R package. *J. Stat. Softw.* **2010**, *33*, 1–22. [CrossRef]
32. Paes, G.P.; Viana, J.M.S.; Mundim, G.B. Linkage disequilibrium, SNP frequency change due to selection, and association mapping in popcorn chromosome regions containing QTLs for quality traits. *Genet. Mol. Biol.* **2016**, *39*, 97–110. [CrossRef]
33. Yan, J.; Shah, T.; Warburton, M.L.; Buckler, E.S.; McMullen, M.D.; Crouch, J. Genetic characterization and linkage disequilibrium estimation of a global maize collection using SNP markers. *PLoS ONE* **2009**, *4*, e8451. [CrossRef] [PubMed]
34. Xu, J.; Liu, Y.; Liu, J.; Cao, M.; Wang, J.; Lan, H.; Xu, Y.; Lu, Y.; Pan, G.; Rong, T. The genetic architecture of flowering time and photoperiod sensitivity in maize as revealed by QTL review and meta-analysis. *J. Integr. Plant. Biol.* **2012**, *54*, 358–373. [CrossRef] [PubMed]
35. Zhang, X.; Tang, B.; Liang, W.; Zheng, Y.; Qiu, F. Quantitative genetic analysis of flowering time, leaf number and photoperiod sensitivity in maize (*Zea mays* L.). *J. Plant. Breed. Crop. Sci.* **2011**, *3*, 168–184.
36. Chardon, F.; Virlon, B.; Moreau, L.; Falque, M.; Joets, J.; Decousset, L.; Murigneux, A.; Charcosset, A. Genetic architecture of flowering time in maize as inferred from quantitative trait loci meta-analysis and synteny conservation with the rice genome. *Genetics* **2004**, *168*, 2169–2185. [CrossRef] [PubMed]
37. Liu, S.; Zenda, T.; Wang, X.; Liu, G.; Jin, H.; Yang, Y.; Dong, A.; Duan, H. Comprehensive Meta-Analysis of Maize QTLs Associated with Grain Yield, Flowering Date and Plant Height Under Drought Conditions. *J. Agric. Sci.* **2019**, *11*, 1–19. [CrossRef]
38. Coles, N.D.; McMullen, M.D.; Balint-Kurti, P.J.; Pratt, R.C.; Holland, J.B. Genetic control of photoperiod sensitivity in maize revealed by joint multiple population analysis. *Genetics* **2010**, *184*, 799–812. [CrossRef]
39. Salvi, S.; Castelletti, S.; Tuberosa, R. An updated consensus map for flowering time QTLs in maize. *Maydica* **2009**, *54*, 501–512.
40. Frey, F.P.; Presterl, T.; Lecoq, P.; Orlik, A.; Stich, B. First steps to understand heat tolerance of temperate maize at adult stage: Identification of QTL across multiple environments with connected segregating populations. *Theor. Appl. Genet.* **2016**, *129*, 945–961. [CrossRef]
41. Buckler, E.S.; Holland, J.B.; Bradbury, P.J.; Acharya, C.B.; Brown, P.J.; Browne, C.; Ersoz, E.; Flint-Garcia, S.; Garcia, A.; Glaubitz, J.C.; et al. The genetic architecture of maize flowering time. *Science* **2009**, *325*, 714–718. [CrossRef]
42. Song, Y.H.; Shogo, I.; Takato, I. Flowering time regulation: Photoperiod-and temperature-sensing in leaves. *Trends Plant Sci.* **2013**, *18*, 575–583. [CrossRef]

43. Pin, P.A.; Nilsson, O. The multifaceted roles of FLOWERING LOCUS T in plant development. *Plant. Cell Environ.* **2012**, *35*, 1742–1755. [CrossRef] [PubMed]

44. Liu, Y.Y.; Yang, K.Z.; Wei, X.X.; Wang, X.Q. Revisiting the phosphatidylethanolamine-binding protein (PEBP) gene family reveals cryptic *FLOWERING LOCUS T* gene homologs in gymnosperms and sheds new light on functional evolution. *New Phytol.* **2016**, *212*, 730–744. [CrossRef] [PubMed]

45. Kikuchi, R.; Kawahigashi, H.; Ando, T.; Tonooka, T.; Handa, H. Molecular and functional characterization of PEBP genes in barley reveal the diversification of their roles in flowering. *Plant. Physiol.* **2009**, *149*, 1341–1353. [CrossRef]

46. Wickland, D.P.; Hanzawa, Y. The *FLOWERING LOCUS T/TERMINAL FLOWER 1* gene family: Functional evolution and molecular mechanisms. *Mol. Plant* **2015**, *8*, 983–997. [CrossRef] [PubMed]

4

Improvement of a RD6 Rice Variety for Blast Resistance and Salt Tolerance through Marker-Assisted Backcrossing

Korachan Thanasilungura [1], Sukanya Kranto [2], Tidarat Monkham [1], Sompong Chankaew [1] and Jirawat Sanitchon [1,*]

[1]   Department of Agronomy, Faculty of Agriculture, Khon Kaen University, Khon Kaen 40002, Thailand; Chun.sn@gmail.com (K.T.); tidamo@kku.ac.th (T.M.); somchan@kku.ac.th (S.C.)

[2]   Ratchaburi Rice Research Center, Muang, Ratchaburi 70000, Thailand; sukanya.me@hotmail.com

*   Correspondence: jirawat@kku.ac.th

Abstract: RD6 is one of the most favorable glutinous rice varieties consumed throughout the north and northeast of Thailand because of its aroma and softness. However, blast disease and salt stress cause decreases in both yield quantity and quality during cultivation. Here, gene pyramiding via marker-assisted backcrossing (MAB) using combined blast resistance QTLs (*qBl 1, 2, 11*, and *12*) and *Saltol* QTL was employed in solving the problem. To pursue our goal, the RD6 introgression line (RGD07005-12-165-1), containing four blast-resistant QTLs, were crossed with the Pokkali salt tolerant variety. Blast resistance evaluation was thoroughly carried out in the fields, from $BC_2F_{2:3}$ to BC_4F_4, using the upland short-row and natural field infection methods. Additionally, salt tolerance was validated in both greenhouse and field conditions. We found that the RD6 "BC_4F_4 132-12-61" resulting from our breeding programme successfully resisted blast disease and tolerated salt stress, while it maintained the desirable agronomic traits of the original RD6 variety. This finding may provide a new improved rice variety to overcome blast disease and salt stress in Northeast Thailand.

Keywords: gene pyramiding; aroma; QTL; chromosome; selection; introgression line

1. Introduction

Rice (*Oryza sativa* L.) is consumed as a staple food in Asia, especially in the southeast region. In Thailand, the indica rice variety RD6 developed from KDML105 through gamma irradiation is one of the most favorable glutinous rice consumed throughout the northeast of Thailand [1,2]. Because of its cooking quality, aroma, and softness, production demand has increased over time. However, its yield of 4.16 ton/ha fails to meet its potential, due to biotic and abiotic stress.

Rice blast disease caused by the fungus *Pyricularia grisea* (Cooke) Sacc. leads to crop losses up to 85% of total yield [3]. Disease symptoms occur in all stages of plant growth, beginning with blast discoloration and wilting of the foliage [4]. Neck blast can be found at the flowering stage, accelerating plant death [5]. Severe damage was also observed within areas of intensive planting with high doses of nitrogen application [6]. Development of new rice varieties resistant to blast fungus is an alternative approach to diminish or control the invasion of this pathogen. The resistance quantitative trait loci (QTL) have been investigated to achieve parental varieties, which are further used for gene pyramiding in breeding programmes. Currently, more than 100 blast-resistant genes have been identified, of which 22 genes structures have been cloned [7]. In Thailand, few studies of blast resistant genes have been conducted. Noenplab et al. [8] studied the relationship of leaf blast and neck blast of resistant genes in the Jao Hom Nil (JHN) variety, in which the resistant QTLs were detected on chromosomes 1 and 11. The resistant QTLs conferred resistance to both leaf blast and neck blast. Suwannual et al. [9]

pyramided four blast-resistant QTLs, individually, on chromosomes 2 and 12 within the P0489 variety, and on chromosomes 1 and 11 carried by the JHN variety, resulting in the creation of new RD6 introgression lines. Their results demonstrated that the RD6 introgression lines carrying a high number of QTLs (achieved through pyramiding) reached a broader spectrum of blast resistance to the blast pathogens prevalent in the region.

In addition to rice blast fungus, salt stress is a crucial constraint for RD6 production. Thailand's northeast region is the country's largest rice-producing area, and it is comprised of two basins: Sakon Nakhon and Nakhon Ratchasima. In those basins containing an understructure of accumulated salt rock, the salt-affected range covers approximately 1.84 Mha [10]. Evaporation during the dry season tends to raise salinity from the subsoil to the surface, thereby increasing salinity intensity and increasing salt stress from 2–4 dS/m to 8–16 dS/m [10–12]. Rice is a salt-sensitive crop, capable of tolerating salinity at moderate levels of electrical conductivity (4–8 dS/m) [13]. Therefore, rice produced under rain-fed, lowland conditions is usually exposed to high levels of soil salinity. The RD6 variety, which is well known, was identified as a geographical indication (GI) within the Tung Gula Rong Hai of the Northeast, Thailand. Specifically, RD6 requires optimal soil salinity to enhance rice seed aroma [14]. However, an abundance of salinity can reduce rice plant growth, tiller number, and seed set-up [15], and the stress caused by excessive salt can significantly reduce total crop yield and result in plant death [16].

The Pokkali variety, derived from the International Rice Research Institute (IRRI), has become a well-known source of salinity tolerance worldwide, attributed to the salt-tolerant QTL located on rice chromosome 1 (*Saltol*) [17–20]. Therefore, several researchers have attempted to develop salt-tolerant rice varieties using the *Saltol* QTL [21–26].

The marker-assisted backcrossing (MAB) method has been employed to obtain beneficial QTLs from donor parents via introgression between the qualitative and quantitative traits from landraces and wild relatives [27] due to the precision method with shortened time frame in both foreground and background selection. MAB provides effective gene selection and/ or QTLs for pyramiding multi-genes/QTLs within the rice population. These benefits further support breeding practices for improved resistance and tolerance [28–32]. The objective of this study was to determine the blast resistance and salt tolerance levels within the RD6 introgression lines by pyramiding four blast-resistant and one salt-tolerant QTL into the RD6 rice variety in both greenhouse and field conditions.

2. Materials and Methods

2.1. Plant Materials and Marker-Assisted Backcrossing Selection (MABS)

Three parental varieties/lines were used to generate the BC_4F_4 population, representing a pseudo-backcrossing approach to increasing the recurrent genetic background of a pyramiding population, comprised of the RD6 (recurrent parent), Pokkali (obtained for the *saltol* QTL present on chromosome 1), and RGD07005-12-165-1 (the RD6 near-isogenic line obtained from the Rice Gene Discovery Unit, Kasetsart University, Thailand). The RGD07005-12-165-1 obtained four blast-resistant QTLs from the JHN and P0489 varieties on chromosomes 1 and 11, and chromosomes 2 and 12, respectively. The breeding program was subsequently improved within the population through MAB, by crossing the RGD07005-12-165-1 with the Pokkali variety to improve salt tolerance. The F_1 was then backcrossed with RGD07005-12-165-1through BC_1F_1, whereas BC_1F_2 was utilized as a marker-assisted selection (MAS) in the blast-resistant and salt-tolerant QTLs. In this step, the flanking marker RM3412/RM10748 was used for the selected *Saltol* QTL [33], RM319/RM212 and RM114/RM224 were used for selected blast-resistant QTLs on chromosomes 1 and 11, respectively [8], and RM48/RM207 and RM313/RM277 were used for selected blast-resistant QTLs on chromosomes 2 and 12, respectively [9], as shown in Figure 1.

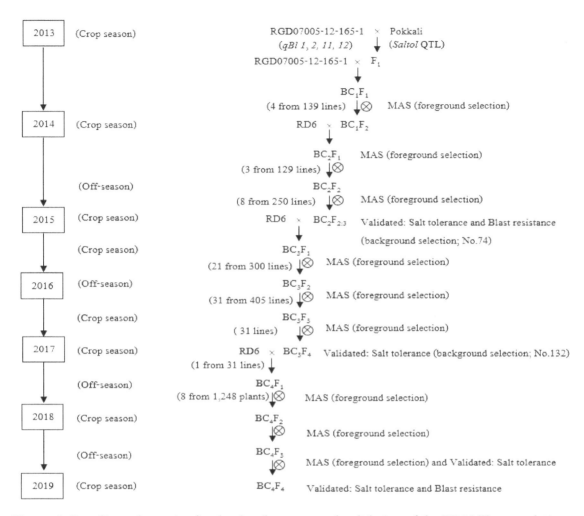

Figure 1. Breeding schematics for the development and validation of the RD6 NILs populations.

Total genomic DNA from young leaves of individual plants, lines, and their parents was extracted according to the method described by Dellaporta et al. [34] with slight modifications. The PCR reactions for SSR markers were carried out in a volume of 10 μL, containing 25 ng of genomic DNA, 1 × PCR buffer, 1.8 mM MgCl2, 0.2 mM dNTP, 0.2 μM forward and reverse primer, and 0.05 unit Taq DNA polymerase. DNA amplification was performed in a DNA thermal cycle for five minutes at 95 °C, followed by 35 cycles of 30 s at 95 °C, 30 s at 55 °C, and two min at 72 °C, with a final extension of seven minutes at 72 °C. The amplification products were separated via 4.5% polyacrylamide gel electrophoresis [32]. The selected lines in the BC_1, BC_2, and BC_3 generations were backcrossed with the RD6 variety, using MAS for *saltol* and blast-resistant QTL selection. Trait qualities; such as glutinous type, aromatic character, and gelatinization temperature (GT) in the BC_2F_2 populations were fixed through MAS using glutinous 23 primer on chromosome 6, badh2 on chromosome 8, and RM190 on chromosome 6 (Table S1). Each backcross generation within the BC_4F_4 populations was evaluated for salt tolerance and blast resistance (Figure 1).

2.2. Evaluation of Salt Tolerance and Blast Resistance in the $BC_2F_{2:3}$ Populations (Exp. 1)

The evaluation of salt tolerance and blast resistance of the $BC_2F_{2:3}$ lines, as well as the parental and check varieties, were conducted at the Department of Agronomy, Faculty of Agriculture, Khon Kaen University, Khon Kaen, Thailand. The salt tolerance evaluation was performed through two methods, salt solution and artificial soil salinity. The salt solution method was laid out in a completely randomized design (CRD), with four replications. Seedlings were transplanted at seven days to 50 × 57 cm^2 Styrofoam sheets with 1.5 cm diameter holes [17]. Fertilizer was applied with

Yoshida nutrient solution [35] at three to twenty-one days of age. NaCl was then added to reach the electrical conductivity (EC) of 4 dS/m. EC was subsequently increased at three-day intervals, reaching EC 8 and 12 dS/m. When the susceptible checks (IR29) presented salt injury symptoms, salt tolerance data were recorded following the standard evaluation score (SES) [36]. The artificial soil salinity method was also laid out in CRD with four replications. Seedlings were transplanted at seven days and transferred from the spent soil trays to the water trays. At 14 days after transplanting, the experimental trays were treated with NaCl, which adjusted the water solution to EC 8dS/m, which increased to EC 12dS/m after two days. Salt tolerance data were recorded similarly in both methods.

Blast resistance was also evaluated via the upland short-row method at the Sakon Nakhon Rice Research Center, Sakon Nakhon, Thailand. The experiment was laid out in CRD with three replications. Seeds of each $BC_2F_{2:3}$ line were sown in rows (approximately) 50 cm long and 10 cm apart. A susceptible KDML105 variety was planted alternately with every two testing varieties. Blast resistance scores were recorded following the SES method [36].

2.3. Evaluation of Salt Tolerance in the BC_3F4 Populations (Exp. 2)

The BC_3F_4 lines and parental varieties were evaluated for salt tolerance in field conditions. The experiment was conducted at the Ban Daeng Village, Ban Fhang, Khon Kaen, Thailand. The experiment was laid out in a randomized complete block design (RCBD) with three replications. Germinated seeds were sown on seedbeds; then, at thirty days, the seedlings were transplanted to the field. Plot sizes were 1×1.5 m^2, in three rows, spaced 25×25 cm between and within rows. The RD6 variety was planted between every five plots within the test lines to ensure that salinity occurred uniformly. Fertilizer (23.44 kg/ha of N, P_2O_5, and K_2O) was applied at four days after transplanting, and hand weeding and chemical application for disease and insect control were performed as needed. When the susceptible check (RD6) presented salt injury symptoms, salt tolerance data were recorded following SES [37]. Moreover, the agronomic traits including 1000/seed weight, seed length, seed width, and seed shape were recorded.

2.4. Evaluation of Salt Tolerance and Blast Resistance Evaluations in the BC_4F_3 Populations (Exp. 3)

The salt tolerance evaluation of the BC_4F_3 lines was conducted in greenhouse conditions at the Department of Agronomy, Faculty of Agriculture, Khon Kaen University, Khon Kaen, Thailand. The experiment was laid out in CRD with three replications, as described in the salt solution method of Exp. 1. The experiment for blast resistance of the BC_4F_3 lines was conducted at the Department of Agronomy, Faculty of Agriculture, Khon Kaen University, Khon Kaen, Thailand through field and upland short-row experiments. The upland short-row method was conducted similarly to Exp. 1. The field experiment was laid out in RCBD with three replications, in plots 0.5×1 m^2, in three rows, spaced 25×25 cm between and within rows. Seedlings were transplanted at 30 days of age. Symptoms of natural blast infection were identified when seedlings began to show signs of infection, following the protocol of the SES [37], in which both leaf and neck blast symptoms were recorded.

Agronomic trait data, such as plant height (PH) and panicle length (PL), were recorded at pre-harvest; whereas post-harvest data recorded 4/panicle seed weight (SW4P), 1000/seed weight (1000SW), total dry weight (TDW), total seed weight (TSW), harvest index (HI), seed length (SL), seed width (SW), and seed shape (SS) (ratio of SL/SW). Additionally, seed qualities of the BC_4F_4 seeds, such as seed morphology, including SL, SW, SS and seed color of brown and paddy rice and aromatic traits, were evaluated and compared with the RD6 variety. The seed aromatic evaluation of each line was achieved through the quantitative determination of 2-acetyl-1-pyrroline (2AP) content using automated headspace gas chromatography following the methods as prescribed by Sriseadka et al. [38]. In brief, polished seed (1.00 g) were ground and then placed in a 20 mL headspace vial. The headspace vials were immediately sealed with PTFE/silicone septa and aluminum caps prior to analysis by static headspace-gas chromatography. A static headspace (Model 7697A, Agilent Technologies, Santa Clara, CA, USA) coupled to an Agilent 7890B Series GC system equipped with an Agilent

5977B GC/MSD system was used. A series of 2AP standard solutions with concentrations of 1.22, 2.45, 4.90, 9.79, and 19.67 ppm in isopropanol were prepared, which was added to headspace vials containing 1.00 g of non-aromatic rice seed (cv. Chai Nat 1) used as the external standard. The optimum headspace operating conditions were oven temperature 110 °C, loop temperature 120 °C, transfer line temperature 130 °C, vial equilibration time 10 min with high speed shaking, pressurizing time 0.15 min, loop equilibration time 0.40 min, and inject time 0.50 min. The headspace volatiles were separated using an HP-5 (25 m × 250 μm × 0.25 μm film thickness) column (J&W Scientific, Folsom, CA, USA). The optimum GC conditions were achieved using an HP-5 column with a splitless injection at 210 °C. The column temperature programme began at 50 °C and increased to 200 °C at 10 °C/min. Purified helium was used as the GC carrier gas at a flow rate of 1.2 mL/min. A calibration curve for 2AP analysis by headspace was generated by spiking known concentrations of 2AP into a non-fragrant rice variety (Chai Nat 1). Samples were run in triplicate, and the concentration of 2AP was calculated based upon the relative peak area of external standard.

2.5. Evaluation of Salt Tolerance in the BC_4F_4 Populations (Exp. 4)

The salt tolerance evaluations of the BC_4F_4 lines were conducted in greenhouse conditions at the Department of Agronomy, Faculty of Agriculture, Khon Kaen University, Khon Kaen, Thailand, via the salt solution method. Laid out in CRD with four replications, the planting methods and experiment protocol were similar to those described in Exp1, except that the EC in Exp. 4 was adjusted to 18 dS/m. The leaf, stem, root, and total dry weight data were recorded. Dry weight of the seedlings was determined by oven-drying the seedlings at 80 °C for 3 days, and the percentages of Na+ and K+ in the leaves, stems, and roots also were determined, according to the flame photometric method [39]. In addition, the percentages of Na+ and K+ in the leaves, stems, and roots also were determined, according to the flame photometric method [39]. In brief, pounding sample volume 0.5 g of each tissues were digest by 10 mL nitric acid (HNO_3) and 5 mL perchloric acid ($HClO_4$), then incubated at 200 °C. The contents were covered to reflux acid fumes generated during digestion until digest appeared translucent. After cooling down, 100 mL deionized water was added to each digestion tube. Contents were then vortexed and passed through qualitative cellulose filter paper (Whatman No.1, Sigma-Aldrich®, St. Louis, MO, USA) and measurement the K+ (768 nm wavelength) and Na+ (589 nm wavelength) by flame photometer (Model 410 Flame Photometer, Sherwood Scientific Limited, Cambridge, UK). The K+ and Na+ concentrations of samples were compared with the known standard solutions of 0.0, 5.0, 10.0, 15, and 20 ppm from a calibration curve with a correlation coefficient (r^2) = 0.999. Finally, the amount of K+ and Na+ were transformed to percentage when compared with dry weight of raw materials.

2.6. Data Analysis

Salt tolerance scores, blast resistance scores, and agronomic trait data were analyzed via the STATISTIC 10© program (1985–2013) (Analytical Software, Tallahassee, FL, USA). Means were compared by the least significant difference (LSD) at $p < 0.05$.

3. Results

3.1. Development of Populations Through MABS

The F_1 population (RGD07005-12-165-1 × Pokkali) was backcrossed with RGD07005-12-165-1 using MAS through BC_2F_2, in which the MAS contained genes of the glutinous type, aromatic, and gelatinization temperature (GT); and the $BC_2F_{2:3}$ populations were then evaluated for salt tolerance and blast resistance. One $BC_2F_{2:3}$ line (no. 74) obtained all QTLs required to obtain the target traits necessary for generation advancement (Table S2). In developing the BC_3 generation, $BC_2F_{2:3}$ (no. 74) was crossed back to RD6 to produce the BC_3F_1, in which the BC_3F_1 was developed through the BC_3F_4 population by MAS. Thirty-one BC_3F_4 lines were evaluated for salt tolerance and agronomic traits within the salted field. Among the BC_3F_4 lines, the BC_3F_4 (no. 132) demonstrated salt tolerance and agronomic characteristics similar to those of the RD6 variety and were subsequently selected for the development of the BC_4F_1 through BC_4F_3. The BC_4F_3-132-12-61 line was successfully developed via the MABS, consisting of one QTL for salt tolerance and four QTLs for blast resistance. BC_4F_3 lines with varied QTL combinations were also validated in the tests.

3.2. Evaluation of Salt Tolerance and Blast Resistance in the $BC_2F_{2:3}$ Populations

Eight of the $BC_2F_{2:3}$ lines were evaluated for salt tolerance in the seedling stage through the salt solution and artificial soil salinity methods. The results concluded that the $BC_2F_{2:3}$ lines presented as highly significant at EC 12 dS/m. $BC_2F_{2:3}$ (nos. 23, 67, and 74), demonstrated tolerance (T) in both methods, similar to that of the tolerant check (Pokkali); while the remaining five $BC_2F_{2:3}$ lines showed moderate tolerance (MT) under both the salt solution and artificial soil salinity tests (Table 1). The $BC_2F_{2:3}$ lines were evaluated for blast resistance via the upland short-row method, through which the blast resistance reaction (R) was similar to that of the resistance checks (JHN and P0489). Their resistance proved greater than that of the susceptible check (RD6), due to the presence of blast-resistant QTLs, and superior to that of the original RD6 (Table 1).

Table 1. Salt tolerance scores at EC 12 Ds/m in salt solution, artificial soil salinity, and blast resistance within the $BC_2F_{2:3}$ populations.

$BC_2F_{2:3}$ Populations	Salt Solution		Artificial Soil Salinity		Blast Resistance	
	Score	Reaction	Score	Reaction	Score	Reaction
No. 74	4.3 c	T	4.5 cd	T	2.0 c	R
No. 42	5.5 b	MT	4.8 cd	T	1.0 c	R
No. 44	5.5 b	MT	4.5 cd	T	2.0 c	R
No. 23	4.8 bc	T	4.5 cd	T	1.0 c	R
No. 56	5.0 bc	MT	4.5 cd	T	2.0 c	R
No. 67	4.0 c	T	4.0 d	T	2.0 c	R
No. 33	5.5 b	MT	5.0 cd	MT	2.0 c	R
No. 7	4.0 c	T	5.8 bc	MT	2.0 c	R
Pokkali (tolerant check)	4.0 c	T	4.0 d	T	6.0 a	MS
IR29 (susceptible check)	9.0 a	HS	9.0 a	HS	4.0 b	MS
RD6 (recurrent parent)	9.0 a	HS	9.0 a	HS	7.0 a	S
RGD07005-12-165-1	8.0 a	S	7.0 b	S	2.0 c	R
Jao Hom Nil (resistant check)	-	-	-	-	1.0 c	R
P0489 (resistant check)	-	-	-	-	2.0 c	R
F-Test	**		**		**	
C.V. (%)	11.22		13.23		24.12	

T = tolerance, MT = moderate tolerance, S = susceptible, HS = highly susceptible, R = resistant, MR = moderate resistance, MS = moderately susceptible. ** = significant different at $p < 0.01$. Different letters after the mean within a column showed a significant difference. CV = the coefficient of variation.

3.3. Evaluation of Salt Tolerance in the BC_3F_4 Populations

$BC_2F_{2:3}$ (No.74) was selected for the next backcross cycle with the RD6, and MAS was performed through the BC_3F_4 population. Thirty-one BC_3F_4 lines were screened for salt tolerance and agronomic performance. The results showed that the 21 BC_3F_4 lines were salt-tolerant (T), whereas the 10 BC_3F_4 lines proved only moderately tolerant (MT). Agronomic traits of the BC_3F_4 lines were similar to those of the RD6 variety, which was in accordance with our objectives (Table 2). Five BC_3F_4 lines (nos.22, 36, 115, 129, and 132) demonstrated superior tolerance (T) and agronomic traits (1000/SW, SL, SW, and SS), again, similar to those of the original RD6, and were selected as donor parents for the development of the BC_4. The results showed that BC_3F_4 (no. 132) produced the best performance for pollination, and was therefore selected for development of the BC_4 population.

Table 2. Salt tolerance scores and agronomic traits within the BC_3F_4 populations.

Line	STS	1000SW (g)	SL (cm)	SW (cm)	SS	Line	STS	1000SW (g)	SL (cm)	SW (cm)	SS
BC_3F_4 no.2	4.33 b–d	26.80 a–g	9.95 h–l	2.62 c–h	3.81 b–g	BC_3F_4 no.124	3.67 cd	27.83 a–d	10.64 a–d	2.70 b–h	3.97 a–e
BC_3F_4 no.14	4.33 b–d	23.93 hi	10.06 f–l	2.64 c–h	3.83 b–g	BC_3F_4 no.126	3.67 cd	27.60 a–d	10.36 b–i	2.64 b–h	3.93 a–f
BC_3F_4 no.22	3.67 cd	24.40 f–i	9.86 i–m	2.58 f–h	3.83 b–g	BC_3F_4 no.129	3.00 d	26.23 b–h	10.18 d–k	2.68 b–h	3.81 b–g
BC_3F_4 no.26	5.00 a–c	24.17 g–i	9.79 j–m	2.59 f–h	3.79 b–g	BC_3F_4 no.132	3.67 cd	26.47 a–h	10.39 b–i	2.73 b–g	3.81 b–g
BC_3F_4 no.35	4.33 b–d	24.47 e–i	10.14 d–k	2.72 b–h	3.74 c–h	BC_3F_4 no.136	4.33 b–d	24.23 g–i	9.88 h–m	2.65 b–h	3.73 c–h
BC_3F_4 no.36	3.67 cd	26.00 c–h	9.76 k–m	2.56 gh	3.82 b–g	BC_3F_4 no.151	6.33 a	25.97 c–h	10.02 g–l	2.75 b–f	3.68 d–h
BC_3F_4 no.40	5.00 a–c	24.30 f–i	9.93 h–l	2.64 b–h	3.47 gh	BC_3F_4 no.152	5.67 ab	27.13 a–f	9.92 h–l	2.79 bc	3.56 f–h
BC_3F_4 no.55	5.67 ab	24.40 f–i	9.35 mm	2.76 b–e	3.40 h	BC_3F_4 no.153	5.00 a–c	25.07 d–i	10.05 g–l	2.73 b–g	3.69 c–h
BC_3F_4 no.58	5.00 a–c	24.50 e–i	9.57 l–n	2.62 c–h	3.66 d–h	BC_3F_4 no.195	5.00 a–c	27.40 a–d	10.59 a–f	2.55 h	4.15 ab
BC_3F_4 no.67	6.33 a	22.90 i	9.04 n	2.68 b–h	3.38 h	BC_3F_4 no.235	3.67 cd	25.80 d–h	10.32 b–j	2.61 d–h	4.00 a–d
BC_3F_4 no.69	5.67 ab	27.67 a–d	10.76 a–c	2.73 b–g	3.97 a–e	BC_3F_4 no.241	3.67 cd	27.27 a–e	10.37 b–i	2.74 b–g	3.81 b–g
BC_3F_4 no.89	4.33 b–d	27.57 a–d	10.40 b–h	2.72 b–h	3.84 b–g	BC_3F_4 no.254	3.00 d	29.07 ab	10.71 a–c	2.64 b–h	4.06 a–c
BC_3F_4 no.93	3.67 cd	26.40 a–h	10.53 a–g	2.73 b–g	3.86 b–f	BC_3F_4 no.265	3.67 cd	27.50 a–d	10.83 ab	2.58 f–h	4.25 a
BC_3F_4 no.109	4.33 b–d	29.13 a	10.97 a	2.62 c–h	3.91 a–f	BC_3F_4 no.298	3.67 cd	26.47 a–h	10.28 c–k	2.64 b–h	3.91 a–f
BC_3F_4 no.113	4.33 b–d	27.63 a–d	10.73 a–c	2.68 b–h	4.01 a–d	Pokkali	4.33 b–d	13.77 j	8.47 o	3.42 a	2.48 i
BC_3F_4 no.115	3.00 d	28.80 a–c	10.64 a–e	2.81 b	3.79 b–g	RD6	6.33 a	24.20 g–i	9.98 h–l	2.78 b–d	3.61 e–h
BC_3F_4 no.118	3.67 cd	25.63 d–i	10.10 e–k	2.59 e–h	3.92 a–f	BC_2F_5 no.74	5.67 ab	24.53 e–i	9.88 h–m	2.68 b–h	3.70 c–h
F-test	**	**	**	**	**	F-test	**	**	**	**	**
C.V. (%)	25.4	6.8	3.96	3.2	6.18	C.V. (%)	25.4	6.8	3.96	3.2	6.18

STS = salt tolerance score, 1000SW = 1000 seeds weight, SL = seed length, SW = seed width, SS = seed shape (ratio of SL/SW). ** = significantly different at $p < 0.01$. Different letters after the mean within a column showed a significant difference. CV = the coefficient of variation.

3.4. Evaluation of Salt Tolerance and Blast Resistance in the BC₄F₃ Populations

This experiment evaluated eight BC$_4$F$_3$ lines obtained via *Saltol* QTL from MAS, specifically, the combination lines, donor parents, and tolerance, and susceptible checks were screened for salt tolerance via the salt solution method (EC 12 dS/m). The results showed highly significant scores within the BC$_4$F$_3$ population and the tolerant check (Pokkali), with salt tolerance scores less than 5.0, indicated as tolerant (T). The recurrent parent (RD6) presented moderate tolerance (MT) with a score of 6.5 (Table 3). The results indicated that the Saltol QTL within the BC$_4$F$_3$ population demonstrated superior performance in salt tolerance to that of the recurrent parent (RD6).

The BC$_4$F$_3$ populations were evaluated for blast resistance in field conditions through natural infection. Since blast infection in the field was not severe, leaves with severe symptoms from the surrounding trap plants were collected and stored in a bag, under dark conditions, for twelve hours to induce spore formation. The natural inoculum was added with water, and spore suspension was then sprayed over the field. The BC$_4$F$_3$ populations and their parents showed a high level of blast resistance. A total of eight BC$_4$F$_3$ lines demonstrated high resistance (HR), similar to those of both the donor parents, whereas the recurrent parent (RD6) presented moderate susceptibility (MS) (Table 3). Importantly, some introgression lines were greater in resistance than the recurrent parent (RD6). The second peak of bimodal rain, which occurred in the flowering stage, initiated significant signs of neck blast. The results found that eight BC$_4$F$_3$ lines showed resistance (R) similar to that of the donor parent, whereas the RD6 was moderately susceptibility (MS) to neck blast (Table 3).

Additionally, blast resistance evaluation was also conducted via the upland short-row method in the seven BC4F3 lines, which presented highly significant blast-resistant levels. The BC$_4$F$_3$ lines showed resistance abilities similar to both donor parents, whereas the RD6 recurrent parent presented moderate resistance (MR). Notably, the blast-resistant genes in the BC$_4$F$_3$ populations provided blast resistance in the seedling stage similar to that of the tilling and grain filling stage, and evidenced greater resistance than that of the RD6 recurrent parent (Table 3).

The agronomic traits were also evaluated in the blast field experiments. The results were highly significant within the BC$_4$F$_3$ populations and their parents for ten traits: PH; PL; SW4P; 1000/SW; TDW; TSW; HI; SL; SW; and SS (Table 4). The BC$_4$F$_3$ 132-12 maintained agronomic traits similar to those of the recurrent parent (RD6) for nine traits, except for the 1000/SW (1000 seed weight), in which the RD6 presented moderate susceptibility (MS) for leaf and neck blast, resulting in low grain filling (Table 4). The results indicate that the newly developed MAB population had greater resistance than that of the original RD6 variety while maintaining its desirable agronomic traits and satisfying consumer demand. The BC$_4$F$_3$ lines also showed improvements in seed length (SL), seed width (SW), and seed shape (SS) as long and slender type. Additionally, the color of paddy rice was straw yellow, similar to the original RD6 varieties (Figure 2).

Table 3. Blast resistance scores of leaf blast neck blast and salt tolerant evaluations in the field and upland short-row methods in the BC$_4$F$_3$ populations.

Line	QTLs	Leaf Blast Resistance	Leaf Blast Mean	Neck Blast Resistance	Neck Blast Mean	Upland Short Row Method Resistance	Upland Short Row Method Mean	Salt Score Tolerance	Salt Score Mean
BC$_4$F$_3$ 132-12-61	*Saltol, qBl* (1 2 11 12)	HR	0.00 c	R	1.53 b	HR	0.00 c	T	3.67 b
BC$_4$F$_3$ 132-25-18	*Saltol, qBl* (1 11 12)	HR	0.00 c	R	1.27 b	HR	0.00 c	T	3.45 b
BC$_4$F$_3$ 132-14-43	*Saltol, qBl* (1 11 12)	HR	0.00 c	R	1.67 b	HR	0.00 c	T	3.67 b
BC$_4$F$_3$ 132-98-87	*Saltol, qBl* (11 12)	HR	0.00 c	R	1.67 b	HR	0.00 c	T	3.67 b
BC$_4$F$_3$ 132-174-54	*Saltol, qBl* (1 11)	HR	0.00 c	R	1.93 b	-	-	T	3.89 b
BC$_4$F$_3$ 132-51-17	*Saltol,* qBl (12)	HR	0.87 c	R	1.00 b	R	1.00 c	T	3.67 b
BC$_4$F$_3$ 132-167-58	*Saltol, qBl* (11)	HR	0.23 c	R	2.53 b	HR	0.33 c	T	3.67 b
BC$_4$F$_3$ 132-276-84	*Saltol, qBl* (2)	HR	0.43 c	R	1.13 b	HR	0.67 c	T	3.89 b
RD6 (recurrent)	-	MS	6.77 a	MS	5.80 a	MR	4.33 b	MS	6.56 a
JHN (resistant check)	*qBl* (1 11)	MR	3.67 b	R	1.00 b	HR	0.00 c	-	-
P0489 (resistant check)	*qBl* (2 12)	HR	0.00 c	R	1.00 b	HR	0.00 c	-	-
KDML105 (susceptible check)	-	-	-	-	-	S	7.67 a	-	-
IR 29 (susceptible check)	-	-	-	-	-	-	-	S	7.45 a
F-test			**		**		**		**
C.V. (%)			48.2		60.73		31.03		17.20

HR = high resistance, R = resistance, MR = moderate resistance, MS = moderate susceptible, S = susceptible. ** = significantly different at $p < 0.01$. Different letters after the mean within a column show a significant difference. CV = the coefficient of variation.

Table 4. Evaluation of agronomic traits and 2-acetyl-1-pyrroline (2AP) content from the blast fields of the BC$_4$F$_3$ populations.

Line	PH (cm)	PL (cm)	SW4P (g)	1000SW (g)	TDW (g)	TSW (g)	HI	SL (cm)	SW (cm)	SS	2AP Content (ppm)
BC$_4$F$_3$ 132-12-61	161.33 ab	27.10 a–c	20.98 a–c	27.90 c–e	354.66 a–c	146.05 a–c	0.42 a	1.09 b	0.26 bc	4.11 cd	3.63 b–d
BC$_4$F$_3$ 132-25-18	165.90 a	27.67 a	20.33 a–c	29.07 ab	367.31 a	157.76 ab	0.43 a	1.09 b	0.26 c	4.20 bc	3.15 d–f
BC$_4$F$_3$ 132-14-43	163.90 ab	27.30 ab	22.03 ab	29.10 ab	330.40 ab	146.54 a–c	0.44 a	1.07 b	0.27 ab	4.06 c–e	4.05 b
BC$_4$F$_3$ 132-98-87	163.00 ab	25.60 de	20.81 a–c	26.93 e	325.53 ab	147.19 a–c	0.45 a	1.02 c	0.26 bc	3.88 ef	4.68 a
BC$_4$F$_3$ 132-174-54	164.43 ab	25.70 c–e	20.80 a–c	28.50 bc	345.89 ab	151.41 a–c	0.44 a	1.05 bc	0.26 bc	4.06 c–f	3.44 c–e
BC$_4$F$_3$ 132-51-17	166.10 a	27.00 a–d	19.82 a–c	28.17 bc	291.88 bc	122.30 bc	0.42 a	1.08 b	0.26 bc	4.08 cd	3.73 bc
BC$_4$F$_3$ 132-167-58	162.90 ab	27.93 a	23.27 a	27.97 cd	334.45 ab	146.77 a–c	0.44 a	1.09 b	0.26 c	4.22 bc	3.84 bc
BC$_4$F$_3$ 132-276-84	165.10 a	26.73 a–d	20.91 a–c	28.53 bc	368.80 a	164.07 a	0.44 a	1.05 bc	0.27 a	3.86 f	3.06 ef
RD6 (recurrent)	167.57 a	25.87 b–e	18.46 bc	25.10 f	295.79 a–c	115.33 cd	0.39 ab	1.02 c	0.26 bc	3.88 ef	4.62 a
JHN (resistant check)	109.80 d	25.03 e	12.99 d	21.80 h	224.05 c	60.30 e	0.26 c	1.07 b	0.24 d	4.43 a	-
P0489 (resistant check)	137.70 c	25.53 de	17.36 c	23.30 g	232.65 c	82.13 de	0.34 b	0.95 d	0.23 d	4.05 c–f	-
F-test	**	**	**	**	**	**	**	**	**	**	**
C.V. (%)	2.46	3.33	11.28	2.16	13.95	16.74	10.15	2.79	1.70	2.85	3.72

PH = plant height, PL = panicle length, SW4P = seed weight 4 panicles, 1000SW = 1000 seeds weight, TDW = total dry weight 4 plants, TSW = total seed weight 4 plants, HI = harvest index, SL = seed length, SW = seed width, SS = seed shape (ratio of SL/SW), 2AP 2-acetyl-1 pyrroline. ** = significantly different at $p < 0.01$. Different letters after the mean within a column show a significant difference. CV = the coefficient of variation.

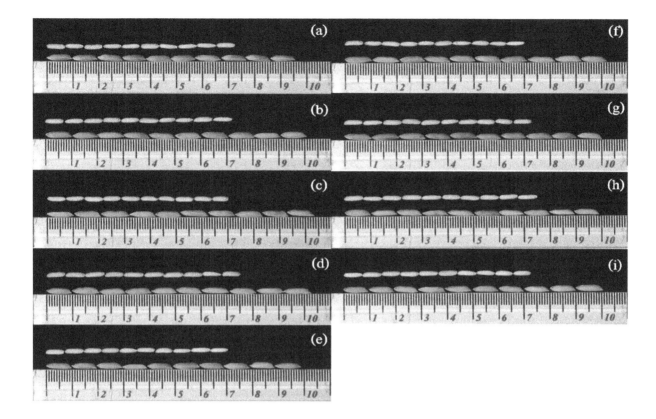

Figure 2. Seed quality, seed length, seed shape, and seed color of 10 seeds of the BC_4F_3 populations compared with the RD6 variety. (**a**) RD6, (**b**) BC_4F_3 133-12-61, (**c**) BC_4F_3 133-25-18, (**d**) BC_4F_3 132-14, (**e**) BC_4F_3 132-98, (**f**) FC_4F_3 132-174, (**g**) BC_4F_3 132-51, (**h**) BC_4F_3 132-167, (**i**) BC_4F_3 132-276. The top row is brown rice and bottom row is paddy rice with straw color. The small scale is in millimeters.

Eight BC_4F_3 lines and the RD6 (recurrent parent) were evaluated for seed aroma through the determination of 2AP content via the automated headspace gas chromatography method. The results showed a significant difference among the BC_4F_3 lines and RD6 variety, with mean values of 2AP content of the BC_4F_3 exceeding 3.00 ppm (Table 4). BC_4F_3 132-98-87 presented the highest 2AP content (4.68 ppm), similar to that of the RD6 (Table 4). The BC_4F_3 lines and RD6 were determined to be similar in fragrance.

3.5. Evaluation of Salt Tolerance in the BC_4F_4 Population

Two BC_4F_4 lines, BC_4F_4 132-12 and BC_4F_4 132-167, presented as tolerant (T) in salt tolerance evaluations, similar to that of the Pokkali, whereas the remaining five BC4F4 lines and the recurrent parent (RD6) showed moderate tolerance (MT) (Table 5). Statistically, seven of the BC_4F_4 lines demonstrated EC values up to 18 dS/m. The dry weights of the seven BC4F4 lines, their parent, and KDML105 check varieties presented as highly significant, in which the tolerant check (Pokkali) presented the highest LDW, SDW, RDW, and TDW evident in the weights of leaf stems and roots, whereas the BC_4F_4 lines were similar to that of the recurrent parent (RD6) (Table 5). The results indicate that salinity significantly affected dry weight.

Table 5. Leaf, stem, root, and total dry weights in the experiment 4.

Line	LDW (g)		SDW (g)		RDW (g)		TDW (g)		Salt Score
BC4F4 132-12-61	0.56	bc	0.47	b	0.27	b	1.3	bc	5.2 c–e
BC4F4 132-25-18	0.47	b–d	0.33	bc	0.22	b	1.02	bc	5.9 b–d
BC4F4 132-14-43	0.55	bc	0.41	bc	0.25	b	1.21	bc	6.0 b–e
BC4F4 132-51-17	0.5	bc	0.35	bc	0.22	b	1.07	bc	5.2 c–e
BC4F4 132-167-58	0.49	bc	0.42	bc	0.21	bc	1.12	bc	4.9 de
BC4F4 132-276-84	0.44	b–d	0.39	bc	0.23	b	1.06	bc	5.5 b–e
RD6 (recurrent)	0.39	cd	0.27	cd	0.2	bc	0.85	cd	6.2 a–c
Pokkali (tolerant check)	1.21	a	1.18	a	0.47	a	2.86	a	3.6 f
IR29 (susceptible check)	0.23	d	0.11	d	0.13	c	0.47	d	7.4 a
KDML105	0.63	b	0.51	b	0.25	b	1.39	b	6.5 ab
F-test	**		**		**		**		**
CV (%)	22.24		21.68		17.1		18.88		11.35

LDW = leaf dry weight, SDW = stem dry weight, RDW = root dry weight, TDW = total dry weight. ** = significantly different at $p < 0.01$. Different letters after the mean within a column show a significant difference. CV = the coefficient of variation.

Additionally, the percentages of Na+ and K+ in leaf stems and roots were also recorded and proved highly significant in all traits. The tolerant checks (Pokkali) presented the lowest Na+ in leaves (1.94), stems (1.75), and roots (4.94), whereas K+ in leaf stems were non-significant within the BC$_4$F$_4$ lines. Moreover, Pokkali also displayed low Na+-to-K+ ratios in rice shoots. The BC$_4$F$_4$ 132-12 also showed low levels of Na+ in both stems and leaves, as well as low Na+-to-K+ ratios in rice shoots comparable to those of the Pokkali variety, yet lower than those of the RD6. We may infer that transferring the saltol QTL from the Pokkali line created greater tolerance in subsequent breeding lines that that of RD6 recurrent parent (Figure 3). Interestingly, the BC$_4$F$_4$ lines presented Na+ levels in leaves and stems similar to those in Pokkali, and significantly different in RD6. This confirmed the Pokkali saltol QTL's ability to exclude Na+ from the leaf blade, expressing salt tolerance through the salt-excluder method. Moreover, the salt tolerance scores were negatively correlated to the leaf, stem, root, and total dry weights (r = −0.7899 **, r = −0.8136 **, r = −0.8140 **, r = −0.8065 **, respectively) and positively correlated with leaf and stem Na+ (r = 0.7670 ** and r = 0.8917 **, respectively) (Table 6). The results indicated that salt stress decreased plant growth in susceptible varieties, while growth was maintained in salt-tolerant varieties and that the accumulation of Na+ in leaves and stems was related to salt susceptibility (Table 6).

Table 6. The correlation between leaf dry weight, stem dry weight, root dry weight, total dry weight, leaf Na+, leaf K+, stem Na+, stem K+, root Na+, root K+, and the salt score of seeds in the BC$_4$F$_4$ populations and check varieties in the experiment 4.

	LDW	SDW	RDW	TDW	Leaves Na$^+$	Leaves K$^+$	Stem Na$^+$	Stem K$^+$	Root Na$^+$	Root K$^+$
SDW	0.9926 **									
RDW	0.9863 **	0.9864 **								
TDW	0.9978 **	0.9981 **	0.9911 **							
Leaves Na$^+$	−0.7490 *	−0.7762 **	−0.7513 *	−0.7665 **						
Leaves K$^+$	−0.0157	−0.0096	0.0313	−0.0044	−0.4369					
Stem Na$^+$	−0.6957 *	−0.6999 *	−0.7268 *	−0.7056 *	0.8532 **	−0.6924 *				
Stem K$^+$	−0.0578	−0.0566	−0.0188	−0.0500	−0.3733	0.9680 **	−0.6326 *			
Root Na$^+$	−0.9140 **	−0.9037 **	−0.8716 **	−0.9054 **	0.5259	0.3967	0.3537	0.4257		
Root K$^+$	0.1762	0.1411	0.2168	0.1666	−0.1362	0.3735	−0.4801	0.2748	0.0495	
Salt score	−0.7899 **	−0.8136 **	−0.8140 **	−0.8065 **	0.7670 **	−0.4516	0.8917 **	−0.4183	0.5386	−0.3405

LDW = leaf dry weight, SDW = stem dry weight, RDW = root dry weight, TDW = total dry weight. * = significantly different at $p < 0.05$. ** = significantly different at $p < 0.01$.

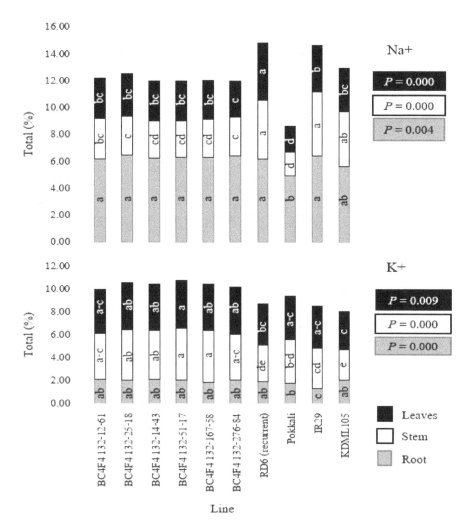

Figure 3. Na+ and K+ on leaves, stems, and roots in the experiment 4. Different letters within a color bar show a mean significant difference of each line.

4. Discussion

Since its release in 1977, the RD6 glutinous rice variety has remained a staple food crop for domestic consumption in Thailand's north and northeast regions. Comprising 83% of total glutinous rice production in these areas, consumers have developed a preference for its superior characteristics. However, the RD6 variety suffers from several production constraints, including biotic stress responsible for both rice blast [40] and bacterial blight disease [41]. Current research has attempted to eliminate sustainable infection-resistant production practices by pyramiding multiple resistant genes [42]. To date, an RD6 introgression line capable of resisting both biotic and abiotic stress has yet to be developed. Thailand's salt rock basins of Sakon Nakhon and Nakhon Ratchasima have demonstrated that consistent levels of salinity can enhance the fragrance of the RD6 rice variety [14] and increase production.

This study proposes the successful introgression of blast-resistant QTLs (*qBL 1, 2, 11*, and *12*) from RGD07005-12-165-1 and the *Saltol* QTL (Pokkali, chromosome 1) to improve the RD6 rice variety through the MAB method within BC$_4$F$_4$ populations. Trait evaluations were completed for the validation of progenies with desirable traits in each advanced population, based on the introgression of the genetic foregrounds and maintenance of the genetic backgrounds, respectively.

Salt salinity was absent in several areas of northeast Thailand, due to high levels of NaCl [43], and such factors as precipitation, soil type, and field management. In past research, the evaluation of salt tolerance was typically conducted through salt screening, hydroponic culture, and soil culture, as well as through pot and field methods [33]. The current study assessed salt tolerance within the breeding populations studied through salt solution, artificial salt culture, and field condition evaluations. Based on the results, salt evaluation under field conditions produced the lowest capability among the tested rice lines (Tables 1–3 and 5), due to the inherent difficulties and uncertainties present under field conditions. Kranto et al. [33] reported that effective alternative screening approaches must be proven to correlate with results produced within the early phases of growth in both greenhouse and field conditions. Within the present study, visual symptom scores of salt stress generated through the salt solution method proved to be the most appropriate method with which to confirm tolerance abilities within a breeding population (Table 5), suggesting that the salt solution method could, therefore, substitute salt tolerance score analysis in field conditions (Table 5). However, we acknowledge the necessity to evaluate RD6 plant types, yield performances, and agronomic traits within the field. The introgression lines developed within our study were evaluated for similarity with the original RD6 agronomic traits, namely plant height, panicle length, 4/panicle seed weight, 1000/seed weight, total dry weight, total seed weight, harvest index, seed length, seed width, and seed shape, as well as seed qualities, such as seed morphology (Tables 2 and 4).

The *Saltol* QTL on chromosome 1 from the Pokkali rice variety has been commonly used for rice improvement in several studies [21–26]. In our results, the *Saltol* QTLs from the Pokkali variety produced the greatest salt tolerance within the RD6 introgression lines (Table 1, Table 2, Table 3, and Table 5). This *saltol* QTL also contributed to the maintenance of low Na+, high K+, and low Na+/K+ homeostasis levels in rice stems, further resulting in increased salt tolerance [24,44] (Figure 3). The Pokkali variety was classified to balance the influx of Na+ and K+ for dilution in the mechanism, creating the ability to exclude Na+ from leaf blades and stems [45,46]. As the water up-take mechanisms in rice accept both nutrients and salt together, the Pokkali variety thereby demonstrated the highest and most significant differences in leaf, stem, root, and total dry weights when compared with other breeding lines (Table 5). However, the BC4F4 lines presented the agronomic traits (above) more closely matched to the RD6 than to the Pokkali (Table 5), due to the advance generation and visual selection of the trait performances (Tables 2 and 4). RD6 performance is very important for farmer acceptance and crop adaptation in our test areas. For example, excessively tall RD6 rice plants present problems in the grain filling stages as a result of heavy wind or rain [47]. Visual selection may explain the differences in percentages of Na+ of the RD6 introgression lines with those of Pokkali (Table 5, Figure 3).

As a photosensitive rice variety, the RD6 grows once a year, during Thailand's rainy season from late May to November [48]. These bimodal rain patterns produce favorable conditions for the occurrence of blast disease, causing damage in all stages of growth. Leaf blast generally occurs during the seedling and tilling stages, whereas neck blast usually occurs during the reproductive phase [4]. In our study, introgression lines were evaluated for blast disease in both the field and upland short-row evaluations.

In this study, the upland short-row method displayed greater incidences of blast disease, due to the favorable microclimate and moisture contents around the experimental plots (Tables 1 and 3) [49]. The experimental field was influenced by bimodal rain, capable of inducing leaf and neck blast symptoms (Table 3), further indicating the resistance of the QTLs [42,50]. Noenplab et al. [8] also reported that the blast QTL on chromosome 11 in the JHN variety successfully contributed to leaf and neck blast resistance. Pyramiding of four blast-resistant QTLs through MAS achieved high levels of blast resistance and broad-spectrum resistance to pathogens prevalent in the region [9]. Moreover, the testing of RD6 introgression lines for durable blast resistance and no-yield penalties were observed [42]. The results, herein, further demonstrated that neck blast disease caused direct yield loss during the grain filling phase [51], as well as lower 1000/SW within the original RD6 variety compared with those of the RD6 introgression lines (Table 4).

The resistance/tolerance abilities of the RD6 introgression lines represent the foreground genetics capable of enhancing plant breeding programs. However, maintaining the background of the original

RD6 variety is also desirable; therefore, the quality and performance of the RD6 within the QTL introgression was also a consideration. The BC_4F_4 populations, herein, were achieved through the introgression of blast-resistant QTLs (*qBL 1, 2, 11*, and *12*) from RGD07005-12-165-1 and *Saltol* QTL (Pokkali) and improved the RD6 rice variety through MAB. Consequently, the performance of the RD6 introgression lines was similar to that of the original RD6 variety (Table 4, Figure 2). The results indicate that foreground and background selection, together with visual selection, accurately depicts the efficiency of MAB.

5. Conclusions

Improvement of the RD6 rice variety for salt tolerance and blast resistance was successfully achieved utilizing the *Saltol* QTL and *qBl* (*1, 2, 11*, and *12*) through marker-assisted backcrossing, together with phenotypic selection. The resulting BC_4F_4 132-12 introgression line exhibited superior salt tolerance, blast resistance, and reduced neck blast and was capable of maintaining higher qualities and agronomic performances than that of the original RD6 variety.

Author Contributions: K.T. and S.K. conceived the study. T.M., S.C., and J.S. designed the experiments. K.T. and S.K. performed the experiments. J.S. and S.C. supervised the study. K.T., T.M., S.C., and J.S. wrote the manuscript. All authors have read and agreed to the published version of the manuscript.

Acknowledgments: This research was supported by The Plant Breeding Research Center for Sustainable Agriculture and The Salt-Tolerance Rice Research Group, Khon Kaen University, Khon Kaen, Thailand. Our gratitude is also extended to the Sakon Nakhon Rice Research Center for their support in our field experiments and Ubon Ratchathani Rice Research Center for their support in 2AP analysis.

References

1. Keeratipibul, S.; Luangsakul, N.; Lertsatchayarn, T. The effect of Thai glutinous rice cultivars, grain length and cultivating locations on the quality of rice cracker (arare). *Food Sci. Technol.* **2008**, *41*, 1934–1943. [CrossRef]

2. Chairote, E.; Jannoey, P.; Chairote, G. Improvement of Monacolin K and minimizing of citrinin content in Korkor 6 (RD 6) red yeast rice. *World Acad. Sci. Eng. Technol.* **2015**, *9*, 43–46.

3. Raj, S.V.; Saranya, R.S.; Kumar, D.S.; Chinnadurai, M. Farm-level economic impact of rice blast: A Bayesian approach. *Agric. Econ. Res. Rev.* **2018**, *31*, 141. [CrossRef]

4. Mentlak, T.A.; Kombrink, A.; Shinya, T.; Ryder, L.S.; Otomo, I.; Saitoh, H.; Terauchi, R.; Nishizawa, Y.; Shibuya, N.; Thomma, B.P.H.J.; et al. Effector-mediated suppression of chitin-triggered immunity by magnaporthe oryzae is necessary for rice blast disease. *Plant Cell* **2012**, *24*, 322–335. [CrossRef]

5. Chumpol, A.; Chankaew, S.; Saepaisan, S.; Monkham, T.; Sanitchon, J. New sources of rice blast resistance obtained from Thai indigenous upland rice germplasm. *Euphytica* **2018**, *214*, 183. [CrossRef]

6. Greer, C.A.; Webster, R.K. Occurrence, distribution, epidemiology, cultivar reaction, and management of rice blast disease in California. *Plant Dis.* **2001**, *85*, 1096–1102. [CrossRef]

7. Wang, D.; Guo, C.; Huang, J.; Yang, S.; Tian, D.; Zhang, X. Allele-mining of rice blast resistance genes at AC134922 locus. *Biochem. Biophys. Res. Commun.* **2014**, *446*, 1085–1090. [CrossRef]

8. Noenplab, A.; Vanavichit, A.; Toojinda, T.; Sirithunya, P.; Tragoonrung, S.; Sriprakhon, S.; Vongsaprom, C. QTL mapping for leaf and neck blast resistance in Khao Dawk Mali105 and Jao Hom Nin recombinant inbred lines. *Sci. Asia* **2006**, *32*, 133–142. [CrossRef]

9. Suwannual, T.; Chankaew, S.; Monkham, T.; Saksirirat, W.; Sanitchon, J. Pyramiding of four blast resistance QTLs into Thai rice cultivar RD6 through marker-assisted selection. *Czech J. Genet. Plant Breed.* **2017**, *53*, 1–8. [CrossRef]

10. Arunin, S.; Pongwichian, P. Salt-affected soils and management in Thailand. *Bull. Soc. Sea Water Sci.* **2015**, *69*, 319–325.

11. Wongsomsak, S. Salinization in Northeast Thailand. *Southeast Asian Stud.* **1986**, *24*, 133–153.

12. Arunin, S. Salt effect soil in Southeast Asia. In Proceedings of the International Symposium on Salt Affected Lagoon Ecosystem, Valencia, Spain, 18–25 September 1995.

13. Akbar, M. Breeding for salinity resistance in rice. In *Prospects for Bio-Saline Research*; Ahmed, R., Pietro, A.S., Eds.; Department of Botany, University of Karachi: Karachi, Sindh, Pakistan, 1986; pp. 37–55.

14. Summart, J.; Thanonkeo, P.; Panichajakul, S.; Prathepha, P.; McManus, M.T. Effect of salt stress on growth, inorganic ion and proline accumulation in Thai aromatic rice, KhaoDawk Mali 105, callus culture. *Afr. J. Biotechnol.* **2010**, *9*, 145–152.

15. Kumar, K.; Kumar, M.; Kim, S.-R.; Ryu, H.; Cho, Y.-G. Insights into genomics of salt stress response in rice. *Rice* **2013**, *6*, 27. [CrossRef] [PubMed]

16. Kobayashi, N.I.; Yamaji, N.; Yamamoto, H.; Okubo, K.; Ueno, H.; Costa, A.; Tanoi, K.; Matsumura, H.; Fujii-Kashino, M.; Horiuchi, T.; et al. OsHKT1;5 mediates Na+ exclusion in the vasculature to protect leaf blades and reproductive tissues from salt toxicity in rice. *Plant J.* **2017**, *91*, 657–670. [CrossRef] [PubMed]

17. Gregorio, G.B.; Senadhira, D.; Mendoza, R.D. Screening rice for salinity tolerance. *IRRI Discuss. Pap. Ser.* **1997**, *22*, 2–23.

18. Bhowmik, S.K.; Islam, M.M.; Emon, R.M.; Begum, S.N.; Siddika, A.; Sultana, S. Identification of salt tolerant rice cultivars via phenotypic and marker-assisted procedures. *Pak. J. Boil. Sci.* **2007**, *10*, 4449–4454.

19. Kavitha, P.G.; Miller, A.J.; Mathew, M.K.; Maathuis, F.J.M. Rice cultivars with differing salt tolerance contain similar cation channels in their root cells. *J. Exp. Bot.* **2012**, *63*, 3289–3296. [CrossRef]

20. Ferreira, L.J.; Azevedo, V.S.; Marôco, J.P.; Oliveira, M.M.; Santos, A.P. Salt tolerant and sensitive rice varieties display differential methylome flexibility under salt stress. *PLoS ONE* **2015**, *10*, e0124060. [CrossRef]

21. Mohammadi-Nejad, G.; Arzani, A.; Rezai, A.M.; Singh, R.K.; Gregorio, G.B. Assessment of rice genotypes for salt tolerance using microsatellite markers associated with the saltol QTL. *Afr. J. Biotechnol.* **2008**, *7*, 730–736.

22. Huyen, L.T.N.; Cuc, L.M.; Ham, L.H.; Khanh, T.D. Introgression the Saltol QTL into Q5BD, the elite variety of Vietnam using marker assisted selection (MAS). *Am. J. Biosci.* **2013**, *1*, 80–84. [CrossRef]

23. Ali, S.; Gautam, R.K.; Mahajan, R.; Krishnamurthy, S.L.; Sharma, S.K.; Singh, R.K. Stress indices and selectable traits in Saltol QTL introgressed rice genotypes for reproductive stage tolerance to sodicity and salinity stresses. *Field Crops Res.* **2013**, *154*, 65–73. [CrossRef]

24. Waziri, A.; Kumar, P.; Purty, R.S. Saltol QTL and their role in salinity tolerance in rice. *Austin. J. Biotechnol. Bioeng.* **2016**, *3*, 1067.

25. Ganie, S.A.; Borgohain, M.J.; Kritika, K.; Talukdar, A.; Pani, D.R.; Mondal, T.K. Assessment of genetic diversity of Saltol QTL among the rice (Oryza sativa L.) genotypes. *Physiol. Mol. Boil. Plants* **2016**, *22*, 107–114. [CrossRef] [PubMed]

26. Aala, W.F.; Gregorio, G.B. Morphological and molecular characterization of novel salt-tolerant rice germplasms from the philippines and bangladesh. *Rice Sci.* **2019**, *26*, 178–188. [CrossRef]

27. Tanksley, S.; Nelson, J. Advanced backcross QTL analysis: A method for simultaneous discovery and transfer of valuable QTLs from unadapted germplasm into elite breeding lines. *Theor. Appl. Genet.* **1996**, *92*, 191–203. [CrossRef]

28. Steele, K.; Price, A.H.; Shashidhar, H.E.; Witcombe, J.R. Marker-assisted selection to introgress rice QTLs controlling root traits into an Indian upland rice variety. *Theor. Appl. Genet.* **2005**, *112*, 208–221. [CrossRef]

29. Neeraja, C.N.; Maghirang-Rodriguez, R.; Pamplona, A.; Heuer, S.; Collard, B.C.; Septiningsih, E.M.; Vergara, G.; Sanchez, D.; Xu, K.; Ismail, A.M.; et al. A marker-assisted backcross approach for developing submergence-tolerant rice cultivars. *Theor. Appl. Genet.* **2007**, *115*, 767–776. [CrossRef]

30. Sundaram, R.M.; Vishnupriya, M.R.; Biradar, S.K.; Laha, G.S.; Reddy, G.A.; Rani, N.S.; Sarma, N.P.; Sonti, R.V. Marker assisted introgression of bacterial blight resistance in Samba Mahsuri, an elite indica rice variety. *Euphytica* **2007**, *160*, 411–422. [CrossRef]

31. Iftekharuddaula, K.M.; Newaz, M.A.; Salam, M.A.; Ahmed, H.U.; Mahbub, M.A.A.; Septiningsih, E.M.; Collard, B.C.; Sanchez, D.L.; Pamplona, A.M.; Mackill, D. Rapid and high-precision marker assisted backcrossing to introgress the SUB1 QTL into BR11, the rainfed lowland rice mega variety of Bangladesh. *Euphytica* **2010**, *178*, 83–97. [CrossRef]

32. Srichant, N.; Chankaew, S.; Monkham, T.; Thammabenjapone, P.; Sanitchon, J. Development of Sakon Nakhon rice variety for blast resistance through marker assisted backcross breeding. *Agronomy* **2019**, *9*, 67. [CrossRef]

33. Kranto, S.; Chankaew, S.; Monkham, T.; Theerakulpisuta, P.; Sanitchon, J. Evaluation for salt tolerance in rice using multiple screening methods. *J. Agric. Sci. Technol.* **2016**, *18*, 1921–1931.

34. Dellaporta, S.L.; Wood, J.; Hicks, J.B. A plant DNA minipreparation: Version II. *Plant Mol. Boil. Rep.* **1983**, *1*, 19–21. [CrossRef]

35. Yoshida, S.; Forno, D.A.; Cock, J.H.; Gomez, K.A. Routine procedure for growing rice plants in cultures solution. *Lab. Man. Physiol. Stud. Rice* **1976**, *17*, 60–65.

36. International Rice Research Institute. *Standard Evaluation System Manual*; International Rice Research Institute: Manila, Philippines, 1996.

37. International Rice Research Institute. *Standard Evaluation System for Rice (SES). Leaf blast*; International Rice Research Institute: Manila, Philippines, 2002; p. 15.

38. Sriseadka, T.; Wongpornchai, S.; Kitsawatpaiboon, P. Rapid method for quantitative analysis of the aroma impact compound, 2-Acetyl-1-pyrroline, in fragrant rice using automated headspace Gas chromatography. *J. Agric. Food Chem.* **2006**, *54*, 8183–8189. [CrossRef] [PubMed]

39. Havre, G.N. The flame photometric determination of sodium, potassium and calcium in plant extracts with special reference to interference effects. *Anal. Chim. Acta* **1961**, *25*, 557. [CrossRef]

40. Wongsaprom, C.; Sirithunya, P.; Vanavichit, A.; Pantuwan, G.; Jongdee, B.; Sidhiwong, N.; Lanceras-Siangliw, J.; Toojinda, T. Two introgressed quantitative trait loci confer a broad-spectrum resistance to blast disease in the genetic background of the cultivar RD6 a Thai glutinous jasmine rice. *Field Crop. Res.* **2010**, *119*, 245–251. [CrossRef]

41. Pinta, W.; Toojinda, T.; Thummabenjapone, P.; Sanitchon, J. Pyramiding of blast and bacterial leaf blight resistance genes into rice cultivar RD6 using marker assisted selection. *Afr. J. Biotechnol.* **2013**, *12*, 4432–4438. [CrossRef]

42. Nan, M.S.A.; Janto, J.; Sribunrueang, A.; Monkham, T.; Sanitchon, J.; Chankaew, S. Field evaluation of RD6 introgression lines for yield performance, blast, bacterial blight resistance, and cooking and eating Qualities. *Agronomy* **2019**, *9*, 825. [CrossRef]

43. Suwanich, K. Geology and geological structure of potash and rock salt deposits in Chalerm Phrakiat District, Nakhon Ratchasima Province in Northeastern Thailand. *Kasetsart J. (Nat. Sci.)* **2010**, *44*, 1058–1068.

44. Chunthaburee, S.; Dongsansuk, A.; Sanitchon, J.; Pattanagul, W.; Theerakulpisut, P. Physiological and biochemical parameters for evaluation and clustering of rice cultivars differing in salt tolerance at seedling stage. *Saudi J. Boil. Sci.* **2015**, *23*, 467–477. [CrossRef]

45. Suriya-arunroj, D.; Supapoj, N.; Vanavichit, A.; Toojinda, T. Screening and selection for physiological characters contributing to salinity tolerance in rice. *Kasetsart J. (Nat. Sci.)* **2005**, *39*, 174–185.

46. Kao, C.H. Mechanisms of salt tolerance in rice plants: Na+ transporters. *Crop Environ. Bioinform.* **2015**, *12*, 113–119.

47. Miyakawa, S. Expansion of an improved variety into rain-fed rice cultivation in Northeast Thailand. *Southeast Asian Stud.* **1995**, *33*, 187–203.

48. Jongdee, B.; Pantuwan, G.; Fukai, S.; Fischer, K. Improving drought tolerance in rainfed lowland rice: An example from Thailand. *Agric. Water Manag.* **2006**, *80*, 225–240. [CrossRef]

49. Vasudevan, K.; Cruz, C.M.V.; Gruissem, W.; Bhullar, N.K. Large scale germplasm screening for identification of novel rice blast resistance sources. *Front. Plant Sci.* **2014**, *5*, 505. [CrossRef]

50. Wang, R.; Fang, N.; Guan, C.; He, W.; Bao, Y.-M.; Zhang, H. Characterization and fine mapping of a blast resistant gene Pi-jnw1 from the japonica rice landrace Jiangnanwan. *PLoS ONE* **2016**, *11*, e0169417. [CrossRef]

51. Goto, K. Estimating losses from rice blast in Japan. In *The Rice Blast Disease*; Johns Hopkins Press: Baltimore, MD, USA, 1965; pp. 195–202.

5

QTL Mapping for Seedling and Adult Plant Resistance to Leaf and Stem Rusts in Pamyati Azieva × Paragon Mapping Population of Bread Wheat

Yuliya Genievskaya [1], Saule Abugalieva [1,2], Aralbek Rsaliyev [3], Gulbahar Yskakova [3] and Yerlan Turuspekov [1,4,*]

[1]   Laboratory of Molecular Genetics, Institute of Plant Biology and Biotechnology, Almaty 050040, Kazakhstan; julia.genievskaya@gmail.com (Y.G.); absaule@yahoo.com (S.A.)
[2]   Department of Biodiversity and Bioresources, Faculty of Biology and Biotechnology, al-Farabi Kazakh National University, Almaty 050040, Kazakhstan
[3]   Laboratory of Phytosanitary Safety, Research Institute of Biological Safety Problems, Gvardeisky 080409, Zhambyl Region, Kazakhstan; aralbek@mail.ru (A.R.); y_gulbahar@mail.ru (G.Y.)
[4]   Agrobiology Faculty, Kazakh National Agrarian University, Almaty 050010, Kazakhstan
*   Correspondence: yerlant@yahoo.com

Abstract: Leaf rust (LR) and stem rust (SR) pose serious challenges to wheat production in Kazakhstan. In recent years, the susceptibility of local wheat cultivars has substantially decreased grain yield and quality. Therefore, local breeding projects must be adjusted toward the improvement of LR and SR disease resistances, including genetic approaches. In this study, a spring wheat segregating population of Pamyati Azieva (PA) × Paragon (Par), consisting of 98 recombinant inbred lines (RILs), was analyzed for the resistance to LR and SR at the seedling and adult plant-growth stages. In total, 24 quantitative trait loci (QTLs) for resistance to rust diseases at the seedling and adult plant stages were identified, including 11 QTLs for LR and 13 QTLs for SR resistances. Fourteen QTLs were in similar locations to QTLs and major genes detected in previous linkage mapping and genome-wide association studies. The remaining 10 QTLs are potentially new genetic factors for LR and SR resistance in wheat. Overall, the QTLs revealed in this study may play an important role in the improvement of wheat resistance to LR and SR per the marker-assisted selection approach.

Keywords: *Triticum aestivum*; QTL; mapping population; leaf rust; stem rust; pathogen races; disease resistance

1. Introduction

Bread wheat (*Triticum aestivum* L.) is one of the major cereal crops in the world. In 2018/2019, the global production of wheat was 734.7 million metric tons, ranking second place amongst the grains after maize [1]. It is used mostly as flour for the production of a large variety of leavened and flat breads and the manufacturing of a wide range of other baked products [2]. In 2018/2019, Kazakhstan was ranked the 12th largest wheat producer in the world [3]. In Kazakhstan, wheat is cultivated on about 13 million hectares annually. The country produces up to 20–25 million tons of bread wheat per year and exports up to 5–7 million tons of the grain [4]. The primary goals of modern wheat breeding programs worldwide include enhancing grain yield and quality and increasing resistance to biotic and abiotic stresses to ensure global food security [5]. Biotic stresses include dangerous fungal diseases and particularly the most common representatives of the *Puccinia* genus: *Puccinia recondita* Rob. ex Desm f. sp. *tritici*, causing leaf rust (LR), and *Puccinia graminis* Pers. f. sp. *tritici* Eriks. & Henn., which is responsible for stem rust (SR) of wheat.

LR generally causes light to moderate yield losses ranging from 1% to 20% over a large area, but when the disease is severe prior to heading time, it may destroy up to 90% of the wheat crop [6]. For example, in Kazakhstan, epiphytotic development of the pathogen on spring wheat during 2000–2001 resulted in 50–100% LR severity on commercial cultivars in the Akmola region (northern Kazakhstan), which is the main wheat-growing region in the country [7].

SR is another important rust disease that is often considered the most devastating of the wheat rust diseases because it can cause complete crop loss over a large area within a short period of time [8]. In 2016, northern Kazakhstan was subjected to an epiphytotic outbreak of SR, resulting in 50% disease development severity in the field, decreasing wheat yield and grain quality [7]. Nowadays, local farmers prefer the usage of fungicides to protect wheat fields from LR and SR; however, this method is harmful to the environment and more expensive than breeding and growing genetically resistant wheat cultivars [9].

LR and SR resistances are controlled by a diverse group of genes, designated as *Lr* and *Sr*, respectively [10]. In the last 100 years, approximately 80 *Lr* resistance genes have been identified and described in bread wheat, durum wheat, and diploid wheat species [10], and the list is still growing. For SR, nearly 60 *Sr* genes have been identified to date in wheat and its wild relatives [10]. Generally, resistance to rust diseases can be broadly categorized into two types. The first is resistance at all growth stages (called seedling resistance), detected at the seedling stage and expressed until the plant dies. This type of resistance is controlled by the R type of genes, and the majority of *Lr* and *Sr* genes belong to this group. The efficacy of the R gene is pathogen-strain-dependent [11]. The second type of resistance is adult plant resistance (APR), where genes are ineffective during the seedling stage but provide robust resistance at maturity [11]. For example, LR resistance genes *Lr12*, *Lr13*, *Lr22a*, *Lr22b*, *Lr34*, *Lr35*, *Lr46*, *Lr48*, *Lr49*, and *Lr67*, and SR resistance genes *Sr2* and *Sr57* are well-characterized APR genes [10]. Durable rust resistance is more likely to be the APR type rather than the seedling type [12]; both types are important for wheat breeding [11].

Two of the most effective methods of quantitative trait locus (QTL) mapping are based on association panels and biparental segregating populations [13]. Both of these methods provide the means to investigate the genome and describe the etiology of complex quantitative traits, including disease resistance [14–17]. Genetic maps are a key tool enabling genetic linkage studies and searches for novel loci responsible for traits. Modern high-throughput sequencing technologies allow for the high-accuracy genotyping of large collections with genetically diverse germplasms [18,19] and segregating mapping populations, such as doubled haploids (DHs), recombinant inbred lines (RILs), F₂, and backcross (BC) populations [20]. Linkage maps were successfully used for the QTL analyses of wheat yield components [21], grain quality traits [22], abiotic [23], and biotic stress factors, including pests [24].

The primary goal of this study was to identify QTLs involved in seedling and adult plant resistance of bread wheat to LR and SR under environmental conditions in southern and southeastern Kazakhstan. To meet this goal, the Pamyati Azieva × Paragon (PA × Par) RILs mapping population (MP) was studied in field and greenhouse (GH) conditions. Previously, this population was successfully used for the analysis of yield-related traits [25] and adult plant resistance to LR and SR in south-east and northern Kazakhstan in 2018 [26,27]. Hence, the current study adds the investigation of seedling resistance in the MP to LR and SR races. In addition to one-year studies in south-east and north Kazakhstan, this work covers the analysis of LR and SR resistance in the MP in southern Kazakhstan in 2018 and 2019.

2. Materials and Methods

2.1. Plant Material and Genotyping

The biparental mapping population PA × Par composed of 98 RILs was developed in greenhouse conditions of the John Innes Centre (Norwich, UK) during 2011–2015 under the ADAPTAWHEAT project [28]. The RIL population was obtained via a single-seed descent method using two parental cultivars: Paragon (elite spring cultivar originated from the U.K.) and Pamyati Azieva (a commercial spring cultivar originating from Russia and registered in Kazakhstan) [29]. Both cultivars were chosen due to their diverse genetic backgrounds and different manifestations of yield traits as well as resistance to diseases. The RIL population was further developed for F_8 generation in the fields of southeastern Kazakhstan.

The RILs and two parental cultivars were genotyped using the Illumina's iSelect 20K single nucleotide polymorphism (SNP) array at the TraitGenetics Company (TraitGenetics GmbH, Gatersleben, Germany). The genotypic data were filtered from markers with >10% missing data and with <0.1 minor allele frequency and consisted of 4595 polymorphic SNP markers.

2.2. Phenotyping of Seedling Resistance in Greenhouse

For the comprehensive study of the PA × Par MP response to LR and SR pathogens, the resistance was evaluated at the seedling and adult plant-growth stages. Race-specific resistance at the seedling stage was assessed in a greenhouse of the Research Institute of Biological Safety Problems (RIBSP, Gvardeisky, Zhambyl region, southern Kazakhstan). For the inoculation of RILs seedlings (7–10 days after sowing) in greenhouse conditions, three races of P. graminis and three races of P. recondita with different levels of virulence to Sr and Lr genes, respectively, were used (Table 1). Inoculated plants were placed in the boxes of the greenhouse with appropriate temperature conditions (22 ± 2 °C for SR, 18 ± 2 °C for LR) and illumination (10,000–15,000 lux, 16 h' light period) [30–32]. RIL reaction was assessed on the 14th day after inoculation, according to the scale reported by Stakman [33]. The experiment was performed in two independent replicates.

Table 1. Virulence/avirulence pattern of pathogen races used in the study.

Disease (Pathogen)	Race	Avirulent (Effective) Genes	Virulent (Ineffective) Genes
LR (*Puccinia recondita* Rob. ex Desm f. sp. *tritici*)	TQKHT	*Lr24, 26, 3ka, 19, 25*	*Lr1, 2a, 2c, 3, 9, 16, 11, 17, 30, 20, 29, 2b, 3bg, 14a, 15*
	TRTHT	*Lr24, 19, 25*	*Lr1, 2a, 2c, 3, 9, 16, 26, 3ka, 11, 17, 30, 20, 29, 2b, 3bg, 14a, 15*
	TQTMQ	*Lr24, 26, 20, 25, 14a, 15*	*Lr1, 2a, 2c, 3, 9, 16, 3ka, 11, 17, 30, 19, 29, 2b, 3bg*
SR (*Puccinia graminis* Pers. f. sp. *tritici* Eriks. & E. Henn.)	TKRTF	*Sr11, 30, 24, 31*	*Sr5, 21, 9e, 7b, 6, 8a, 9g, 36, 9b, 17, 9a, 9d, 10, 38, Tmp, McN*
	PKCTC	*Sr21, 11, 36, 9b, 30, 24, 31, 38*	*Sr5, 9e, 7b, 6, 8a, 9g, 17, 9a, 9d, 10, Tmp, McN*
	RKRTF	*Sr9e, 11, 30, 24, 31*	*Sr5, 21, 7b, 6, 8a, 9g, 36, 9b, 17, 9a, 9d, 10, 38, Tmp, McN*

LR, leaf rust; SR, stem rust.

Races of P. graminis were differentiated in 2018 [34] using the North American nomenclature [35] with the assistance of five sets of SR-differentiating wheat cultivars. Races of P. recondita were also identified in 2018 using 20 Thatcher near-isogenic lines (NILs) sets of Lr genes [36–38]. For the nomenclature of P. recondita races, Virulence Analysis Tools [39] were used.

2.3. Adult Plant Resistance and Yield Components in Field Conditions

APR in the field was tested in two environments: the RIBSP and the Kazakh Research Institute of Agriculture and Plant Industry (KRIAPI, Almalybak, Almaty region, southeastern Kazakhstan) (Table 2).

Table 2. Meteorological data on average temperature and precipitations during the vegetation period in the fields of Research Institute of Biological Safety Problems (RIBSP, southern Kazakhstan) and Kazakh Research Institute of Agriculture and Plant Industry (KRIAPI, southeastern Kazakhstan).

	KRIAPI				
	March	April	May	June	July
Temperature (°C)	8.2	12.8	17.0	22.3	27.0
Precipitation (mm)	27.3	168.4	39.3	72.7	22.6
	RIBSP				
	March	April	May	June	July
Temperature (°C)	13.0	17.0	24.0	29.0	33.0
Precipitation (mm)	1.0	2.3	1.8	0.9	0.1

In RIBSP fields, mixed races of LR and SR urediniospores common in Kazakhstan were applied as inoculum. The inoculum was activated at a temperature of 37–40 °C for 30 min, followed by watering in a humid chamber at a temperature of 18–22 °C for 2 h. At the booting stage, individual plants were treated with an aqueous suspension of leaf and stem rust urediniospores dissolved in Tween 80 detergent. After inoculation, the plots were covered with plastic wrap for 16–18 h. In KRIAPI fields, inoculation occurred with local LR and SR pathogen populations in uncontrolled natural conditions.

Thus, experiments were conducted in three independent environments, including the study of seedling resistance in greenhouse conditions, the study of APR at RIBSP (controlled inoculation), and APR at KRIAPI (uncontrolled inoculation). In both field conditions, phenotyping of APR to LR and SR was performed in two independent replicates at the stage of grain ripening with the maximum level of disease manifestation. Disease assessment was performed using the scale of Stakman for SR [33] and the scale of Mains and Jackson for LR [40]. The severity of rust infection on leaf and stem surfaces was evaluated using the modified Cobb scale [41,42]. To meet the data format required for linkage analysis, the results of LR and SR evaluations at both seedling and adult plant-growth stages were converted to the 0–9 linear disease scale as described by Zhang et al. [43].

To identify the influence of LR and SR severity on the productivity of the studied population, two important yield-related components, thousand kernel weight (TKW, g) and kernel yield per plot (YP, g/m^2), were also evaluated.

2.4. Statistical Analysis of Phenotypic Data and QTL Mapping

Phenotypic data processing, descriptive statistics, and one-tailed correlation tests were performed with SPSS Statistics v. 22 (SPSS Inc., Chicago, IL, USA).

The composite interval mapping (CIM) method with the Kosambi mapping function was used for the detection of QTLs by Windows QTL Cartographer v2.5 [44]. The threshold value for the logarithm of odds (LOD) score was calculated based on 1000 permutations and was 3.0 for all experiments with a walking step of 1 cM. QTLs were detected for each environment and replication separately (seedling resistance in GH, APR at KRIAPI, and APR at RIBSP). QTLs identified in individual environments and/or replications overlapping in 20 cM intervals and associated with the same trait were considered as identical [45]. Genetic maps with QTLs were drawn using MapChart v. 2.32 software [46]. For the markers with the same positions, only one single nucleotide polymorphism (SNP) maker was selected for the map.

All genes present within the interval of 500 kb to the left and 500 kb to the right (1 Mb in total) from the peak marker were identified using the Ensembl Plant database [47]. As a reference, the genome of *T. aestivum* RefSeq v1 was used. The exact position of the peak SNP in the genome was determined using a BLAST tool [48]. Proteins and RNA gene products were identified using the UniProt database [49] via cross-reference from Ensembl Plant.

3. Results

3.1. Phenotyping Variations of Seedling and Adult Plant Resistance in Mapping Population

The values of the resistance to target diseases in parents and 98 RILs are summarized in Table 3.

Table 3. Descriptive statistics for leaf rust (LR) and stem rust (SR) resistance at two plant-growth stages in the Pamyati Azieva × Paragon mapping population.

Env.	Plant-Growth Stage	Race (Disease)	Parents IT [1]		RILs IT					
			PA	Par	Mean	Range	R (%)	MR (%)	MS (%)	S (%)
GH	Seedling	TQTMQ (LR)	6 (3)	5 (3-)	6.1 ± 0.9	5–8	0	0	89.8	10.2
		TQKHT (LR)	8 (4-)	6 (3)	5.9 ± 1.0	4–9	0	3.1	91.8	5.1
		TRTHT (LR)	6 (3)	6 (3)	5.9 ± 0.8	4–8	0	1.0	96.9	2.1
	Seedling	TKRTF (SR)	8 (4-)	8 (4-)	7.0 ± 1.4	5–9	0	0	45.9	54.1
		PKCTC (SR)	5 (3-)	8 (4-)	5.4 ± 1.8	1–9	1.0	22.4	58.2	18.4
		RKRTF (SR)	6 (3)	5 (3-)	6.5 ± 1.4	5–9	0	0	64.3	35.7
RIBSP	Adult	LR	8 (30S)	6 (50MS)	3.8 ± 1.4	0–9	38.8	11.2	29.6	20.4
		SR	9 (80S)	8 (40S)	6.6 ± 2.9	1–9	1.0	23.5	18.4	57.1
KRIAPI		LR	6 (40MS)	6 (40MS)	6.6 ± 1.7	2–9	0	9.2	64.3	26.5
		SR	0 (0)	1 (10R)	1.8 ± 2.3	0–8	56.1	26.5	18.4	1.0

[1]—parents IT scores are given in 0–9 numeric scale, traditional IT scores are given in parentheses. Env., environment; GH, greenhouse conditions; PA, Pamyati Azieva; Par, Paragon; IT, infection type; R, percentage of resistant lines (0–1 on 9-point scale); MR, percentage of moderately resistant lines (2–4 on 9-point scale); MS, percentage of moderately susceptible lines (5–7 on 9-point scale); S, percentage of susceptible lines (8–9 on 9-point scale).

The average seedling resistance of RILs to LR races was between 5.9 and 6.1 points, corresponding to the moderately susceptible (MS) level. The major part of the population belonged to the MS group, with only several lines observed in the susceptible (S) group. Several lines were also in the resistant (R) group to the races TQKHT and TRTHT. Parental cultivars demonstrated an MS level of resistance to studied LR races, except for PA, which was susceptible to the race TQKHT. As for seedling resistance to SR, the average level in the RILs population was MS to all three SR races. However, unlike in the case of LR, the distribution of lines among resistance groups was different. Races TKRTF and RKRTF were divided between MS and S groups with a dominance of the S reaction to TKRTF and MS reaction to RKRTF. R and moderately resistant (MR) levels were detected only in the race PKCTC. Levels of resistance in parental cultivars were similar to races TKRTF (S) and RKRTF (MS), but PA demonstrated higher resistance to the race PKCTC than Par.

At the adult plant stage, the reactions of parents and RILs to LR and SR were significantly different between the studied environments. At RIBSP, the average reaction of RILs to LR was MR, with almost even distribution among all possible reactions observed in the population. At KRIAPI, the average level of resistance was MS with a dominance of MS and S reactions in the population. The parental cultivars demonstrated the same reaction to LR at KRIAPI, but at RIBSP, Par was more resistant than PA. For SR at the adult plant stage at RIBSP, the majority of RILs were in the S group, and the average level was MS. At KRIAPI, the largest part of the population was in the R group and the average level was MR. Parental cultivars also were in the S group at RIBSP and the R group at KRIAPI.

The analysis of variance showed that the resistance of RILs to LR at the seedling growth stage was significantly affected by the RIL genotype ($p < 0.01$) and the race of LR pathogen ($p < 0.05$), but not by genotype × race interaction (Table 4). For SR resistance, all factors had a significant influence ($p < 0.001$) on the resistance at the seedling stage (Table 4).

Table 4. ANOVA of plant genotype (Geno), pathogen race (Race), and plant genotype × pathogen race (Geno: Race) effects on seedling resistance to leaf rust (LR) and stem rust (SR) in the Pamyati Azieva × Paragon mapping population.

Factor	df	LR			SR		
		MeanS	F	p	MeanS	F	p
Geno	97	2.506	1.630	0.001	5.380	2.679	8.39×10^{-11}
Race	2	5.420	3.525	0.031	136.940	68.180	$<2.00 \times 10^{-16}$
Geno: Race	194	1.396	0.908	0.766	4.540	2.262	1.20×10^{-10}
Residuals	294	1.537			2.010		

df, degree of freedom; MeanS, mean square.

3.2. Correlations among SR and LR Seedling and Adult Plant Resistance and Influence of APR on the Yield-Related Traits

Significant positive correlations were found among the reactions to all three LR races at the seedling stage, as well as between APR to LR at KRIAPI and seedling resistance to LR races TQTMQ and TRTHT (Table 5). APR to LR at KRIAPI was also positively correlated with APR to SR at KRIAPI and seedling resistance to SR race TKRTF. For the other SR races, race PKCTC had a positive correlation with APR to SR at KRIAPI, and race RKRTF was negatively correlated with LR race TQTMQ. The only significant correlation of APR to SR at RIBSP was associated with LR race TRTHT.

Table 5. Correlations among race-specific seedling resistance and adult plant resistance (APR) to leaf rust (LR) and stem rust (SR) in the Pamyati Azieva × Paragon mapping population.

	LR TQTMQ	LR TQKHT	LR TRTHT	APR LR RIBSP	APR LR KRIAPI	SR TKRTF	SR PKCTC	SR RKRTF	APR SR RIBSP
LR TQKHT	0.193 *								
LR TRTHT	0.207 *	0.283 **							
APR LR RIBSP	0.008 ns	0.122 ns	0.154 ns						
APR LR KRIAPI	0.251 **	−0.028 ns	0.192 *	−0.124 ns					
SR TKRTF	−0.152 ns	0.000 ns	0.141 ns	0.088 ns	0.228 *				
SR PKCTC	−0.150 ns	0.021 ns	0.058 ns	0.081 ns	−0.009 ns	−0.019 ns			
SR RKRTF	−0.202 *	−0.006 ns	0.091 ns	−0.125 ns	0.069 ns	0.050 ns	0.110 ns		
APR SR RIBSP	−0.071 ns	0.096 ns	0.197 *	0.130 ns	0.038 ns	−0.075 ns	−0.040 ns	−0.029 ns	
APR SR KRIAPI	0.127 ns	0.040 ns	0.117 ns	0.018 ns	0.245 **	−0.006 ns	0.182 *	0.015 ns	−0.043 ns

APR, adult plant resistance; ns not significant; * $p < 0.05$; ** $p < 0.01$.

The negative influence of LR and SR severities at the adult plant stage on the wheat YP and TKW was confirmed by significant negative correlations ($p < 0.01$) between these traits at RIBSP (Table 6). At KRIAPI, the severity of LR at the adult plant stage was negatively correlated with YP only.

Table 6. Correlations between leaf rust (LR) and stem rust (SR) resistance at the adult plant stage and yield-related traits in the Pamyati Azieva × Paragon mapping population.

	TKW RIBSP	YP RIBSP	TKW KRIAPI	YP KRIAPI
LR RIBSP	−0.175 **	−0.252 **	–	–
SR RIBSP	−0.490 **	−0.474 **	–	–
LR KRIAPI	–	–	−0.055 ns	−0.200 *
SR KRIAPI	–	–	0.134 ns	−0.151 ns

TKW, thousand kernel weight (g); YP, yield per plot (g/m^2); ns not significant; * $p < 0.05$; ** $p < 0.01$.

3.3. Identification of QTLs for Seedling and Adult Plant Resistance to LR in the RIL Population

A total of 11 QTLs for resistance to LR were identified at the seedling and adult plant-growth stages. Out of these 11 QTLs, eight QTLs were detected for different LR races at the seedling stage, two QTLs were for APR, and one QTL was observed for both seedling and adult plant resistance (Table 7, Figure 1). QTLs for LR resistance were located on 10 chromosomes of the A, B, and D genomes. The phenotypic variations explained by an individual QTL ranged from 11.6% to 25.7%. Because all QTLs for LR resistance identified in this study explained more than 10% of the phenotypic variation, they were considered major QTLs [50]. The LOD score of QTLs for LR resistance was in the range of 3.2–8.6.

For the LR race TQTMQ, three QTLs identified on chromosomes 4A, 5B, and 7B were revealed. They explained 12.1–15.7% of the phenotypic variations. The alleles of all three QTLs associated with the increase in resistance to LR originated from PA. For the second LR race TQKHT, two QTLs on chromosomes 6A and 7D explained 25.7% and 11.6% of the variations in phenotype, respectively. Both QTLs associated with higher resistance to race TQKHT originated from Paragon. For the third LR race TRTHT, three QTLs on chromosomes 3A, 4D, and 6A were observed. Identified QTLs explained 12.3–16.2% of the variation in resistance to race TQKHT. The alleles of QTLs *QLr.ipbb-3A.2* and *QLr.ipbb-6A.5* increasing resistance were from Paragon, and the allele of *QLr.ipbb-4D.1* was from PA.

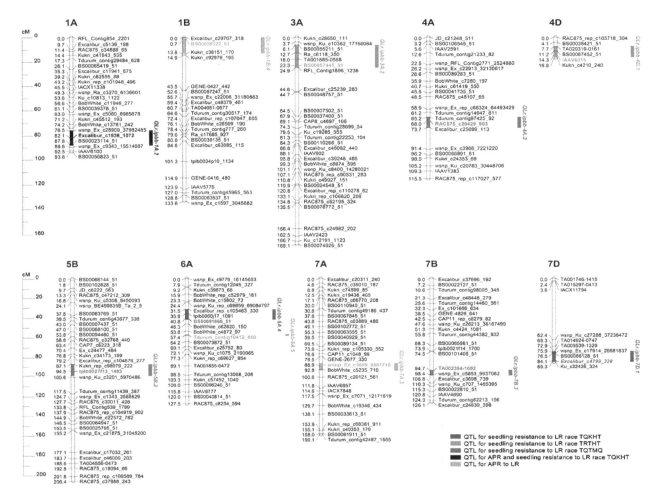

Figure 1. Pamyati Azieva × Paragon genetic map with quantitative trait loci (QTLs) for adult plant resistance (APR) to leaf rust (LR) in two regions and seedling resistance to three LR races. The region containing the QTL is indicated by a vertical bar on the right and followed by the name of the QTL. Single nucleotide polymorphism (SNP) markers are shown on the right and their genetic positions (cM) on the left. The peak marker for each QTL is highlighted in color and bolded. Colors of QTL indicate APR or race-specific seedling resistance.

Table 7. The list of quantitative trait loci (QTLs) for adult plant resistance (APR) and race-specific seedling resistance to leaf rust (LR) identified in the Pamyati Azieva × Paragon recombinant inbred lines (RILs) population.

Race	Env.	QTL	Flanking Markers (Left–Right)	Chr.	CI (cM)	Peak (cM)	Max. LOD	R^2 (%)	Add. Effect[1]	Allele[1]
TQTMQ	GH	QLr.ipbb-4A.2	IAAV7104–Excalibur_c25699_113	4A	65.2–73.9	68.0	3.3	12.1	-0.47	PA
	GH	QLr.ipbb-5B.2	Kukri_rep_c98079_222–BS00075815_51	5B	87.4–99.2	94.5	4.3	15.2	-0.57	PA
	GH	QLr.ipbb-7B.3	BS00023023_51–wsnp_Ex_c5653_9937062	7B	93.5–98.9	94.7	4.9	15.7	-0.79	PA
TQKHT	GH	QLr.ipbb-6A.4	Excalibur_rep_c105463_330–Ku_c37893_495	6A	31.5–42.0	40.8	8.6	25.7	0.63	Paragon
	GH	QLr.ipbb-7D.1	Kukri_c16416_647–BS00062644_51	7D	74.0–84.9	84.1	4.2	11.6	0.41	Paragon
TRTHT	GH	QLr.ipbb-3A.2	Excalibur_c74666_291–RFL_Contig1896_1236	3A	6.2–24.7	23.3	5.1	16.2	0.46	Paragon
	GH	QLr.ipbb-4D.1	TA020319-0161–BS00022436_51	4D	7.1–16.2	14.3	3.7	12.3	-0.39	PA
	GH	QLr.ipbb-6A.5	BobWhite_c30930_192–BS00022992_51	6A	56.9–58.5	57.4	4.8	15.1	0.44	Paragon
APR	RIBSP	QLr.ipbb-1B.4	Excalibur_c29707_318–Kukri_c36151_170	1B	0–13.0	0.7	3.2	13.6	1.29	Paragon
APR	RIBSP	QLr.ipbb-7A.3	Ra_c4601_2417–CAP7_c7296_88	7A	86.4–94.0	88.9	3.4	11.7	1.18	Paragon
TQKHT/APR	GH/KRIAPI	QLr.ipbb-1A.2	RAC875_c12348_720–BS00022824_51	1A	75.9–88.2	82.1	4.6	16.0	0.86	Paragon

CI, confidence interval; LOD, logarithm of odds; $R2$, phenotypic variance explained by the QTL; Chr., chromosome; Add., additive effect. [1]—Additive effect of QTL indicates increasing of the trait expression explained by the allele of one parent (positive for PA and negative for Paragon). In the case of disease resistance, increased expression of the trait is undesired, and the effective allele is taken from the other parent (negative for PA and positive for Paragon).

Two QTLs for APR to LR at RIBSP were identified on chromosomes 1B and 7A and explained 13.6% and 11.7% of phenotypic variation, respectively. For both APR QTLs, alleles associated with increased LR resistance were from Paragon. One QTL for APR to LR at KRIAPI was also detected at the seedling stage for the resistance to LR race TQKHT on chromosome 1A. It explained 16.0% of LR resistance variation. In both cases, alleles associated with higher resistance originated from Paragon.

3.4. QTLs for SR Resistance at Seedling and Adult Plant Stages Identified in PA × Par Mapping Population

A total of 13 QTLs were detected in this study for SR resistance at the seedling and adult plant-growth stages. Among them, seven race-specific QTLs were identified at the seedling stage (three QTLs for race TKRTF, three QTLs for race PKCTC, and one QTL for race RKRTF), three QTLs were observed for APR (two QTLs at KRIAPI and one QTL at RIBSP), and three QTLs were revealed in both the seedling and adult stages (Table 8, Figure 2). The identified QTLs for SR resistance were distributed among nine chromosomes of A, B, and D genomes and explained from 8.9% to 39.1% of the variation in the resistance to SR. In total, 11 out of 13 QTLs for SR resistance had $R^2 > 10\%$ and could be considered major QTLs. The LOD score for the detected QTL varied from 3.0 to 6.8.

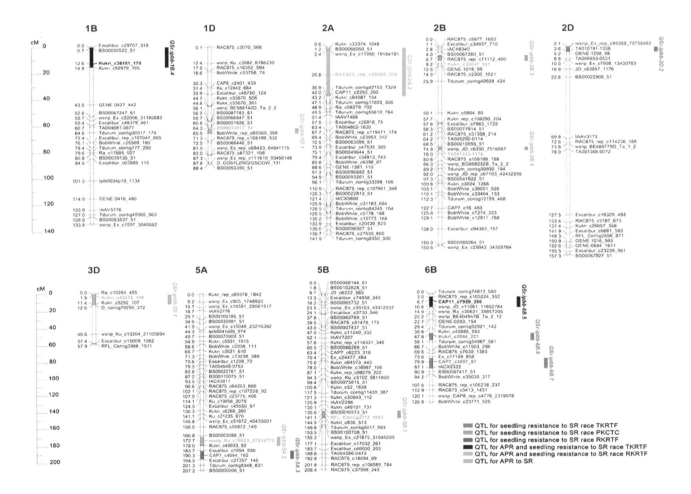

Figure 2. Pamyati Azieva × Paragon genetic map with quantitative trait loci (QTLs) for adult plant resistance (APR) to stem rust (SR) in two regions and seedling resistance to three SR races. The region containing the QTL is indicated by a vertical bar on the right and followed by the name of the QTL. Single nucleotide polymorphism (SNP) markers are positioned on the right and their genetic positions (cM) are shown on the left. The peak marker for each QTL is highlighted in color and bold. Colors of QTL indicate APR or race-specific seedling resistance.

Table 8. The list of quantitative trait loci (QTLs) for APR and race-specific seedling resistance to stem rust (SR) identified in Pamyati Azieva × Paragon recombinant inbred lines (RILs) population.

Race	Env.	QTL	Flanking Markers (Left–Right)	Chr.	CI (cM)	Peak (cM)	Max. LOD	R2 (%)	Add. Effect *	Allele *
TKRTF	GH	QSr.ipbb-2D.2	wsnp_Ex_rep_c80588_75758453-TA010191-1338	2D	0–3.7	0.1	3.0	10.0	0.43	Paragon
	GH	QSr.ipbb-6B.6	wsnp_Ex_c12618_20079758-wsnp_Ex_c1249_2399894	6B	46.0–54.3	47.8	4.2	14.1	−0.57	PA
	GH	QSr.ipbb-6B.7	wsnp_Ku_c43368_50890819-wsnp_Ku_c4910_8793327	6B	75.5–87.3	79.9	4.7	19.2	−0.66	PA
PKCTC	GH	QSr.ipbb-2B.3	RAC875_rep_c71112_400-Kukri_c1175_1577	2B	6.3–10.4	9.2	3.9	12.2	−0.63	PA
	GH	QSr.ipbb-2B.4	BS00041323_51-RAC875_c32503_134	2B	75.9–78.3	78.0	2.9	8.9	0.54	Paragon
	GH	QSr.ipbb-5B.1	RAC875_rep_c114200_428-wsnp_RFL_Contig1548_762547	5B	137.6–146.3	141.1	4.9	14.7	0.96	Paragon
RKRTF	GH	QSr.ipbb-5A.3	BS00036851_51-Excalibur_c27357_146	5A	184.7–194.4	190.3	6.8	24.7	0.72	Paragon
APR	KRIAPI	QSr.ipbb-1D.1	BS0005182 6_51-BobWhite_rep_c65565_359	1D	61.7–65.5	64.3	5.0	18.0	2.24	Paragon
	RIBSP	QSr.ipbb-3D.1	Ra_c10284_405-Kukri_c5252_107	3D	0–11.4	1.9	3.1	11.3	−0.99	PA
	KRIAPI	QSr.ipbb-5A.2	BS00089076_51-CAP11_c2623_196	5A	168.2–178.5	172.7	3.0	10.8	0.72	Paragon
TKRTF/APR	GH/KRIAPI	QSr.ipbb-1B.4	Excalibur_c29707_318-Kukri_c92979_195	1B	0–15.2	12.6	4.2	15.7	0.59/−0.77	Paragon/PA
RKRTF/APR	GH/RIBSP	QSr.ipbb-2A.2	Kukri_c33374_1048-Tdurum_contig42153_5854	2A	0.5–37.2	30.8	5.2	39.1	−1.45	PA
TKRTF/APR	GH/RIBSP	QSr.ipbb-6B.5	RAC875_rep_c105224_352-Kukri_rep_c107077_360	6B	2.7–15.1	6.7	4.6	17.3	−0.56/1.26	PA/Paragon

* Additive effect of QTL indicates increasing of the trait expression explained by the allele of one parent (positive for PA and negative for Paragon). In the case of disease resistance, increased expression of the trait is undesired, and the effective allele is taken from the other parent (negative for PA and positive for Paragon).

Three QTLs for resistance to SR race TKRTF were identified, including two QTLs mapped on chromosome 6B and one QTL on 2D. These QTLs explained 10.0–19.2% of the variation in SR resistance to race TKRTF. QTLs *QSr.ipbb-6B.6* and *QSr.ipbb-6B.7* had alleles increasing SR resistance to race TKRTF carried by PA, whereas resistance allele *QSr.ipbb-2D.2* originated from Paragon. For the second SR race PKCTC, two QTLs for resistance to this race were located on chromosome 2B and one QTL was on 5B. The phenotypic variance conditioned by these QTLs varied from 8.9% to 14.7%. The third SR race RKRTF allowed the identification of one race-specific QTL on chromosome 5A explaining 24.7% of the phenotypic variation. Its allele, associated with an increase in SR resistance, originated from Paragon.

Three QTLs for APR to SR were identified on chromosomes 1D, 3D, and 5A, and explained from 10.8% to 18.0% of SR resistance variation. Two QTLs identified at KRIAPI had alleles increasing resistance to SR originating from Paragon, and the allele of QTL at RIBSP was from PA. The last three QTLs for LR resistance occurred multiple times in the experiment and are located on chromosomes 1B, 2A, and 6B. The QTL *QSr.ipbb-1B.4* was detected as race-specific to TKRTF at the seedling stage and as APR QTL at KRIAPI. It explained 15.7% of SR resistance variation and had alleles increasing resistance originating from Paragon at the seedling stage and PA at the adult plant stage. The QTL *QSr.ipbb-2A.2* was identified as effective against SR race RKRTF at the seedling stage and as APR QTL at RIBSP. This QTL explained 39.1% of the phenotypic variation, and alleles increasing SR resistance at both growth stages were inherited from PA. The QTL *QSr.ipbb-6B.5* was discovered at the seedling stage to race TKRTF and at the adult plant stage at RIBSP. The QTL explained 17.3% of SR resistance variations. Its alleles increasing resistance originated from PA in the case of seedling resistance and from Paragon at the adult growth stage.

3.5. Comparison of Identified QTLs with Previous Works and Gene Identification

The QTLs identified in this study were analyzed in comparison with previously reported QTLs for LR and SR resistance in the PA × Par RILs population [27] and with QTLs for LR and SR resistance at RIBSP identified using genome-wide association study (GWAS) [51]. The location of each identified QTL was compared to the genetic positions of known *Lr* and *Sr* genes (Table 9). In total, four candidate *Lr* genes and four QTLs were found for five QTLs associated with LR resistance in this study. In the analysis of QTLs for SR resistance, we found similarities with the genetic locations of eight previously identified QTLs and/or candidate *Sr* genes.

Table 9. Comparison of quantitative trait loci (QTLs) for leaf rust (LR) and stem rust (SR) resistance identified in this study in a Pamyati Azieva × Paragon mapping population with previously described QTLs and candidate *Lr* and *Sr* genes.

#	Trait	Type of Resistance (Race)	QTL	Reference QTL	Candidate Genes
1		TQKHT/APR	QLr.ipbb-1A.2	QLr.ipbb-1A.1 [51]	-
2		APR	QLr.ipbb-1B.4	-	Lr21 [47]
3	LR	TRTHT	QLr.ipbb-3A.2	-	Lr63, Lr66 [52]
4		TQKHT	QLr.ipbb-6A.4	QLr.ipbb-6A.1 [51]	-
5		TRTHT	QLr.ipbb-6A.5	QLr.ipbb-6A.2 [51]	-
6		TQTMQ	QLr.ipbb-7B.3	QLr.ipbb-7B.1 [51]	Lr14 [52]
1		TKRTF/APR	QSr.ipbb-1B.4	-	Sr31 [53]
2		APR	QSr.ipbb-1D.1	-	Sr18 [54]
3		RKRTF/APR	QSr.ipbb-2A.2	-	Sr32-2A [53]
4		PKCTC	QSr.ipbb-2B.3	-	Sr32-2B, Sr39 [53]
5	SR	PKCTC	QSr.ipbb-2B.4	QSR.IPBB-2B [26]	Sr36 [53]
6		TKRTF	QSr.ipbb-2D.2	-	Sr32-2D [53]
7		TKRTF/APR	QSr.ipbb-6B.5	QSR.IPBB-6B.1 [27], QSr.ipbb-6B.3 [51]	-
8		TKRTF	QSr.ipbb-6B.7	QSR.IPBB-6B.2 [27], QSr.ipbb-6B.4 [51]	Sr11 [54]

The region of each QTL was analyzed for the presence of protein-coding genes in the interval 500 kb upstream and 500 kb downstream from the most significant SNP (Table S1). The analysis of LR-associated QTL regions suggested the presence of 158 genes ranging from 6 (*QLr.ipbb-7B.3* and *QLr.ipbb-7D.1*) to 22 (*QLr.ipbb-1B.4* and *QLr.ipbb-6A.4*) genes per interval. A similar search for SR-associated QTL regions indicated the presence of 226 genes ranging from 5 (*QSr.ipbb-2B.4*) to 29 (*QSr.ipbb-1B.4*) genes per interval. Among these 158 genes identified for QTLs associated with LR, 48.9% coded for proteins with functions known in *T. aestivum*, 48.6% for uncharacterized proteins, and 2.5% for RNAs. For QTLs associated with SR, 56.6% of genes coded for proteins uncharacterized in *T. aestivum*, 41.2% described protein-coding genes, and 2.2% coded for RNAs. Among genes coding for uncharacterized proteins, sequences similar to the 24 QTL regions for LR and 39 QTL regions for SR were identified in other grass species (Table S1). Orthologous genes with their sequence similarity level higher than 70% were selected and are listed.

4. Discussion

4.1. General Resistance of RILs in Studied Environments

At the seedling stage, the majority of RILs and parental cultivars showed MS and S levels of resistance to all races of LR and SR, except for the SR race PKCTC, where several lines were identified as R and MR (Table 3). The ANOVA test showed a more significant influence of pathogen genotype (race) on the resistance of RILs rather than the genotype of wheat lines (Table 4). This result indicated that these genetic factors associated with resistance are race-specific. In the world and in Kazakhstan, breeding programs are mostly focused on the combination of seedling resistance and APR in new cultivars. Pyramiding of seedling gene(s) with slow rusting APR gene(s) usually results in higher resistance of the crop. This agrees with wheat R genes conferring resistance to LR (*Lr1*, *Lr10*, *Lr21*) and SR (*Sr22*, *Sr33*, *Sr35*, *Sr45*, *Sr50*) being cloned and widely used in wheat breeding [55]. However, the significant positive correlations among LR races observed in this study (Table 5) suggested the involvement of genetic factors that are effective against all three races. The presence of strong positive correlations between APR to LR and SR at KRIAPI also indicated that genes conferring LR resistance are either closely linked or may have a pleiotropic effect on genes that control SR resistance [26,56]. Positive correlations were simultaneously observed between seedling resistance to LR races TQTHQ and TKTHT and APR to LR at KRIAPI, as well as between seedling resistance to SR race PKCTC and APR to SR at KRIAPI (Table 5). The relationship between race-specific seedling and broad adult plant resistances could be influenced by the presence of LR and SR races in the fields at KRIAPI. This also suggested that the wheat germplasm growing in this region could be effectively and rapidly screened for resistance to LR and SR at the seedling stage in a greenhouse [57].

LR and SR resistances are complex traits [58]; this was confirmed by the range of reactions to pathogens and the presence of transgressive segregations. Even when parents demonstrated the same level of resistance, such as APR to SR, RILs still showed transgressive phenotypes in the direction of either resistance (RIBSP) or susceptibility (KRIAPI) (Table 3). This phenomenon is not rare; it was previously described for many other quantitatively inherited wheat traits; for example, in studies of grain quality traits [22], grain Zn and Fe concentrations [59], grain yield and plant height [60], and rust diseases [61,62].

4.2. QTL Mapping for Leaf Rust Resistance

Alleles conferring increased resistance of QTLs for LR race TQKHT and APR at RIBSP originated from Par (Table 7). The higher LR resistance of Par in comparison with PA indicated that the U.K. cultivar is a promising source for wheat breeding programs in Kazakhstan. PA was simultaneously found to be a source for QTLs with increased LR resistance to race TQTMQ.

The 11 QTLs for the resistance to LR at the seedling and adult plant-growth stages can be divided into two categories: (1) similar to QTLs previously detected for LR resistance and (2) presumably novel QTLs. The first category consisted of 6 out of 11 QTLs for LR resistance (Table 9). Four of the QTLs for LR resistance with similar genetic positions (*QLr.ipbb-1A.2* (APR at KRIAPI), *QLr.ipbb-6A.4* (seedling resistance to TQKHT), *QLr.ipbb-6A.5* (seedling resistance to TRTHT), and *QLr.ipbb-7B.3* (seedling resistance to TQTMQ) were previously identified in a GWAS study performed at RIBSP in 2018/2019 at the adult plant stage [50]. Hence, multiple occurrences of QTLs associated with the resistance to LR in different conditions and environments indicated the broad stability of these loci. *QLr.ipbb-7B.3* may be associated with the gene *Lr14* located in a similar region of the genome (Table 9). The effectiveness of allele *Lr14a* was described for northern Kazakhstan and *Lr14b* for eastern and western Kazakhstan [7]. *Lr14* was also described as an effective resistance factor to TQTMQ (Table 1). The APR QTL *QLr.ipbb-1B.4* is associated with the gene *Lr21*, positioned in close proximity to the peak of the QTL (Table S1). This gene was described as effective in southeastern Kazakhstan [7]. The last QTL from the first group, *QLr.ipbb-3A.2*, is probably associated with genes *Lr63* and *Lr66* (Table 9). Unfortunately, information is lacking about the role of these genes in the wheat-growing areas of Kazakhstan. However, *Lr63* and *Lr66* are known to condition low to intermediate infection types to most of *P. recondita* isolates [63]. The remaining five QTLs identified for LR resistance are presumably novel genetic factors, since there were no reliable matches between their positions in the genome and previously identified QTLs or genes.

4.3. QTLs for Stem Rust

In 13 QTLs for the resistance to SR identified in this study, alleles presumably increasing resistance originated from both PA and Par (Table 8). Similar to LR resistance, SR-resistance-associated QTLs could be divided into two loci groups, where the first group has similar genetic positions with previously reported QTLs for SR resistance (Table 9), and the second group has none of those matches. The first group includes 8 out of 13 QTLs identified for SR resistance. For three of them—*QSr.ipbb-2B.4* (seedling resistance to PKCTC), *QSr.ipbb-6B.5* (seedling resistance to TKRTF and APR in RIBSP), and *QSr.ipbb-6B.7* (seedling resistance to TKRTF)—QTLs for SR resistance with similar positions in the genome were identified in a previous work involving the PA × Par mapping population [27] and in a GWAS study using resistance data obtained from RIBSP [51]. Similar to the LR study, these findings may indicate the stability of identified QTLs. In addition to the information with QTL similarities, several specific *Sr* genes seem to be associated with QTLs from this study (Table 9). One of the most interesting findings was the identification of three QTLs on distal ends of chromosomes 2A, 2B, and 2D responsible for seedling resistance to SR races RKRTF, PKCTC, and TKRTF, respectively. These QTLs could be associated with the gene *Sr32*, which was mapped in these regions of chromosomes 2A [64], 2B [65], and 2D [66]. The gene was previously reported as effective against Ug99 and related SR races [66]. The other *Sr* genes involved in resistance to SR races in the Ug99 lineage and possibly associated with QTL from this study are *Sr31* (resistant to TTKSF and TTKSP), *Sr36* (all Ug99 lineage races, except TTTSK), and *Sr39* (all Ug99 lineage races) [67]. Among the SR races used in this study, *Sr36* was described as effective against PKCTC (Table 1). The resistance pattern is similar to *QSr.ipbb-2B.4*, which is located in a nearby region of the chromosome. The second group of the genetic factors consisted of the remaining five QTLs that could be novel QTLs associated with resistance to SR.

4.4. QTLs Cluster on Chromosome 1B

The QTLs associated with several traits are common in wheat. It may occur due to pleiotropic effect or their tight linkage. For the resistance to wheat fungal diseases, pleiotropic APR genes *Lr34/Yr18/Pm38/Sr57* [68], *Lr46/Yr29/Pm39/Sr58* [69], and *Lr67/Yr46/Pm46/Sr55* [70] were previously described. Among the QTLs identified for LR and SR resistance in this study, two QTLs (*QLr.ipbb-1B.4* and *QSr.ipbb-1B.4*) occupy the same interval on chromosome 1B (Tables 7 and 8, Figures 1 and 2). In addition, the *QLr.ipbb-1B.4* interval contains the resistance gene *Lr21* less than 500 kb from the significant peak, whereas the interval of *QSr.ipbb-1B.4* has genes for disease resistance proteins and resistance-related kinases next to the peak marker (Table S1). *Lr21* was described as effective for southeastern Kazakhstan [7]. Common markers in these intervals suggest the usefulness for marker-assisted breeding of these QTLs to develop wheat cultivars with durable rust resistance for gene pyramiding [11].

5. Conclusions

Overall, 24 QTLs for the resistance to rust diseases at the seedling and adult plant stages were identified in this study, including 11 QTLs for LR and 13 QTLs for SR. Among the QTLs associated with LR, eight QTLs were race-specific and detected at the seedling stage, two QTLs were at the stage of the adult plant, and one QTL was identified in both stages. The QTLs for LR-resistance explained from 11.6% (*QLr.ipbb-7D.1*) to 25.7% (*QLr.ipbb-6A.4*) of the phenotypic variation and were detected on 10 chromosomes. The increased resistance to LR in TQTMQ race-specific QTLs originated from PA; in QTLs specific for the race TQKHT and APR, alleles were from Par. For TRTHT, the origin of resistance alleles in identified QTLs was both parental cultivars. For SR resistance, seven QTLs were race-specific and detected at the seedling stage, three QTLs were identified at the adult plant stage, and three QTLs were identified at both growth stages. SR-associated QTLs explained from 8.9% (*QSr.ipbb-2B.4*) to 39.1% (*QSr.ipbb-2A.2*) of variation in SR resistance and were mapped on nine chromosomes. The alleles increasing resistance to SR originated from both parents: effective alleles in six QTLs were from Par, in five QTLs from PA, and two QTLs had a different origin of resistance at the seedling and adult plant stages. Among the QTLs from this study, 10 QTLs were putative and 14 matching QTLs were found in previous works involving the PA × Par population, a GWAS study at RIBSP, and possible candidate resistance genes. The cluster of QTLs associated with both LR and SR resistances was identified on chromosome 1B. Thus, the QTLs revealed in this study may play an essential role in the improvement of wheat resistance to LR and SR via marker-assisted selection.

Author Contributions: Conceptualization, S.A. and Y.T.; methodology, A.R. and Y.T.; formal analysis, Y.G. and G.Y.; investigation, Y.G., A.R., and S.A.; resources, Y.T. and A.R.; data curation, Y.G., G.Y., and Y.T.; writing—original draft preparation, Y.G.; writing—review and editing, Y.G., S.A., A.R., G.A., and Y.T.; supervision, Y.T.; project administration, S.A.; funding acquisition, A.R. All authors have read and agreed to the published version of the manuscript.

Acknowledgments: This work was conducted within the framework of the project "Development of new DNA markers associated with the resistance of bread wheat to the most dangerous fungal diseases in Kazakhstan" in Program "Development of the innovative systems for increasing the resistance of wheat varieties to especially dangerous diseases in the Republic of Kazakhstan": BR06249329 supported by the Ministry of Agriculture of the Republic of Kazakhstan.

References

1. Statista. Available online: https://www.statista.com/ (accessed on 12 March 2020).
2. Giraldo-Carbajo, P.; Barzana, M.E.B.; Manzano-Agugliaro, F.; Giménez, E. Worldwide Research Trends on Wheat and Barley: A Bibliometric Comparative Analysis. *Agronomy* **2019**, *9*, 352. [CrossRef]
3. USDA. Available online: https://www.usda.gov/ (accessed on 17 March 2020).
4. Statistics Committee. Ministry of National Economy of the Republic of Kazakhstan. Available online: https://stat.gov.kz/ (accessed on 15 March 2020).
5. Tadesse, W.; Sanchez-Garcia, M.; Assefa, S.G.; Amri, A.; Bishaw, Z.; Ogbonnaya, F.C.; Baum, M. Genetic Gains in Wheat Breeding and Its Role in Feeding the World. *Crop Breed. Genet. Genom.* **2019**, *1*, e190005.
6. Samborski, D.J. Wheat leaf rust. In *The Cereal Rusts Vol. II: Diseases, Distribution, Epidemiology and Control*; Roelfs, A.P., Bushnell, W.R., Eds.; Academic Press: Orlando, FL, USA, 1985; pp. 39–59.
7. Koyshybaev, M. *Wheat Diseases*, 1st ed.; FAO: Ankara, Turkey, 2018; p. 365.
8. Eversmeyer, M.G.; Kramer, C.L. Epidemiology of Wheat Leaf and Stem Rust in the Central Great Plains of the USA. *Annu. Rev. Phytopathol.* **2000**, *38*, 491–513. [CrossRef] [PubMed]
9. Summers, R.W.; Brown, J.K.M. Constraints on breeding for disease resistance in commercially competitive wheat cultivars. *Plant Pathol.* **2013**, *62*, 115–121. [CrossRef]
10. McIntosh, R.A.; Dubcovsky, J.; Rogers, W.J.; Morris, C.F.; Xia, X.C. Catalogue of Gene Symbols for Wheat: 2017 Supplement. Available online: https://shigen.nig.ac.jp/wheat/komugi/genes/macgene/supplement2017.pdf (accessed on 27 July 2020).

11. Ellis, J.G.; Lagudah, E.; Spielmeyer, W.; Dodds, P.N. The past, present and future of breeding rust resistant wheat. *Front. Plant Sci.* **2014**, *5*, 641. [CrossRef]

12. Bariana, H.S.; Hayden, M.J.; Ahmed, N.U.; Bell, J.A.; Sharp, P.J.; McIntosh, R.A. Mapping of durable adult plant and seedling resistances to stripe rust and stem rust diseases in wheat. *Aust. J. Agric. Res.* **2001**, *52*, 1247. [CrossRef]

13. Xu, Y.; Li, P.; Yang, Z.; Yang, Z. Genetic mapping of quantitative trait loci in crops. *Crop J.* **2017**, *5*, 175–184. [CrossRef]

14. Arruda, M.P.; Brown, P.; Brown-Guedira, G.; Krill, A.M.; Thurber, C.; Merrill, K.R.; Foresman, B.J.; Kolb, F.L. Genome-Wide Association Mapping of Fusarium Head Blight Resistance in Wheat using Genotyping-by-Sequencing. *Plant Genome* **2016**, *9*, 1–14. [CrossRef]

15. Li, G.; Xu, X.; Tan, C.; Carver, B.F.; Bai, G.; Wang, X.; Bonman, J.M.; Wu, Y.; Hunger, R.; Cowger, C. Identification of powdery mildew resistance loci in wheat by integrating genome-wide association study (GWAS) and linkage mapping. *Crop J.* **2019**, *7*, 294–306. [CrossRef]

16. Cheng, B.; Gao, X.; Cao, N.; Ding, Y.; Gao, Y.; Chen, T.; Xin, Z.; Zhang, L.Y. Genome-wide association analysis of stripe rust resistance loci in wheat accessions from southwestern China. *J. Appl. Genet.* **2020**, *61*, 37–50. [CrossRef]

17. Ollier, M.; Talle, V.; Brisset, A.-L.; Le Bihan, Z.; Duerr, S.; Lemmens, M.; Goudemand, E.; Robert, O.; Hilbert, J.-L.; Buerstmayr, H. QTL mapping and successful introgression of the spring wheat-derived QTL Fhb1 for Fusarium head blight resistance in three European triticale populations. *Theor. Appl. Genet.* **2020**, *133*, 457–477. [CrossRef] [PubMed]

18. Unamba, C.I.N.; Nag, A.; Sharma, R.K. Next Generation Sequencing Technologies: The Doorway to the Unexplored Genomics of Non-Model Plants. *Front. Plant Sci.* **2015**, *6*, 6. [CrossRef]

19. Turuspekov, Y.; Plieske, J.; Ganal, M.; Akhunov, E.; Abugalieva, S. Phylogenetic analysis of wheat cultivars in Kazakhstan based on the wheat 90 K single nucleotide polymorphism array. *Plant Genet. Resour.* **2015**, *15*, 29–35. [CrossRef]

20. Singh, M.; Upadhyaya, H.D. *Genetic and Genomic Resources for Grain Cereals Improvement*, 1st ed.; Academic Press: Cambridge, MA, USA, 2015; pp. 58–59.

21. Hu, J.; Wang, X.; Zhang, G.; Jiang, P.; Chen, W.; Hao, Y.; Ma, X.; Xu, S.; Jia, J.; Kong, L.; et al. QTL mapping for yield-related traits in wheat based on four RIL populations. *Theor. Appl. Genet.* **2020**, *133*, 917–933. [CrossRef] [PubMed]

22. Goel, S.; Singh, K.; Singh, B.; Grewal, S.; Dwivedi, N.; Alqarawi, A.A.; Allah, E.A.; Ahmad, P.; Singh, N.K. Analysis of genetic control and QTL mapping of essential wheat grain quality traits in a recombinant inbred population. *PLoS ONE* **2019**, *14*, e0200669. [CrossRef] [PubMed]

23. Li, L.; Mao, X.; Wang, J.; Chang, X.; Reynolds, M.; Jing, R. Genetic dissection of drought and heat-responsive agronomic traits in wheat. *Plant Cell Environ.* **2019**, *42*, 2540–2553. [CrossRef]

24. Emebiri, L.C.; Tan, M.K.; El-Bouhssini, M.; Wildman, O.; Jighly, A.; Tadesse, W.; Ogbonnaya, F.C. QTL mapping identifies a major locus for resistance in wheat to Sunn pest (*Eurygaster integriceps*) feeding at the vegetative growth stage. *Theor. Appl. Genet.* **2017**, *130*, 309–318. [CrossRef] [PubMed]

25. Amalova, A.Y.; Yermekbayev, K.A.; Griffiths, S.; Abugalieva, S.I.; Turuspekov, Y.K.; Centre, U.K.J.I. Phenotypic variation of common wheat mapping population Pamyati Azieva x Paragon in south-east of Kazakhstan. *Int. J. Biol. Chem.* **2019**, *12*, 11–17. [CrossRef]

26. Genievskaya, Y.; Amalova, A.; Ydyrys, A.; Sarbayev, A.; Griffiths, S.; Abugalieva, S.; Turuspekov, Y. Resistance of common wheat (*Triticum aestivum* L.) mapping population Pamyati Azieva × Paragon to leaf and stem rusts in conditions of south-east Kazakhstan. *Eurasian J. Ecol.* **2019**, *61*, 14–23.

27. Genievskaya, Y.; Fedorenko, Y.; Sarbayev, A.; Amalova, A.; Abugalieva, S.; Griffiths, S.; Turuspekov, Y. Identification of QTLs for resistance to leaf and stem rusts in bread wheat (*Triticum aestivum* L.) using a mapping population of 'Pamyati Azieva × Paragon'. *Vavilov J. Genet. Breed.* **2019**, *23*, 887–895. [CrossRef]

28. ADAPTAWHEAT. Available online: https://www.jic.ac.uk/adaptawheat/ (accessed on 12 April 2020).

29. Yermekbayev, K.; Turuspekov, Y.; Ganal, M.; Plieske, J.; Griffiths, S. Construction and utilization of the hexaploid map Pamyati Azieva × Paragon. In *Proceedings of PlantGen (Plant Genetics, Genomics, Bioinformatics and Biotechnology), Almaty, Kazakhstan, 29 May–2 June 2017*; Turuspekov, Y., Abugalieva, S., Eds.; Institute of Plant Biology and Biotechnology: Almaty, Kazakhstan, 2017.

30. Chu, C.-G.; Friesen, T.L.; Xu, S.S.; Faris, J.D.; Kolmer, J.A. Identification of novel QTLs for seedling and adult plant leaf rust resistance in a wheat doubled haploid population. *Theor. Appl. Genet.* **2009**, *119*, 263–269. [CrossRef] [PubMed]

31. Pretorius, Z.A.; Jin, Y.; Bender, C.M.; Herselman, L.; Prins, R. Seedling resistance to stem rust race Ug99 and marker analysis for *Sr2*, *Sr24* and *Sr31* in South African wheat cultivars and lines. *Euphytica* **2011**, *186*, 15–23. [CrossRef]

32. Spanic, V.; Rouse, M.N.; Kolmer, J.A.; Anderson, J.A. Leaf and stem seedling rust resistance in wheat cultivars grown in Croatia. *Euphytica* **2014**, *203*, 437–448. [CrossRef]

33. Stakman, E.C.; Stewart, D.M.; Loegering, W.Q. Identification of physiologic races of Puccinia graminis var. tritici. *US Agric. Res. Serv.* **1962**, *617*, 1–53.

34. Rsaliyev, A.; Rsaliyev, S.S. Principal approaches and achievements in studying race composition of wheat stem rust. *Vavilov J. Genet. Breed.* **2019**, *22*, 967–977. [CrossRef]

35. Roelfs, A.P. An International System of Nomenclature for *Puccinia graminis* f. sp. tritici. *Phytopathology* **1988**, *78*, 526. [CrossRef]

36. Long, D.L.; Kolmer, J.A. A North American system of nomenclature for *Puccinia recondita* f. sp. tritici. *Phytopathology* **1989**, *79*, 525–529. [CrossRef]

37. Kolmer, J.A.; Ordoñez, M.E. Genetic Differentiation of Puccinia triticina Populations in Central Asia and the Caucasus. *Phytopathology* **2007**, *97*, 1141–1149. [CrossRef]

38. Kolmer, J.A.; Kabdulova, M.G.; Mustafina, M.A.; Zhemchuzhina, N.S.; Dubovoy, V. Russian populations of *Puccinia triticina* in distant regions are not differentiated for virulence and molecular genotype. *Plant Pathol.* **2014**, *64*, 328–336. [CrossRef]

39. Schachtel, G.A.; Dinoor, A.; Herrmann, A.; Kosman, E. Comprehensive Evaluation of Virulence and Resistance Data: A New Analysis Tool. *Plant Dis.* **2012**, *96*, 1060–1063. [CrossRef]

40. Mains, E.B.; Jackson, H.S. Physiologic specialization in the leaf rust of wheat *Puccinia triticiana* Erikss. *Phytopathology* **1926**, *16*, 89–120.

41. Peterson, R.F.; Campbell, A.B.; Hannah, A.E. A Diagrammatic Scale for Estimating Rust Intensity on Leaves and Stems of Cereals. *Can. J. Res.* **1948**, *26*, 496–500. [CrossRef]

42. Roelfs, A.; Singh, R.; Saari, E.E. *Rust Diseases of Wheat: Concepts and Methods of Disease Management*; CIMMYT: Mexico City, Mexico, 1992; p. 45.

43. Zhang, D.; Bowden, R.; Bai, G. A method to linearize Stakman infection type ratings for statistical analysis. In Proceedings of the Borlaug Global Rust Initiative 2011 Technical Workshop, Saint Paul, MN, USA, 13–16 June 2011.

44. Windows QTL Cartographer V2.5. Available online: https://brcwebportal.cos.ncsu.edu/qtlcart/WQTLCart.htm (accessed on 9 February 2020).

45. Rong, J.; Feltus, F.A.; Waghmare, V.N.; Pierce, G.J.; Chee, P.W.; Draye, X.; Saranga, Y.; Wright, R.J.; Wilkins, T.A.; May, O.L.; et al. Meta-analysis of Polyploid Cotton QTL Shows Unequal Contributions of Subgenomes to a Complex Network of Genes and Gene Clusters Implicated in Lint Fiber Development. *Genetics* **2007**, *176*, 2577–2588. [CrossRef]

46. Voorrips, R.E. MapChart: Software for the graphical presentation of linkage maps and QTLs. *J. Hered.* **2002**, *93*, 77–78. [CrossRef] [PubMed]

47. EnsemblPlant. Triticum Aestivum. Available online: https://plants.ensembl.org/Triticum_aestivum/Info/Index (accessed on 27 February 2020).

48. BLAST Tool. Triticum Aestivum. Available online: https://plants.ensembl.org/Triticum_aestivum/Tools/Blast (accessed on 2 March 2020).

49. UniProt. Available online: https://www.uniprot.org/ (accessed on 2 March 2020).

50. Collard, B.C.; Jahufer, M.Z.Z.; Brouwer, J.B.; Pang, E.C.K. An introduction to markers, quantitative trait loci (QTL) mapping and marker-assisted selection for crop improvement: The basic concepts. *Euphytica* **2005**, *142*, 169–196. [CrossRef]

51. Genievskaya, Y.; Abugalieva, S.; Rsaliyev, A.; Turuspekov, Y. Genome-wide association mapping for resistance to leaf, stem, and yellow rusts of bread wheat in conditions of South Kazakhstan. *PeerJ* **2020**. submitted work under review.

52. Aoun, M.; Breiland, M.; Turner, M.K.; Loladze, A.; Chao, S.; Xu, S.S.; Ammar, K.; Anderson, J.A.; Kolmer, J.A.; Acevedo, M. Genome-Wide Association Mapping of Leaf Rust Response in a Durum Wheat Worldwide Germplasm Collection. *Plant Genome* **2016**, *9*, 1–24. [CrossRef] [PubMed]

53. Yu, L.-X.; Barbier, H.; Rouse, M.N.; Singh, S.; Singh, R.P.; Bhavani, S.; Huerta-Espino, J.; Sorrells, M.E. A consensus map for Ug99 stem rust resistance loci in wheat. *Theor. Appl. Genet.* **2014**, *127*, 1561–1581. [CrossRef]

54. Hart, G.E.; Gale, M.D.; McIntosh, R.A. Linkage maps of *Triticum aestivum* (hexaploid wheat, 2n = 42, genomes A, B & D) and T. tauschii (2n = 14, genome D). In *Progress in Genome Mapping of Wheat and Related Species, Proceedings of the 3rd Publ Wkshp Int Triticeae Mapping Initiative, Mexico City, Mexico, 22–26 September 1992*; Hoisington, D., McNab, A., Eds.; CIMMYT: Mexico City, Mexico, 1993; pp. 32–46.

55. Periyannan, S.; Milne, R.J.; Figueroa, M.; Lagudah, E.; Dodds, P.N. An overview of genetic rust resistance: From broad to specific mechanisms. *PLoS Pathog.* **2017**, *13*, e1006380. [CrossRef]

56. Figlan, S.; Terefe, T.G.; Shimelis, H.; Tsilo, T.J. Adult plant resistance to leaf rust and stem rust of wheat in a newly developed recombinant inbred line population. *S. Afr. J. Plant Soil* **2017**, *35*, 111–119. [CrossRef]

57. Tadesse, W.; Reents, H.J.; Hsam, S.L.K.; Zeller, F.J. Relationship of seedling and adult plant resistance and evaluation of wheat germplasm against tan spot (*Pyrenophora tritici-repentis*). *Genet. Resour. Crop Evol.* **2010**, *58*, 339–346. [CrossRef]

58. Niks, R.E.; Qi, X.; Marcel, T.C. Quantitative Resistance to Biotrophic Filamentous Plant Pathogens: Concepts, Misconceptions, and Mechanisms. *Annu. Rev. Phytopathol.* **2015**, *53*, 445–470. [CrossRef] [PubMed]

59. Crespo-Herrera, L.A.; Govindan, V.; Stangoulis, J.C.R.; Hao, Y.; Singh, R.P. QTL Mapping of Grain Zn and Fe Concentrations in Two Hexaploid Wheat RIL Populations with Ample Transgressive Segregation. *Front. Plant Sci.* **2017**, *8*, 1800. [CrossRef]

60. Asif, M.; Yang, R.C.; Navabi, A.; Iqbal, M.; Kamran, A.; Lara, E.P.; Randhawa, H.; Pozniak, C.; Spaner, D. Mapping QTL, selection differentials, and the effect of *Rht-B1* under organic and conventionally managed systems in the Attila × CDC Go spring wheat mapping population. *Crop Sci.* **2015**, *55*, 1129–1142. [CrossRef]

61. Tian, Z.M.; Zhang, L.P.; Sun, Q.L.; Shan, F.H.; Tian, Z.H. Research review on wheat QTLs. *Inn. Mong. Agric. Sci. Technol.* **2007**, *3*, 68–71.

62. Bokore, F.E.; Knox, R.E.; Cuthbert, R.D.; Pozniak, C.J.; McCallum, B.D.; N'Diaye, A.; Depauw, R.M.; Campbell, H.L.; Munro, C.; Singh, A.; et al. Mapping quantitative trait loci associated with leaf rust resistance in five spring wheat populations using single nucleotide polymorphism markers. *PLoS ONE* **2020**, *15*, e0230855. [CrossRef]

63. Kolmer, J.A.; Anderson, J.A.; Flor, J.M. Chromosome Location, Linkage with Simple Sequence Repeat Markers, and Leaf Rust Resistance Conditioned by Gene *Lr63* in Wheat. *Crop Sci.* **2010**, *50*, 2392–2395. [CrossRef]

64. McIntosh, R.A.; Wellings, C.R.; Park, R.F. *Wheat Rusts, An Atlas of Resistance Genes*; CSIRO Publications: Melbourne, Australia, 1995; pp. 93–99.

65. McIntosh, R.A.; Dubcovsky, J.; Rogers, W.J.; Morris, C.F.; Appels, R.; Xia, X.C. Catalogue of gene symbols for wheat: 2011 supplement. *Annu. Wheat Newsl.* **2011**, *57*, 303–321.

66. Mago, R.; Verlin, D.; Zhang, P.; Bansal, U.; Bariana, H.; Jin, Y.; Ellis, J.; Hoxha, S.; Dundas, I. Development of wheat–*Aegilops speltoides* recombinants and simple PCR-based markers for *Sr32* and a new stem rust resistance gene on the 2S#1 chromosome. *Theor. Appl. Genet.* **2013**, *126*, 2943–2955. [CrossRef]

67. Singh, R. Current status, likely migration and strategies to mitigate the threat to wheat production from race Ug99 (TTKS) of stem rust pathogen. *CAB Rev. Perspect. Agric. Veter-Sci. Nutr. Nat. Resour.* **2006**, *1*, 1–13. [CrossRef]

68. Singh, R.P.; Herrera-Foessel, S.A.; Huerta-Espino, J.; Bariana, H.; Bansal, U.; McCallum, B.; Hiebert, C.; Bhavani, S.; Singh, S.; Lan, C.X.; et al. *Lr34/Yr18/Sr57/Pm38/Bdv1/Ltn1* confers slow rusting, adult plant resistance to *Puccinia graminis tritici*. In Proceedings of the 13th International Cereal Rusts and Powdery Mildews Conference, Beijing, China, 28 August–1 September 2012.

69. Singh, R.P.; Herrera-Foessel, S.A.; Huerta-Espino, J.; Lan, C.X.; Basnet, B.R.; Bhavani, S.; Lagudah, E.S. Pleiotropic gene *Lr46/Yr29/Pm39/Ltn2* confers slow rusting, adult plant resistance to wheat stem rust fungus. In Proceedings of the Borlaug Global Rust Initiative, New Delhi, India, 19–22 August 2013.

70. Herrera-Foessel, S.A.; Singh, R.P.; Lillemo, M.; Huerta-Espino, J.; Bhavani, S.; Singh, S.; Lan, C.; Calvo-Salazar, V.; Lagudah, E.S. Lr67/Yr46 confers adult plant resistance to stem rust and powdery mildew in wheat. *Theor. Appl. Genet.* **2014**, *127*, 781–789. [CrossRef] [PubMed]

6

Genetic Dissection of the Seminal Root System Architecture in Mediterranean Durum Wheat Landraces by Genome-Wide Association Study

Martina Roselló [1], Conxita Royo [1], Miguel Sanchez-Garcia [2] and Jose Miguel Soriano [1,*]

[1]   Sustainable Field Crops Programme, Institute for Food and Agricultural Research and Technology (IRTA), 25198 Lleida, Spain

[2]   International Centre for Agricultural Research in Dry Areas (ICARDA), Rabat 10112, Morocco

*   Correspondence: josemiguel.soriano@irta.cat

Abstract: Roots are crucial for adaptation to drought stress. However, phenotyping root systems is a difficult and time-consuming task due to the special feature of the traits in the process of being analyzed. Correlations between root system architecture (RSA) at the early stages of development and in adult plants have been reported. In this study, the seminal RSA was analysed on a collection of 160 durum wheat landraces from 21 Mediterranean countries and 18 modern cultivars. The landraces showed large variability in RSA, and differences in root traits were found between previously identified genetic subpopulations. Landraces from the eastern Mediterranean region, which is the driest and warmest within the Mediterranean Basin, showed the largest seminal root size in terms of root length, surface, and volume and the widest root angle, whereas landraces from eastern Balkan countries showed the lowest values. Correlations were found between RSA and yield-related traits in a very dry environment. The identification of molecular markers linked to the traits of interest detected 233 marker-trait associations for 10 RSA traits and grouped them in 82 genome regions named marker-train association quantitative trait loci (MTA-QTLs). Our results support the use of ancient local germplasm to widen the genetic background for root traits in breeding programs.

Keywords: durum wheat; landraces; marker-trait association; root system architecture

1. Introduction

Wheat is estimated to have been first cultivated around 10,000 years before present (BP) in the Fertile Crescent region. It spread to the west of the Mediterranean Basin and reached the Iberian Peninsula around 7000 years BP [1]. During this migration, both natural and human selection resulted in the development of local landraces considered to be very well adapted to the regions where they were grown and containing the largest genetic diversity within the species [2]. From the middle of the 20th century, as a consequence of the Green Revolution, the cultivation of local landraces was progressively abandoned and replaced by the improved, more productive, and genetically uniform semi-dwarf cultivars. However, scientists are convinced that local landraces may provide new alleles to improve commercially valuable traits [3]. Introgression of these alleles into modern cultivars can be very useful, especially in breeding for suboptimal environments.

Drought is the most important environmental factor limiting wheat productivity in many parts of the world. Therefore, improving yield under water-limited conditions is one of the major challenges for wheat production worldwide. Breeding for adaptation to drought is extremely challenging due to the complexity of the target environments and the stress-adaptive mechanisms adopted by plants to withstand and mitigate the negative effects of a water deficit [4]. These mechanisms allow the plant to escape (e.g., early flowering date), avoid (e.g., root system), and/or tolerate (e.g., osmolyte accumulation) the negative effects of drought, which plays a role in determining final crop performance [5]. The crop traits to be considered as selection targets under drought conditions must be genetically correlated with yield and should have a greater heritability than yield itself [6,7]. Among these traits, early vigour, leaf area duration, crop water status, radiation use efficiency, and root architecture have been identified to be associated with yield under rainfed conditions (reviewed by Reference [8]).

Root system architecture (RSA) is crucial for wheat adaptation to drought stress. Roots exhibit a high level of morphological plasticity in response to soil conditions, which allows plants to adapt better, which is particularly under drought conditions. However, evaluating root architecture in the field is very difficult, expensive, and time-consuming, especially when a large number of plants need to be phenotyped. Several studies have reported a correlation of RSA in the early stages of development with RSA in adult plants [9], Manschadi et al. [10] reported that adult root geometry is strongly related to seminal root angle (SRA). Wasson et al. [11] described a relationship of root vigor between plants grown in the field and controlled conditions. Several systems have been adopted to enable early screening of the RSA in wheat [12].

Identifying quantitative trait loci (QTLs) and using marker-assisted selection is an efficient way to increase selection efficiency and boost genetic gains in breeding programs. However, while numerous studies have reported QTLs for RSA in bi-parental crosses [13], very few of them were based on association mapping [12,14–18]. Association mapping is a complementary approach to bi-parental linkage analysis and provides broader allelic coverage with higher mapping resolution. Association mapping is based on linkage disequilibrium, defined as the non-random association of alleles at different loci, and is used to detect the relationship between phenotypic variation and genetic polymorphism.

The main objectives of the present study were a) to identify differences in RSA among genetic subpopulations of durum wheat Mediterranean landraces, b) to find correlations of RSA with yield-related traits in different rainfed Mediterranean environments, and c) to identify molecular markers linked to RSA in the old Mediterranean germplasm through a genome-wide association study.

2. Materials and Methods

2.1. Plant Material

The germplasm used in the current study consisted of a set of 160 durum wheat landraces from 21 Mediterranean countries and 18 modern cultivars from a previously structured collection [2,19]. The landraces were classified into four genetic subpopulations (SPs) that matched their geographical origin as follows: the eastern Mediterranean (19 genotypes), the eastern Balkans and Turkey (20 genotypes), the western Balkans and Egypt (31 genotypes), the western Mediterranean (71 genotypes), and 19 genotypes that remained as admixed (Supplementary Materials Table S1).

2.2. Phenotyping

Eight uniform seeds per genotype were cultured following the paper roll method [20,21] in two replicates of four seeds. The seeds were placed at the top of a filter paper (420 × 520 mm) with the embryo facing down and sprayed with a 0.4% sodium hypochlorite solution. Subsequently, the papers were folded in half to obtain a 210 × 520 mm rectangle with the seeds fixed at the top. The papers were misted with deionized water and rolled by hand. The rolls were placed in plastic pots with deionized water at the bottom that was regularly checked to ensure it did not evaporate. The experiment was conducted in a growth chamber at 25 °C and darkness conditions. One week after sowing, the seeds were transferred to a black surface to take digital images that were processed by SmartRoot software [22] (Figure 1). Nine traits for the seminal root system architecture (RSA) were measured: total root number (TRN), primary root length (PRL, cm), total lateral root length (LRL, cm), primary root surface (PRS, cm^2), total lateral root surface (LRS, cm^2), primary root volume (PRV, cm^3), total lateral root volume (LRV, cm^3), primary root diameter (PRD, cm), and mean lateral root diameter (LRD, cm).

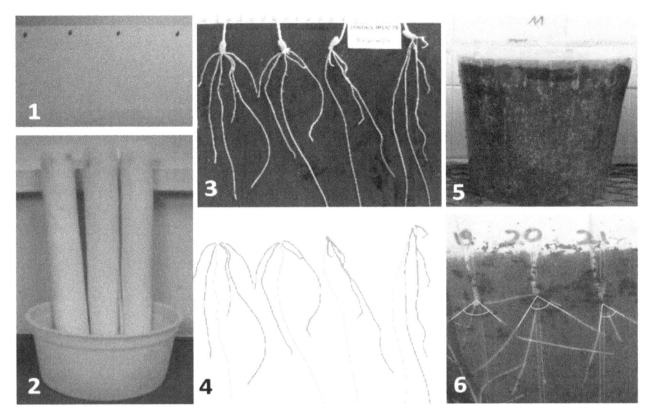

Figure 1. Experimental setup for root system architecture analysis. First, seeds were placed on humid filter paper (**1**) and rolled. Paper rolls were placed in plastic pots with deionized water at the bottom for root growth (**2**). One week after sowing, the seeds were transferred to a black surface for digital imaging (**3**) that were processed by SmartRoot software [22] (**4**). The seminal root angle was measured using the clear pots (**5,6**).

Additionally, the SRA (°) was measured at the facilities of the International Center for Agricultural Research in the Dry Areas (ICARDA) in Rabat (Morocco) using the clear pot method described by Richard et al. [23] (Figure 1). Using a randomized complete block design, eight seeds per genotype were grown in 4 L clear pots filled with peat. The seeds were placed with the embryo facing down and close to the pot wall to facilitate root growth along the transparent wall. The pots were then watered, placed inside 4 L black pots, and kept at 20 °C and darkness conditions in a growth chamber. Five days after sowing, digital images were taken and processed with ImageJ software [24].

Data from field experiments conducted under rainfed conditions during two years of contrasting water input from sowing to physiological maturity (285 mm in 2008 and 104 mm in 2014) in Lleida, North-eastern Spain [25] were used to assess the relationships between RSA traits and yield-related traits.

The experiments were carried out in a non-replicated modified augmented design with three replicated checks (the cultivars 'Claudio,' 'Simeto,' and 'Vitron') and plots of 6 m^2 (8 rows, 5 m long with a 0.15 m spacing). Sowing density was adjusted to 250 viable seeds m^{-2} and the plots were maintained free of weeds and diseases.

2.3. Statistical Analysis

Combined analyses of variance (ANOVA) were performed for the RSA traits of the structured accessions (141 landraces and 18 modern cultivars), considering the accessions and the replicate as random effects. The sum of squares of the cultivar effect was partitioned into differences between SPs and differences within them. The Kenward-Roger correction was used due to the unbalanced number of genotypes within the SPs. Since the experiment was divided into six sets with one check, least squared means were calculated using Simeto as a check and compared using the Tukey test [26] at $p < 0.01$.

Raw field data were fitted to a linear mixed model with the check cultivars as fixed effects and the row number, column number, and genotype as random effects [27]. Restricted maximum likelihood was used to estimate the variance components and to produce the best linear unbiased predictors (BLUPs) for yield and yield components. The relationships between RSA traits and yield-related traits were assessed through correlation analyses. All calculations were carried out using the SAS statistical package [28].

2.4. Genotyping

DNA isolation was performed from leaf samples following the method reported by Doyle and Doyle [29]. High throughput genotyping was performed at Diversity Arrays Technology Pty Ltd. (Canberra, Australia) (http://www.diversityarrays.com) with the genotyping by sequencing (GBS) DArTseq platform [30]. A total of 46,161 markers were used to genotype the association mapping panel, including 35,837 presence/absence variants (PAVs) and 10,324 single nucleotide polymorphisms (SNPs). Markers were ordered according to the consensus map of wheat v4 available at https://www.diversityarrays.com/.

2.5. Linkage Disequilibrium

Linkage disequilibrium (LD) among markers was calculated for the A and B genomes using markers with a map position on the wheat v4 consensus map, and a minor allele frequency greater than 5%, using TASSEL 5.0 [31]. Pair-wise LD was measured using the squared allele frequency correlations r^2 and the values for genomes A and B were plotted against the genetic distance to determine how fast the LD decays. A LOESS curve was fitted to the plot using the JMP v12Pro statistical package (SAS Institute Inc, Cary, NC, USA).

2.6. Genome-Wide Association Study

A genome-wide association study (GWAS) was performed with 160 landraces for the mean of measured traits with TASSEL 5.0 software [31]. A mixed linear model was conducted using the population structure determined by Soriano et al. [19] as the fixed effect and a kinship (K) matrix as the random effect (Q + K) at the optimum compression level. A false discovery rate threshold [32] was established at $-\log_{10}p > 4.6$ ($p < 0.05$), using 2135 markers according to the results of the LD decay, to consider a marker-trait association (MTA) significant. Moreover, a second, less restrictive threshold was established at $-\log_{10}p > 3$. To simplify the MTA information, those associations located within LD blocks were considered to belong to the same QTL and were named marker-trait association quantitative trait loci (MTA-QTLs). Graphical representation of the genetic position of MTA-QTLs was carried out using MapChart 2.3 [33].

2.7. Gene Annotation

Gene annotation for the target region of significant MTAs was performed using the gene models for high-confidence genes reported for the wheat genome sequence [34] available at https://wheat-urgi.versailles.inra.fr/Seq-Repository/.

3. Results

3.1. Phenotypic Analyses

The ANOVA showed that, for all traits, the phenotypic variability was mainly explained by the cultivar effect, since it accounted for 63.41% (PRD) to 90.57% (LRD) of the total sum of squares (Table 1). A summary of the genetic variation of the RSA traits is shown in Supplementary Materials

Table S2. The partitioning of the sum of squares of the cultivar effect into differences between and within SPs revealed that the variability induced by the genotype was mainly explained by differences within SPs on a range from 70.1% for TRN to 91.5 for PRV (Table 1). Differences between SPs were statistically significant for all traits, accounting for 8.5% (PRV) to 30.5% (TRN) of the sum of squares of the genotype effect (Table 1). Western Mediterranean landraces showed the highest number of seminal roots and the narrowest root angle, whereas the eastern Balkans and Turkey SP showed the widest angle (Table 2). The highest values for root size–related traits (length, surface and volume) in both primary and lateral roots were recorded in the eastern Mediterranean landraces. The western Balkans and Egypt subpopulation showed the largest root diameter (Table 2). The comparison of mean values of eastern Balkans and Turkish landraces revealed that the Turkish ones had high values for all traits except TRN, LRL, and root diameter (Supplementary Materials Table S3). The modern cultivars showed intermediate values for all RSA traits (Table 2).

Table 1. Percentage of the sum of squares of the ANOVA model for the seminal root system architecture traits in a set of 159 Mediterranean durum wheat genotypes structured into five genetic subpopulations by Soriano et al. [19].

Source of Variation	df	TRN	SRA	PRL	LRL	PRS	LRS	PRV	LRV	PRD	LRD
Genotype	158	84.1 ***	69.4 ***	87.0 ***	86.8 ***	85.3 ***	86.1 ***	82.9 ***	86.8 ***	63.41 ***	90.57 ***
Between subpopulations	4	30.5 ***	16.6 ***	15.4 ***	22.7 ***	11.6 ***	18.5 ***	8.5 **	12.7 ***	10.44 **	17.96 ***
Within subpopulations	154	70.1 ***	83.4 ***	85.0 ***	77.8 ***	88.6 ***	82.0 ***	91.5 ***	87.6 ***	89.33 **	81.87 ***
Replicate	1	0.001	1.83 **	0.00	0.43 *	0.36 *	0.36 *	0.86 **	0.33 **	1.35 *	0.12
Error	157	15.9	28.8	13.0	12.9	14.3	13.6	16.1	12.9	35.25	9.33
Total	316										

TRN, total root number. SRA, seminal root angle. PRL, primary root length. LRL, total lateral root length. PRS, primary root surface. LRS, total lateral root surface. PRV, primary root volume. LRV, total lateral root volume. PRD, primary root diameter. LRD, mean lateral root diameter. * $p < 0.05$. ** $p < 0.01$. *** $p < 0.001$.

Table 2. Means comparison of seminal root system architecture traits measured in a set of 159 Mediterranean durum wheat genotypes structured into five genetic subpopulations [19]. Means within columns with different letters are significantly different at $p < 0.01$ following a Tukey test.

	TRN	SRA	PRL	LRL	PRS	LRS	PRV	LRV	PRD	LRD
EM	4.8 b	94.7 ab	13.8 a	25.1 a	2.5 a	4.6 a	38.3 a	67.2 a	0.57 b	0.57 b
EB + T	4.8 b	98.2 a	10.3 c	17.2 b	1.9 c	3.2 b	28.5 b	47.6 b	0.57 b	0.58 b
WB + E	4.3 c	87.6 bc	10.4 c	16.5 b	2.1 bc	3.2 b	33.6 ab	52.3 b	0.61 a	0.62 a
WM	5.2 a	84.5 c	11.8 ab	23.5 a	2.2 bc	4.3 a	33.0 ab	63.9 a	0.58 b	0.58 b
Modern	4.5 bc	93.9 ab	12.8 ab	20.8 ab	2.4 ab	3.7 ab	35.2 ab	54.5 ab	0.56 b	0.57 b

TRN, total root number. SRA, seminal root angle. PRL, primary root length. LRL, total lateral root length. PRS, primary root surface. LRS, total lateral root surface. PRV, primary root volume. LRV, total lateral root volume. PRD, primary root diameter. LRD, mean lateral root diameter. EM, Eastern Mediterranean. EB + T, Eastern Balkans and Turkey. WB + E, Western Balkans and Egypt. WM, Western Mediterranean.

Correlation coefficients between RSA traits and yield-related traits were calculated for two field experiments with contrasting water input (285 and 104 mm of rainfall from sowing to physiological maturity). Whereas, for the rainiest environment, only the relationship between SRA and number of spikes per square meter (NSm^2) was statistically significant ($p = 0.043$, $r^2 = 0.16$)). For the driest environment, 14 correlations involving all the yield-related traits and RSA traits except root diameter were statistically significant (Figure 2) (r^2 between 0.17 for NSm^2 and PRL and PRS to 0.30 for TKW and TRN). Most of the significant correlations were positive. Only the relationship between SRA and thousand kernel weight (TKW) was negative.

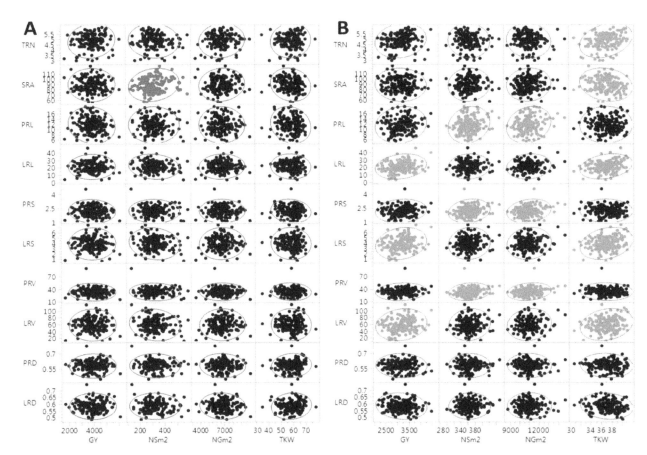

Figure 2. Correlations between seminal root system architecture traits and yield-related traits determined in field experiments receiving high (density ellipse in red, **A**) and low (density ellipse in green, **B**) water input from sowing to physiological maturity. Significant correlation coefficients ($p < 0.05$) are indicated with red and green points. TRN, total root number. SRA, seminal root angle. PRL, primary root length. LRL, total lateral root length. PRS, primary root surface. LRS, total lateral root surface. PRV, primary root volume. LRV, total lateral root volume. PRD, primary root diameter. LRD, mean lateral root diameter. GY, grain yield. NSm^2, number of spikes per square meter. NGm^2, number of grains per square meter. TKW, thousand kernel weight.

3.2. Marker-Trait Associations

A total of 46,161 DArTseq markers, including PAVs and SNPs, were used to genotype the set of 160 durum wheat landraces. To reduce the risk of false positives, markers and accessions were analyzed for the presence of duplicated patterns and missing values. Of 35,837 PAVs, 24,188 were placed on the wheat v4 consensus map. Of these, those with more than 30% of missing data and those with a minor allele frequency lower than 5% were removed from the analysis, leaving 19,443 PAVs. A total of 6957 SNPs were mapped, leaving a total of 4686 SNPs after marker filtering as before. Additionally, 413 markers were duplicated between PAVs and SNPs, so the corresponding PAVs were eliminated. A total of 23,716 markers remained for the subsequent analysis.

Linkage disequilibrium was estimated for locus pairs in genomes A and B using a sliding window of 50 cM. A total of 471,319 and 681,389 possible pair-wise loci were observed for genomes A and B, respectively. Of these locus pairs, 52% and 43% showed significant linkage disequilibrium at $p < 0.01$ and $p < 0.001$, respectively. Mean r^2 was 0.12 for genome A and 0.11 for genome B. These means were used as a threshold for estimating the intercept of the LOESS curve to determine the distance at which LD decays in each genome. Markers were in LD in a range from less than 1 cM in genome B to 1 cM in genome A (Supplementary Materials Table S4).

Results of the GWAS are reported in Figure 3 and in Supplementary Materials Table S5. Using a restrictive threshold based on a false discovery rate at $p < 0.05$ ($-\log_{10}p > 4.6$) and the LD decay, only 12 MTAs corresponding to seven markers were significant. Using a common threshold of $-\log_{10}p > 3$, as previously reported by other authors [35–38], a total of 233 MTAs involving 176 markers were identified. MTAs were equally distributed in both genomes (50.2% in the A genome and 49.8% in the B genome). Chromosomes 2B and 7A harbored the highest number of MTAs (39 and 32 respectively), carrying 30% of the total number of MTAs, whereas chromosomes 4B and 7B harbored the lowest number of MTAs, 8 and 6 MTAs, respectively (Figure 3A). Root volume was the trait showing the highest number of MTAs (77), followed by root surface (46), root diameter (37), root length and number (26), and SRA (21) (Figure 3B). The mean percentage of phenotypic variance explained (PVE) per MTA was similar for all traits, ranging from 0.09 to 0.11 (Figure 3C). Most of the MTAs showed low PVE, in agreement with the quantitative nature of the analyzed traits. The percentage of MTAs with a PVE lower than 0.1 was 71%, whereas that of MTAs with a PVE lower than 0.15 was 98% (Figure 3D).

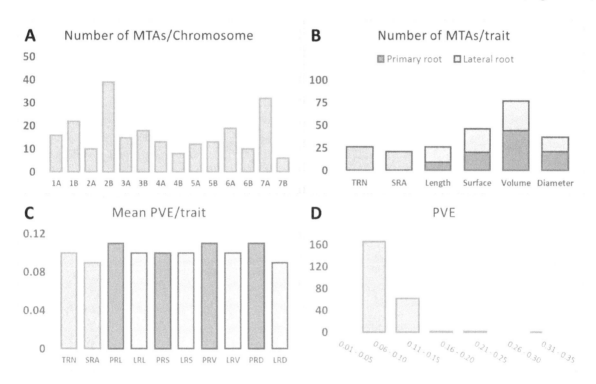

Figure 3. Summary of marker trait associations (MTA). (**A**) Number of MTAs per chromosome. (**B**) Number of MTAs per trait. (**C**) Mean PVE per trait. (**D**) PVE. TRN, total root number. SRA, seminal root angle. PRL, primary root length. LRL, total lateral root length. PRS, primary root surface. LRS, total lateral root surface. PRV, primary root volume. LRV, total lateral root volume. PRD, primary root diameter. LRD, mean lateral root diameter.

To simplify the MTA information, those MTAs located within a region of 1 cM, as reported by the LD decay, were considered part of the same QTL. Thus, the 233 associations were restricted to 81 MTA-QTLs (Figure 4 and Table 3). Of the 82 MTA-QTLs, 33 had only one MTA, whereas, for the remaining 49, the number of MTAs per MTA-QTL ranged from 2 in 19 MTA-QTLs to 15 in mtaq-7A.1. When several consecutive pairs of MTAs were separated for a distance of 1 cm, the whole block was considered as the same MTA-QTL. The genomic distribution of MTA-QTLs showed that chromosome 1A, 4A, and 5B harbored 8 MTA-QTLs, chromosomes 1B, 3A, 3B, and 5A 7 MTA-QTLs, 6A 6 MTA-QTLs, 7A 5 MTA-QTLs, 2A, 2B, 4B, and 6B 4 MTA-QTLs and chromosome 7B harbored 3 MTA-QTLs. For the 48 MTA-QTLs with more than one MTA, 10 were related to one trait. Of these, mtaq-1A.5, mtaq-3A.1, mtaq-4A.4, and mtaq-4A.5 carried associations related to root volume, mtaq-2B.1, and mtaq-3B.7 to root diameter, mtaq-3B.1, and mtaq-7A.5 to root number, and mtaq-4A.3 and mtaq-6A.5 to the root angle.

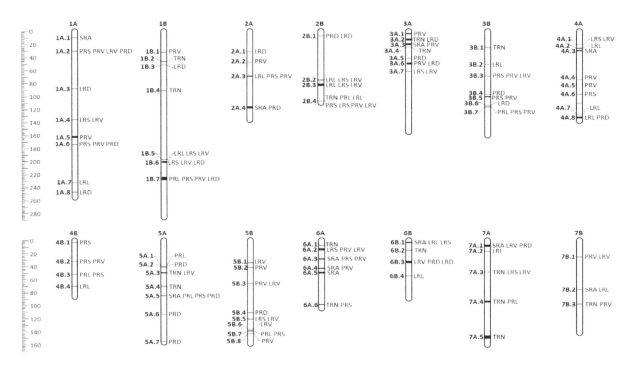

Figure 4. MTA-QTL map. MTA-QTLs are indicated in bold on the left side of the chromosome and traits involved in each MTA-QTL are on the right side. The rule on the left indicates genetic distance in cM. TRN, total root number. SRA, seminal root angle. PRL, primary root length. LRL, total lateral root length. PRS, primary root surface. LRS, total lateral root surface. PRV, primary root volume. LRV, total lateral root volume. PRD, primary root diameter. LRD, mean lateral root diameter.

Table 3. MTA-QTLS.

MTA-QTLs	Chromosome	Position (cM)	MTAs	Trait
mtaq-1A.1	1A	9.24	1	SRA
mtaq-1A.2	1A	29.71	4	PRS PRV LRV PRD
mtaq-1A.3	1A	88.15	1	LRD
mtaq-1A.4	1A	135.37	2	LRS LRV
mtaq-1A.5	1A	160.75–163.11	3	PRV
mtaq-1A.6	1A	173.41	3	PRS PRV PRD
mtaq-1A.7	1A	231.76	1	LRL
mtaq-1A.8	1A	246.3	1	LRD
mtaq-1B.1	1B	31.69	1	PRV
mtaq-1B.2	1B	45.68	1	TRN
mtaq-1B.3	1B	51.29	1	LRD
mtaq-1B.4	1B	90.37	1	TRN
mtaq-1B.5	1B	196.56	3	LRL LRS LRV
mtaq-1B.6	1B	199.9–201.49	3	LRS LRV LRD
mtaq-1B.7	1B	223.51–227.36	12	PRL PRS PRV LRD
mtaq-2A.1	2A	31.13	1	LRD
mtaq-2A.2	2A	46.78	1	PRV
mtaq-2A.3	2A	68.39–68.96	4	LRL PRS PRV
mtaq-2A.4	2A	115.8–118.32	4	SRA PRD
mtaq-2B.1	2B	6.7	2	PRD LRD
mtaq-2B.2	2B	75.09–75.13	13	LRL LRS LRV
mtaq-2B.3	2B	80.79–83.84	16	LRL LRS LRV
mtaq-2B.4	2B	106.98–107.03	8	TRN PRL LRL PRS LRS PRV LRV
mtaq-3A.1	3A	3.32–3.58	3	PRV
mtaq-3A.2	3A	11.88–12.93	2	TRN LRD
mtaq-3A.3	3A	18.37–20.39	3	SRA PRV
mtaq-3A.4	3A	23.99	1	TRN

Table 3. *Cont.*

MTA-QTLs	Chromosome	Position (cM)	MTAs	Trait
mtaq-3A.5	3A	40.97	1	PRD
mtaq-3A.6	3A	48.06–49.67	3	PRV LRD
mtaq-3A.7	3A	61.57	2	LRS LRV
mtaq-3B.1	3B	24.98–25	2	TRN
mtaq-3B.2	3B	50.7	1	LRL
mtaq-3B.3	3B	68.36	4	PRS PRV LRV
mtaq-3B.4	3B	96.48	1	PRD
mtaq-3B.5	3B	100.07–101.44	3	PRS PRV
mtaq-3B.6	3B	112.86	4	LRD
mtaq-3B.7	3B	115.61	3	PRL PRS PRV
mtaq-4A.1	4A	20.42–26.03	2	LRS LRV
mtaq-4A.2	4A	26.03	1	LRL
mtaq-4A.3	4A	28.85–28.87	2	SRA
mtaq-4A.4	4A	74.09	2	PRV
mtaq-4A.5	4A	96.08	2	PRV
mtaq-4A.6	4A	109.72	1	PRS
mtaq-4A.7	4A	127.56	1	LRL
mtaq-4A.8	4A	131.42–132.72	2	LRL PRD
mtaq-4B.1	4B	2.79	1	PRS
mtaq-4B.2	4B	31.93	4	PRS PRV
mtaq-4B.3	4B	51.22	3	PRL PRS
mtaq-4B.4	4B	70.04	1	LRL
mtaq-5A.1	5A	38.83	1	PRL
mtaq-5A.2	5A	40.51	1	PRD
mtaq-5A.3	5A	48.57–48.65	2	TRN LRV
mtaq-5A.4	5A	69.82	1	TRN
mtaq-5A.5	5A	84.51	5	SRA PRL PRS PRD
mtaq-5A.6	5A	112.96	1	PRD
mtaq-5A.7	5A	155.41	1	PRD
mtaq-5B.1	5B	33.99	1	LRV
mtaq-5B.2	5B	40.83	1	PRV
mtaq-5B.3	5B	65.51	2	PRV LRV
mtaq-5B.4	5B	111.15	1	PRD
mtaq-5B.5	5B	120.34	2	LRS LRV
mtaq-5B.6	5B	135.45	1	LRV
mtaq-5B.7	5B	138.69	4	PRL PRS
mtaq-5B.8	5B	142.12	1	PRV
mtaq-6A.1	6A	7.11	1	TRN
mtaq-6A.2	6A	11.95–14.24	8	LRS PRV LRV
mtaq-6A.3	6A	27.82–28.69	3	SRA PRS PRV
mtaq-6A.4	6A	42.36	3	SRA PRV
mtaq-6A.5	6A	48.39–50.08	2	SRA
mtaq-6A.6	6A	98.51–98.82	2	TRN PRS
mtaq-6B.1	6B	2.41–3.31	5	SRA LRL LRS
mtaq-6B.2	6B	14.26	1	TRN
mtaq-6B.3	6B	31.49–33.46	3	LRV PRD LRD
mtaq-6B.4	6B	53.66	1	LRL
mtaq-7A.1	7A	5.7–9.43	15	SRA LRV PRD
mtaq-7A.2	7A	16.28	1	LRL
mtaq-7A.3	7A	47.85	3	TRN LRS LRV
mtaq-7A.4	7A	92.69–94.34	2	TRN PRL
mtaq-7A.5	7A	145.94–150.31	11	TRN
mtaq-7B.1	7B	24.48	2	PRV LRV
mtaq-7B.2	7B	74.86–75.24	2	SRA LRL
mtaq-7B.3	7B	97.45	2	TRN PRV

TRN, total root number. SRA, seminal root angle. PRL, primary root length. LRL, total lateral root length. PRS, primary root surface. LRS, total lateral root surface. PRV, primary root volume. LRV, total lateral root volume. PRD, primary root diameter. LRD, mean lateral root diameter.

Among all significant MTAs, markers with different alleles between extreme genotypes for each trait (i.e., the upper and lower 10th percentile) were identified except for PRL (Table 4, Figure 5). Frequency of the most common allele among genotypes from the upper 10th percentile ranged from 67% for LRD to 90% for PRV, whereas, for the lower 10th percentile, they ranged from 74% for TRN to 93% for LRD (Figure 5).

Table 4. Selected significant markers from the GWAS with different allele composition for the upper (UP) and lower (LOW) 10th percentile of genotypes. Different letters on the UP and LOW 10th phenotype indicate that means are significantly different at $p < 0.01$ following a Tukey test.

Trait	Phenotype			Marker	Chromosome	Position	R^3	Most Frequent Allele			
	Mean	UP 10th	LOW 10th					UP	Frequency	LOW	Frequency
TRN (N)	4.9	5.8 [a]	3.7 [b]	2260740_SNP	7A	148.38	0.09	T	0.80	C	0.81
				1252655_PAV	7B	97.45	0.11	1	0.94	0	0.67
SRA (°)	88.5	111.0 [a]	67.1 [b]	1125557_PAV	2A	115.80	0.09	0	1.00	1	1.00
				1117775_PAV	2A	118.32	0.10	1	0.75	0	0.71
LRL (cm)	21.8	36.5 [a]	9.3 [b]	4408432_PAV	6B	3.31	0.09	1	0.88	0	0.73
				4408958_PAV	6B	3.31	0.09	1	0.88	0	0.73
				1098568_PAV	6B	53.66	0.08	1	0.77	0	0.86
PRS (cm²)	2.2	3.2 [a]	1.3 [b]	4406631_PAV	4B	31.93	0.09	0	0.71	0	0.86
				4406980_PAV	4B	31.93	0.09	1	0.71	1	0.86
LRS (cm²)	4.0	6.5 [a]	1.8 [b]	1201756_PAV	2B	107.03	0.15	1	1.00	0	0.73
				987263_PAV	3A	61.57	0.10	0	0.92	1	0.88
				4408432_PAV	6B	3.31	0.09	1	0.81	0	0.79
				4408958_PAV	6B	3.31	0.09	1	0.81	0	0.79
PRV (mm³)	33.7	49.8 [a]	20.2 [b]	997799_SNP	1B	31.69	0.12	A	0.86	G	0.77
				1201756_PAV	2B	107.03	0.11	1	0.87	0	0.71
				4406631_PAV	4B	31.93	0.09	0	0.93	1	0.87
				4406980_PAV	4B	31.93	0.09	0	0.93	1	0.87
LRV (mm³)	60.5	99 [a]	27.7 [b]	1201756_PAV	2B	107.03	0.15	1	0.94	0	0.73
				987263_PAV	3A	61.57	0.10	0	0.93	1	0.81
				1126050_SNP	5B	33.99	0.07	A	0.81	M	0.81
				1149356_PAV	7B	24.48	0.08	0	0.87	1	0.81
PRD (mm)	0.58	0.66 [a]	0.48 [b]	1113225_SNP	5A	84.51	0.09	G	0.87	C	0.92
				1864057_SNP	6B	33.46	0.07	C	0.81	M	0.81
LRD (mm)	0.58	0.66 [a]	0.51 [b]	4005012_PAV	1B	51.29	0.10	0	0.67	1	0.93

TRN, total root number. SRA, seminal root angle. PRL, primary root length. LRL, total lateral root length. PRS, primary root surface. LRS, total lateral root surface. PRV, primary root volume. LRV, total lateral root volume. PRD, primary root diameter. LRD, mean lateral root diameter.

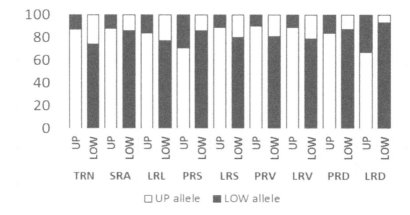

Figure 5. Marker allele frequency means from landraces within the upper and lower 10th percentile for the analyzed traits. All significant markers shown in Table 4 are included. TRN, total root number. SRA, seminal root angle. PRL, primary root length. LRL, total lateral root length. PRS, primary root surface. LRS, total lateral root surface. PRV, primary root volume. LRV, total lateral root volume. PRD, primary root diameter. LRD, mean lateral root diameter.

3.3. Gene Annotation

Of the 176 markers showing significant associations, 31 were identified in the reference sequence of the wheat genome [34] (Table 5). Eight of them were positioned within gene models, whereas, for the rest, the closest gene model to the corresponding marker was taken into consideration. The gene models described in Table 5 included molecules related to abiotic stress resistance, seed formation, carbohydrate remobilization, disease resistance proteins, and other genes involved in different cellular metabolic pathways.

Table 5. Gene models within MTA-QTL positions. Only MTAs with markers mapped against the genome sequence are included. Genome position of the gene model is indicated in Mb.

DArTseq Marker	MTA-QTL	Gene Model	Position	Description
1109244_SNP	mtaq-1A.5	TraesCS1A01G363600	540.1	Jacalin lectin family protein
1210090_SNP	mtaq-1A.7	TraesCS1A01G424800	579.8	Cellulose synthase
997799_SNP	mtaq-1B.1	TraesCS1B01G022500	10.1	Protein trichome birefringence
1003552_SNP	mtaq-1B.7	TraesCS1B01G430400	654.8	F-box domain protein
1085277_SNP *	mtaq-2A.3	TraesCS2A01G250600	378.4	9-cis-epoxycarotenoid dioxygenase
1083104_SNP	mtaq-2A.3	TraesCS2A01G281000	469.4	Dynamin-like family protein
1117775_PAV	mtaq-2A.4	TraesCS2A01G541700	752.9	LEA hydroxyproline-rich glycoprotein family
1075469_SNP	mtaq-2B.1	TraesCS2B01G004500	2.4	Cytochrome P450 family protein
1256467_PAV	mtaq-3A.1	TraesCS3A01G018600	11.5	F-box domain protein
1082068_PAV	mtaq-3A.2	TraesCS3A01G034100	19.3	Receptor-like kinase
1130621_PAV	mtaq-3A.5	TraesCS3A01G132300	108.9	Blue copper protein
987263_PAV *	mtaq-3A.7	TraesCS3A01G393600	641.6	Pectin lyase-like superfamily protein
1101009_SNP	mtaq-3B.4	TraesCS3B01G516800	759.9	Ribosomal protein S4
3034109_PAV	mtaq-4A.6	TraesCS4A01G419000	688.9	Histone acetyltransferase of the CBP family 5
1250077_PAV *	mtaq-4B.3	TraesCS4B01G345800	639.4	Basic helix-loop-helix DNA-binding protein
1240561_PAV	mtaq-6A.3	TraesCS6A01G041500	21.7	Transmembrane protein 97
1047867_PAV	mtaq-6A.3	TraesCS6A01G415600	615.3	Cobyric acid synthase
1105573_PAV	mtaq-6A.5	TraesCS6A01G242300	453.9	50S ribosomal protein L19
989287_PAV *	mtaq-6A.6	TraesCS6A01G417400	615.8	F-box domain protein
1129380_PAV *	mtaq-6B.1	TraesCS6B01G000200	0.1	NBS-LRR resistance-like protein
1864057_SNP *	mtaq-6B.3	TraesCS6B01G335600	590.9	Hexosyltransferase
1098568_PAV *	mtaq-6B.4	TraesCS6B01G399700	675.2	bZIP transcription factor family protein
1130796_PAV	mtaq-7A.1	TraesCS7A01G015100	0.0	Mitochondrial pyruvate carrier
2253648_PAV	mtaq-7A.1	TraesCS7A01G016700	7.3	Transmembrane protein DUF594
1139027_PAV	mtaq-7A.1	TraesCS7A01G015400	6.7	Signal peptidase complex catalytic subunit SEC11
1076865_PAV	mtaq-7A.1	TraesCS7A01G024800	9.7	WAT1-related protein
1059554_SNP *	mtaq-7A.3	TraesCS7A01G100600	61.8	GDSL esterase/lipase
1665955_PAV	mtaq-7A.4	TraesCS7A01G442400	636.7	BTB/POZ domain
1149356_PAV	mtaq-7B.1	TraesCS7B01G058300	60.6	Glutamate receptor
1075278_SNP	mtaq-7B.2	TraesCS7B01G378200	642.6	Receptor-like kinase
1252655_PAV	mtaq-7B.3	TraesCS7B01G421300	690.2	NBS-LRR resistance-like protein

* Markers located within gene models.

4. Discussion

Roots exhibit a high level of morphological plasticity in response to soil conditions, which allows plants to better adapt, particularly under drought conditions. Several authors have reported the role of RSA traits in response to drought stress [39,40]. Wasson et al. [11] suggested that a deep root system with the appropriate density along the soil profile would confer an advantage on wheat grown in rainfed agricultural systems. Therefore, identifying new alleles for improving root architecture under drought conditions and introgressing them into adapted phenotypes is a desirable approach for breeding purposes. The current study analyzed a collection of durum wheat landraces representative of the variability existing within the Mediterranean Basin in an attempt to broaden the genetic background present in commercial cultivars.

Evaluating root architecture in the field is a difficult, expensive, and time-consuming assignment, especially when a large number of plants need to be phenotyped. It has been reported that the root geometry of adult plants is strongly related to the seminal root angle (SRA), with deeply rooted wheat genotypes showing a narrower SRA [10]. Different systems have been adopted to enable early screening of the root system architecture in wheat, assuming that genotypes that differ in root architecture at an early developmental stage would also differ in the field at stages when nutrient and/or water capture become critical for grain yield [12].

4.1. Phenotypic Variation

The germplasm analyzed in the present study, including mostly durum wheat landraces from the Mediterranean Basin, showed wide variability in RSA traits. The variability found was higher than that observed in other studies using elite accessions [12,14] or even landraces, as reported by Ruiz et al. [41] analyzing a collection of Spanish durum wheat landraces. These results, and the intermediate values obtained for all traits in modern cultivars, support the use of ancient local germplasm for widening the genetic background in breeding programs.

Means comparison of phenotypic traits revealed large differences among SPs associated with their geographical origin. Eastern Mediterranean landraces, collected in the area closest to the origin of tetraploid wheat, showed the largest root size in terms of length, surface, and volume, and the widest root angle. The wheat-growing areas of this region, which comprises Syria, Jordan, Israel, and Egypt, are the warmest and driest within the Mediterranean Basin [42]. In addition, when SRA traits were analyzed separately for the two components of the eastern Balkans and Turkey subpopulation, large differences appeared between them, with Turkish landraces being much more similar to the eastern Mediterranean ones than to the eastern Balkan ones, since the latter showed the lowest values for root length, surface, and volume. Turkish landraces also showed a wide root angle, as did the eastern Mediterranean ones. The differences found in SRA between the eastern Balkans and Turkish landraces are sustained by two lines of evidence. One is the contrasting environmental conditions of the wheat-growing areas of northern Balkan countries and Turkey, since the analysis of long-term climate data demonstrated less rainfall and higher temperatures and solar radiation in the latter [42]. The other is that the northern Balkan landraces likely originated in the steps of southern Russia and the Volga region [2,43], which also suggests contrasting environmental conditions in the zones of origin of the eastern Balkan and Turkish landraces. The phenotypic analysis carried out in the current study revealed that landraces from regions where drought stress is prevalent have a larger root size and a wider root angle. This architecture should allow a larger proportion of the soil to be covered for more efficient water capture, and this hypothesis is supported by correlations between RSA and yield traits. Although low, likely due to the very early stage when the root traits were measured, differences in the number of significant correlations were observed between the two environments with the highest and lowest water input reported by Roselló et al. [25]. Root size–related traits were positively correlated with the number of grains and spikes per unit area (primary roots) and with grain yield and grain weight (lateral roots) in the driest environment. SRA was negatively correlated with TKW, as reported previously by Canè et al. [12], who concluded that it was due to the influence of the root angle on the distribution of roots in the soil layers, which affects the water uptake from deeper layers. In our study, the genotypes with the narrowest angle corresponded to those from the western Mediterranean countries, which Royo et al. [42] and Soriano et al. [19] reported to have heavier grains.

4.2. Marker-Trait Associations

The current study attempts to dissect the genetic architecture controlling the seminal root system in a collection of landraces from the Mediterranean Basin by association analysis. A mixed linear model accounting for the genetic relatedness between cultivars (random effect) and their population structure (fixed effect) (K + Q model) was used in order to reduce the number of spurious associations.

A total of 233 significant associations were identified for the 10 RSA traits underlying the complex genetic control of RSA. However, in order to simplify this information and to integrate closely linked MTAs in the same QTL, those MTAs located within LD blocks were considered as belonging to the same MTA-QTL. As a result, the number of genome regions involved in RSA was reduced to 82. The relationships between RSA and yield-related traits was also suggested by the presence of pleiotropic

MTA-QTLs. The comparison of the genome regions identified in the current study with those related to yield and yield components by Roselló et al. [44] showed that 45% of the RSA MTA-QTLs were located with yield-related trait MTA-QTLs. These results are in agreement with the findings of Canè et al. [12], who found that 30% of the RSA-QTLs affected agronomic traits, which provided evidence of the implications of RSA in field performance of durum wheat at early growth stages.

In the last few years, GWAS for RSA have been limited in comparison with QTL mapping for root traits based on bi-parental populations (see Soriano and Álvaro [13] for a review). A comparison with previous studies reporting MTAs for RSA resulted in several common regions with the current study. Three common regions were found with the study of Canè et al. [12], but different traits were included for MTAs in those QTLs (mtaq-3A.3, mtaq-3A.5, mtaq-3A.6, and mtaq-6B.2). Two MTAs were in common with those reported by Ayalew et al. [15], who identified five significant associations with root length under stress (2) and non-stress (3) conditions. The MTA reported under stress conditions in chromosome 2B may correspond with mtaq-2B.2, which also shows an association with LRL. However, the association on chromosome 3B, although in a common region with mtaq-3B.4, differed in RSA. When MTA-QTLs were compared with QTLs from bi-parental populations, twelve genomic regions were located within the meta-QTL positions defined by Soriano and Álvaro [13] after the compilation of 754 QTLs from 30 studies.

Candidate genes at the MTA peak were sought using the high-confidence gene annotation from the wheat genome sequence [34]. Among these genes, those involved in plant growth and development as well as tolerance to abiotic stresses may be of special interest. On chromosome 1A, the marker 1210090_SNP in mtaq-1A.7 is located close to a cellulose synthase gene. This type of gene is involved in plant cell growth and structure [45]. A trichome birefringence (TB) protein was identified in mtaq-1B.1. According to Zhu et al. [46], the TB-like27 protein mutants in *Arabidopsis* increased aluminium accumulation in cell walls, which inhibited root elongation through structural and functional damage. Three peaks corresponded with F-box domains located in mtaq-1B.7, mtaq-3A.1, and mtaq-6A.6. According to Hua et al. [47], this is the protein subunit of E3 ubiquitin ligases involved in the response to abiotic stresses. Li et al. [48] overexpressed the F-box *TaFBA1* in transgenic tobacco to improve heat tolerance, and one of the results was increased root length in the transgenic plants. 9-cis-epoxycarotenoid dioxygenase (NCED) is a key enzyme in the biosynthesis of ABA in higher plants, which regulates the response to various environmental stresses [49]. This enzyme is located within mtaq-2A.3. In mtaq-2A.4, the marker 1117775_PAV corresponded with a late embryogenesis abundant (LEA) hydroxyproline-rich glycoprotein. These proteins play a role in the response to abiotic stresses. They are mainly accumulated in seeds, but have been found in roots during the whole developmental cycle [50]. The marker 1098568_PAV, in mtaq-6B.4, is located within a gene coding a bZIP transcription factor family protein. This type of transcription factor is involved in abiotic stresses [51]. Zhang et al. [52] observed that the root growth of transgenic plants overexpressing the gene *TabZIP14-B* was hindered more severely than that of the control plants. Another gene involved in abiotic stress tolerance is the mitochondrial pyruvate carrier (MPC) located in mtaq-7A.1 [53]. This gene is involved in cadmium tolerance in *Arabidopsis*, which prevents its accumulation. Roots are the predominant plant tissue for cadmium absorption or exclusion. He et al. [53] found that the root length of mutant plants of *Arabidopsis* for MPC genes was substantially shorter than the wild-type plants. A protein related to WAT1 (WALLS ARE THIN1) involved in secondary cell wall thickness [54] is located in the peak of mtaq-7A.1.

5. Conclusions

Including local landraces in breeding programs is a useful approach to broadening the genetic variability of crops [3]. The variability for root system architecture traits found in Mediterranean landraces and the high number of genome regions controlling them—most of them not reported previously—makes this germplasm a valuable source for root architecture improvement. The identification of extreme genotypes for root architecture traits can help identify parents for the development of new mapping populations to tackle a map-based cloning approach to the genes of interest. In the present study, we identified the molecular markers linked to these genotypes with different allele composition that facilitate the introgression of the corresponding traits through marker-assisted breeding.

Supplementary Materials
Table S1: Accessions included in the study, Table S2: Statistics of the seminal RSA traits, Table S3: Means comparison of seminal root system architecture traits for the eastern Balkans (EB) and Turkish durum wheat landraces. Means within columns with different letters are significantly different at $p < 0.05$ following a Tukey test, Table S4: Linkage disequilibrium decay plots. (A) Genome A. (B) Genome B. The LOESS curve is represented in blue. The horizontal red line corresponds to the r^2 mean for each genome, Table S5: Significant markers associated with seminal root system architecture traits obtained in 160 durum wheat Mediterranean landraces.

Author Contributions: Conceptualization, M.R. and J.M.S.; Methodology, M.R., M.S.-G., and J.M.S.; Formal Analysis, M.R.; Investigation, M.R., M.S.-G., J.M.S.; Resources, C.R.; Data Curation, M.R., C.R.; Writing—Original Draft Preparation, M.R., J.M.S.; Writing—Review & Editing, C.R., M.S.-G., and J.M.S.; Visualization, J.M.S.; Supervision, C.R., M.S.-G., and J.M.S.; Project Administration, C.R. and J.M.S.; Funding Acquisition, C.R. and J.M.S.

Acknowledgments: Projects AGL-2012-37217 (C.R.) and AGL2015-65351-R (J.M.S. and C.R.) of the Spanish Ministry of Science, Innovation and Universities (http://www.ciencia.gob.es/) funded this study. The authors acknowledge the contribution of the CERCA Program (Generalitat de Catalunya). M.R. is a recipient of a PhD grant from the Instituto Nacional de Investigación y Tecnología Agraria y Alimentaria (INIA). J.M.S. was hired by the INIA-CCAA program funded by INIA and the Generalitat de Catalunya.

Abbreviations

BP	Before Present
DArTseq	Diversity Arrays Technology sequencing
EB + T	Eastern Balkans and Turkey
EM	Eastern Mediterranean
FDR	False Discovery Rate
GWAS	Genome Wide Association Study
GY	Grain Yield
LRD	Lateral Roots Diameter
LRL	Lateral Roots Length
LRS	Lateral Roots Surface
LRV	Lateral Roots Volume
MTA	Marker-Trait Association
NGm^2	Number of Grains per square meter
NSm^2	Number of Spikes per square meter
PAV	Presence/Absence Variants
PRD	Primary Root Diameter
PRL	Primary Root Length
PRS	Primary Root Surface
PRV	Primary Root Volume
PVE	Phenotypic Variance Explained
QTL	Quantitative Trait Loci
RSA	Root System Architecture
SNP	Single Nucleotide Polymorphism
SP	Subpopulation
SRA	Seminal Root Angle
TKW	Thousand Kernel Weight
TRN	Total Root Number
WB + E	Western Balkans and Egypt
WM	Western Mediterranean

References

1. Feldman, M. Origin of cultivated wheat. In *The World Wheat Book: A History of Wheat Breeding*; Bonjean, A.P., Angus, W.J., Eds.; Lavoisier Publishing: Paris, France, 2001; pp. 3–56. ISBN 1898298726.

2. Nazco, R.; Villegas, D.; Ammar, K.; Peña, R.J.; Moragues, M.; Royo, C. Can Mediterranean durum wheat landraces contribute to improved grain quality attributes in modern cultivars? *Euphytica* **2012**, *185*, 1–17. [CrossRef]

3. Lopes, M.S.; El-Basyoni, I.; Baenziger, P.S.; Singh, S.; Royo, C.; Ozbek, K.; Aktas, H.; Ozer, E.; Ozdemir, F.; Manickavelu, A.; et al. Exploiting genetic diversity from landraces in wheat breeding for adaptation to climate change. *J. Exp. Bot.* **2015**, *66*, 3477–3486. [CrossRef] [PubMed]

4. Reynolds, M.P.; Mujeeb-Kazi, A.; Sawkins, M. Prospects for utilising plant-adaptive mechanisms to improve wheat and other crops in drought- and salinity-prone environments. *Ann. Appl. Biol.* **2005**, *146*, 239–259. [CrossRef]

5. Maccaferri, M.; Sanguineti, M.C.; Demontis, A.; El-Ahmed, A.; García del Moral, L.; Maalouf, F.; Nachit, M.; Nserallah, N.; Ouabbou, H.; Rhouma, S.; et al. Association mapping in durum wheat grown across a broad range of water regimes. *J. Exp. Bot.* **2011**, *62*, 409–438. [CrossRef] [PubMed]

6. Royo, C.; García del Moral, L.; Slafer, G.A.; Nachit, M.; Araus, J.L. Selection tools for improving yield-associated physiological traits. In *Durum Wheat Breeding: Current Approaches and Future Strategies*; Royo, C., Nachit, M.M., Di Fonzo, N., Araus, J., Pfeiffer, W., Slafer, G., Eds.; Food Products Press: New York, NY, USA, 2005; pp. 563–598.

7. Royo, C.; Villegas, D. Field measurements of canopy spectra for biomass assessment of small-grain cereals. In *Biomass—Detection, Production and Usage*; Darko, M., Ed.; InTech Open: London, UK, 2011; pp. 27–52. ISBN 978-953-307-492-4.

8. Tuberosa, R. Phenotyping for drought tolerance of crops in the genomics era. *Front. Physiol.* **2012**, *3*, 347. [CrossRef] [PubMed]

9. Løes, A.-K.; Gahoonia, T.S. Genetic variation in specific root length in Scandinavian wheat and barley accessions. *Euphytica* **2004**, *137*, 243–249. [CrossRef]

10. Manschadi, A.M.; Hammer, G.L.; Christopher, J.T.; Manschadi, A.M.; Hammer, G.L.; Christopher, J.T. Genotypic variation in seedling root architectural traits and implications for drought adaptation in wheat (*Triticum aestivum* L.). *Plant Soil* **2008**, *303*, 115–129. [CrossRef]

11. Wasson, A.P.; Richards, R.A.; Chatrath, R.; Misra, S.C.; Prasad, S.V.S.; Rebetzke, G.J.; Kirkegaard, J.A.; Christopher, J.; Watt, M. Traits and selection strategies to improve root systems and water uptake in water-limited wheat crops. *J. Exp. Bot.* **2012**, *63*, 3485–3498. [CrossRef]

12. Canè, M.A.; Maccaferri, M.; Nazemi, G.; Salvi, S.; Francia, R.; Colalongo, C.; Tuberosa, R. Association mapping for root architectural traits in durum wheat seedlings as related to agronomic performance. *Mol. Breed.* **2014**, *34*, 1629–1645. [CrossRef]

13. Soriano, J.M.; Álvaro, F. Discovering consensus genomic regions in wheat for root-related traits by QTL meta-analysis. *Sci. Rep.* **2019**, in press.

14. Sanguineti, M.C.; Li, S.; Maccaferri, M.; Corneti, S.; Rotondo, F.; Chiari, T.; Tuberosa, R. Genetic dissection of seminal root architecture in elite durum wheat germplasm. *Ann. Appl. Biol.* **2007**, *151*, 291–305. [CrossRef]

15. Ayalew, H.; Liu, H.; Börner, A.; Kobiljski, B.; Liu, C.; Yan, G. Genome-wide association mapping of major root length QTLs under PEG induced water stress in wheat. *Front. Plant Sci.* **2018**, *9*, 1759. [CrossRef] [PubMed]

16. Alahmad, S.; El Hassouni, K.; Bassi, F.M.; Dinglasan, E.; Youssef, C.; Quarry, G.; Aksoy, A.; Mazzucotelli, E.; Juhász, A.; Able, J.A.; et al. A major root architecture QTL responding to water limitation in durum wheat. *Front. Plant Sci.* **2019**, *10*, 436. [CrossRef] [PubMed]

17. Beyer, S.; Daba, S.; Tyagi, P.; Bockelman, H.; Brown-Guedira, G.; IWGSC; Mohammadi, M. Loci and candidate genes controlling root traits in wheat seedlings—A wheat root GWAS. *Funct. Integr. Genom.* **2019**, *19*, 91–107. [CrossRef] [PubMed]

18. Li, L.; Peng, Z.; Mao, X.; Wang, J.; Chang, X.; Reynolds, M.; Jing, R. Genome-wide association study reveals genomic regions controlling root and shoot traits at late growth stages in wheat. *Ann. Bot.* **2019**, 1–14. [CrossRef]

19. Soriano, J.; Villegas, D.; Aranzana, M.; García del Moral, L.; Royo, C. Genetic structure of modern durum wheat cultivars and mediterranean landraces matches with their agronomic performance. *PLoS ONE* **2016**, *11*, e0160983. [CrossRef] [PubMed]

20. Rahnama, A.; Munns, R.; Poustini, K.; Watt, M. A screening method to identify genetic variation in root growth response to a salinity gradient. *J. Exp. Bot.* **2011**, *62*, 69–77. [CrossRef] [PubMed]

21. Watt, M.; Moosavi, S.; Cunningham, S.C.; Kirkegaard, J.A.; Rebetzke, G.J.; Richards, R.A. A rapid, controlled-environment seedling root screen for wheat correlates well with rooting depths at vegetative, but not reproductive, stages at two field sites. *Ann. Bot.* **2013**, *112*, 447–455. [CrossRef] [PubMed]

22. Lobet, G.; Pagès, L.; Draye, X. A Novel image-analysis toolbox enabling quantitative analysis of root system architecture. *Breakthr. Technol.* **2011**, *157*, 29–39. [CrossRef] [PubMed]

23. Richard, C.A.I.; Hickey, L.T.; Fletcher, S.; Jennings, R.; Chenu, K.; Christopher, J.T. High-throughput phenotyping of seminal root traits in wheat. *Plant Methods* **2015**, *11*, 13. [CrossRef] [PubMed]

24. Abramoff, M.D.; Magalhaes, P.J.; Ram, S.J. Image Processing with ImageJ. *Biophotonics Int.* **2004**, *11*, 36–42.

25. Roselló, M.; Villegas, D.; Álvaro, F.; Soriano, J.M.; Lopes, M.S.; Nazco, R.; Royo, C. Unravelling the relationship between adaptation pattern and yield formation strategies in Mediterranean durum wheat landraces. *Eur. J. Agron.* **2019**, *107*, 43–52. [CrossRef]

26. Tukey, J.W. Comparing individual means in the analysis of variance. *Biometrics* **1949**, *5*, 99–114. [CrossRef] [PubMed]

27. Little, R.; Milliken, G.; Stroup, W.; Wolfinger, R. SAS system for mixed models. *Technometrics* **1997**, *39*, 344.

28. SAS Institute. *SAS Enterprise Guide*; SAS Institute: Cary, NC, USA, 2014.

29. Doyle, J.; Doyle, J. A rapid DNA isolation procedure for small quantities of fresh leaf tissue. *Phytochem. Bull.* **1987**, *19*, 11–15.

30. Sansaloni, C.; Petroli, C.; Jaccoud, D.; Carling, J.; Detering, F.; Grattapaglia, D.; Kilian, A. Diversity arrays technology (DArT) and next-generation sequencing combined: Genome-wide, high throughput, highly informative genotyping for molecular breeding of Eucalyptus. *BMC Proc.* **2011**, *5*, 54. [CrossRef]

31. Bradbury, P.J.; Zhang, Z.; Kroon, D.E.; Casstevens, T.M.; Ramdoss, Y.; Buckler, E.S. TASSEL: Software for association mapping of complex traits in diverse samples. *Bioinformatics* **2007**, *23*, 2633–2635. [CrossRef] [PubMed]

32. Benjamini, Y.; Hochberg, Y. Controlling the false discovery rate: A practical and powerful approach to multiple testing. *J. R. Stat. Soc.* **1995**, *57*, 289–300. [CrossRef]

33. Voorrips, R.E. MapChart: Software for the graphical presentation of linkage maps and QTLs. *J. Hered.* **2002**, *93*, 77–78. [CrossRef]

34. IWGSC. Shifting the limits in wheat research and breeding using a fully annotated reference genome. *Science* **2018**, *361*, eaar7191. [CrossRef]

35. Mwadzingeni, L.; Shimelis, H.; Rees, D.J.G.; Tsilo, T.J. Genome-wide association analysis of agronomic traits in wheat under drought-stressed and non-stressed conditions. *PLoS ONE* **2017**, *12*, e0171692. [CrossRef] [PubMed]

36. Wang, S.-X.; Zhu, Y.-L.; Zhang, D.-X.; Shao, H.; Liu, P.; Hu, J.-B.; Zhang, H.; Zhang, H.-P.; Chang, C.; Lu, J.; et al. Genome-wide association study for grain yield and related traits in elite wheat varieties and advanced lines using SNP markers. *PLoS ONE* **2017**, *12*, e0188662. [CrossRef] [PubMed]

37. Mangini, G.; Gadaleta, A.; Colasuonno, P.; Marcotuli, I.; Signorile, A.M.; Simeone, R.; De Vita, P.; Mastrangelo, A.M.; Laidò, G.; Pecchioni, N.; et al. Genetic dissection of the relationships between grain yield components by genome-wide association mapping in a collection of tetraploid wheats. *PLoS ONE* **2018**, *13*, e0190162. [CrossRef] [PubMed]

38. Sukumaran, S.; Reynolds, M.P.; Sansaloni, C. Genome-wide association analyses identify QTL hotspots for yield and component traits in durum wheat grown under yield potential, drought, and heat stress environments. *Front. Plant Sci.* **2018**, *9*, 81. [CrossRef] [PubMed]

39. Christopher, J.; Christopher, M.; Jennings, R.; Jones, S.; Fletcher, S.; Borrell, A.; Manschadi, A.M.; Jordan, D.; Mace, E.; Hammer, G. QTL for root angle and number in a population developed from bread wheats (*Triticum aestivum*) with contrasting adaptation to water-limited environments. *Theor. Appl. Genet.* **2013**, *126*, 1563–1574. [CrossRef] [PubMed]

40. Paez-García, A.; Motes, C.; Scheible, W.-R.; Chen, R.; Blancaflor, E.; Monteros, M. Root traits and phenotyping strategies for plant improvement. *Plants* **2015**, *4*, 334–355. [CrossRef] [PubMed]

41. Ruiz, M.; Giraldo, P.; González, J.M. Phenotypic variation in root architecture traits and their relationship with eco-geographical and agronomic features in a core collection of tetraploid wheat landraces (*Triticum turgidum* L.). *Euphytica* **2018**, *214*, 54. [CrossRef]

42. Royo, C.; Nazco, R.; Villegas, D. The climate of the zone of origin of Mediterranean durum wheat (*Triticum durum* Desf.) landraces affects their agronomic performance. *Genet. Resour. Crop Evol.* **2014**, *61*, 1345–1358. [CrossRef]

43. Dedkova, O.S.; Badaeva, E.D.; Amosova, A.V.; Martynov, S.P.; Ruanet, V.V.; Mitrofanova, O.P.; Pukhal'skiy, V.A. Diversity and the origin of the European population of *Triticum dicoccum* (Schrank) Schuebl. As revealed by chromosome analysis. *Russ. J. Genet.* **2009**, *45*, 1082–1091. [CrossRef]

44. Roselló, M.; Royo, C.; Sansaloni, C.; Soriano, J.M. GWAS for yield and related traits under rainfed mediterranean environments revealed different genetic architecture in pre- and post-green revolution durum wheat collections. *Front. Plant Sci.* **2019**, in press.

45. Lei, L.; Li, S.; Gu, Y. Cellulose synthase complexes: Composition and regulation. *Front. Plant Sci.* **2012**, *3*, 75. [CrossRef] [PubMed]

46. Zhu, X.F.; Sun, Y.; Zhang, B.C.; Mansoori, N.; Wan, J.X.; Liu, Y.; Wang, Z.W.; Shi, Y.Z.; Zhou, Y.H.; Zheng, S.J. *Trichome Birefringence-Like27* affects aluminum sensitivity by modulating the *O*-acetylation of xyloglucan and aluminum-binding capacity in *Arabidopsis*. *Plant Physiol.* **2014**, *166*, 181–189. [CrossRef] [PubMed]

47. Hua, Z.; Vierstra, R.D. The cullin-RING ubiquitin-protein ligases. *Annu. Rev. Plant Biol.* **2011**, *62*, 299–334. [CrossRef] [PubMed]

48. Li, Q.; Wang, W.; Wang, W.; Zhang, G.; Liu, Y.; Wang, Y.; Wang, W. Wheat f-box protein gene *TaFBA1* is involved in plant tolerance to heat stress. *Front. Plant Sci.* **2018**, *9*, 521. [CrossRef] [PubMed]

49. Zhang, S.J.; Song, G.Q.; Li, Y.L.; Gao, J.; Liu, J.J.; Fan, Q.Q.; Huang, C.Y.; Sui, X.X.; Chu, X.S.; Guo, D.; et al. Cloning of 9-*cis*-epoxycarotenoid dioxygenase gene (*TaNCED1*) from wheat and its heterologous expression in tobacco. *Biol. Plant.* **2014**, *58*, 89–98. [CrossRef]

50. Gao, J.; Lan, T. Functional characterization of the late embryogenesis abundant (LEA) protein gene family from *Pinus tabuliformis* (*Pinaceae*) in *Escherichia coli*. *Sci. Rep.* **2016**, *6*, 19467. [CrossRef]

51. Sornaraj, P.; Luang, S.; Lopato, S.; Hrmova, M. Basic leucine zipper (bZIP) transcription factors involved in abiotic stresses: A molecular model of a wheat bZIP factor and implications of its structure in function. *Biochim. Biophys. Acta* **2016**, *1860*, 46–56. [CrossRef]

52. Zhang, W.; Chen, S.; Abate, Z.; Nirmala, J.; Rouse, M.N.; Dubcovsky, J. Identification and characterization of *Sr13*, a tetraploid wheat gene that confers resistance to the Ug99 stem rust race group. *Proc. Natl. Acad. Sci. USA* **2017**, *114*, E9483–E9492. [CrossRef]

53. He, L.; Jing, Y.; Shen, J.; Li, X.; Liu, H.; Geng, Z.; Wang, M.; Li, Y.; Chen, D.; Gao, J.; et al. Mitochondrial pyruvate carriers prevent cadmium toxicity by sustaining the TCA cycle and glutathione synthesis. *Plant Physiol.* **2019**, *180*, 198–211. [CrossRef]

54. Ranocha, P.; Dima, O.; Nagy, R.; Felten, J.; Corratgé-Faillie, C.; Novák, O.; Morreel, K.; Lacombe, B.; Martinez, Y.; Pfrunder, S.; et al. *Arabidopsis* WAT1 is a vacuolar auxin transport facilitator required for auxin homoeostasis. *Nat. Commun.* **2013**, *4*, 2625. [CrossRef]

7

An SNP-Based High-Density Genetic Linkage Map for Tetraploid Potato using Specific Length Amplified Fragment Sequencing (SLAF-Seq) Technology

Xiaoxia Yu [†], Mingfei Zhang [†], Zhuo Yu *, Dongsheng Yang, Jingwei Li, Guofang Wu and Jiaqi Li

Agricultural College, Inner Mongolia Agricultural University, Hohhot 010000, China; yuxiaoxia1985@sina.com (X.Y.); zhangmingfei0207@163.com (M.Z.); yangdongsheng007@163.com (D.Y.); ljw2016409@163.com (J.L.); wuguofang25@163.com (G.W.); lijiaqi1127@sina.com (J.L.)

* Correspondence: yuzhuo58@sina.com

† These authors contributed equally to this work.

Abstract: Specific length amplified fragment sequencing (SLAF-seq) is a recently developed high-resolution strategy for the discovery of large-scale de novo genotyping of single nucleotide polymorphism (SNP) markers. In the present research, in order to facilitate genome-guided breeding in potato, this strategy was used to develop a large number of SNP markers and construct a high-density genetic linkage map for tetraploid potato. The genomic DNA extracted from 106 F_1 individuals derived from a cross between two tetraploid potato varieties YSP-4 × MIN-021 and their parents was used for high-throughput sequencing and SLAF library construction. A total of 556.71 Gb data, which contained 2269.98 million pair-end reads, were obtained after preprocessing. According to bioinformatics analysis, a total of 838,604 SLAF labels were developed, with an average sequencing depth of 26.14-fold for parents and 15.36-fold for offspring of each SLAF, respectively. In total, 113,473 polymorphic SLAFs were obtained, from which 7638 SLAFs were successfully classified into four segregation patterns. After filtering, a total of 7329 SNP markers were detected for genetic map construction. The final integrated linkage map of tetraploid potato included 3001 SNP markers on 12 linkage groups, and covered 1415.88 cM, with an average distance of 0.47 cM between adjacent markers. To our knowledge, the integrated map described herein has the best coverage of the potato genome and the highest marker density for tetraploid potato. This work provides a foundation for further quantitative trait loci (QTL) location, map-based gene cloning of important traits and marker-assisted selection (MAS) of potato.

Keywords: tetraploid potato; SNP markers; SLAF-seq technology; high-density genetic linkage map

1. Introduction

Potato, *Solanum tuberosum* L., is the fourth most important food crop in the world behind maize, wheat, and rice, with a total production of more than 388 million tons in 2017 [1]. Nevertheless, cultivated potato is a highly heterozygous outcrossing autotetraploid (2n = 4x = 48), which causes complexities in genetic or genomic studies, and provides many challenges for breeding. Therefore, more breeding efforts have been focused on improving important traits, such as processing quality, nutritional value, as well as disease/pest resistance.

A high-density genetic linkage map can provide a large amount of information that facilitates map-based cloning, QTL identification, and comparative genomic researches, establishing a general tool for marker-assisted selection breeding (MAS). However, the construction of linkage maps in autopolyploids always has more difficulties than that in polyploids as well as allopolyploid species, due to their complicated segregation patterns and chromosomal pairing [2–5]. Over the past two decades, multiple linkage maps have been constructed for potato (both diploid and autotetraploid potato) for the purpose of better understanding the potato genome, facilitating map-based cloning, and developing markers for MAS [6–11]. Gebhardt et al. (1991) [6] reported the first potato map in the world, including 135 restriction fragment length polymorphism (RFLP) molecular markers and defining 12 distinct linkage groups, which was drawn from segregation data derived from the interspecific cross of diploid potato ($2n = 2x = 24$), *S. phureja* × (*S. tuberosum* × *S. chacoense*). Yamanaka et al. (2005) [10] constructed an integrated genetic linkage map of diploid potato, using 106 F_1 individuals from a cross of two wild and landrace germplasm 86.61.26 × 84.194.30 as the mapping population. This map included 13 newly developed P450-based analogue (PBA), 27 random amplified polymorphic DNA (RAPD), 4 inter-simple sequence repeat (ISSR), 22 simple sequence repeat (SSR), 9 restriction fragment length polymorphism-sequence-tagged sites (RFLP-STSs), and 7 RFLP markers, with a coverage of 673 cM and an average marker distance of 8.2 cM. Van Os et al. (2006) [11] constructed an ultradense map of potato with more than 10,000 amplified fragment length polymorphism (AFLP) markers from a heterozygous diploid potato population. It is also the densest meiotic recombination map ever constructed.

With the rapid development of next-generation sequencing technologies, single nucleotide polymorphism (SNP) markers have been developed to construct high-density genetic linkage maps for many important crop species, such as maize [12,13], rice [14,15], and wheat [16,17]. For potato, Xun et al. (2011) [18] used a homozygous double-monoploid potato clone to sequence and assemble 86% of the 844-megabase genome, which bridged the gap between genomics and applied breeding with an in-depth understanding of the structure and function of the potato genome, and provided an effective tool and data to develop potato SNP markers. To date, several high-density genetic linkage maps based on SNP markers have been reported with the accomplishment and subsequent development of the potato's whole genome sequence. Felcher et al. (2012) [19] first used SNP markers and two diploid potato populations to create two linkage maps, where over 4400 markers were mapped, including 787 markers common to both populations, and the map sizes were 965 and 792 cM, respectively. Hackett et al. (2013) [20] constructed a high-density SNP map of tetraploid potato based on obtained Infinium 8300 Potato SNP Array data, which included 1130 markers with a coverage of 1087.5 cM, using a mapping population of 190 progenies from a cross between the breeding clone 12601ab1 and the cultivar stirling. Endelman et al. (2016) [21] first used a diploid inbred line-based F_2 population to construct a genetic linkage map of diploid potato with 2264 SNP markers. To sum up, most potato linkage maps are generated from diploid populations of wild species and primitive cultivars. Linkage mapping in tetraploid potato species is still a challenge despite the recent advances in mapping methodology, genotyping, and molecular marker technology.

Due to the advances in next generation sequencing (NGS) technologies, new high-throughput genotyping methods hold promise for the detection of a large number of SNPs in a short time, which include genotyping-by-sequencing (GBS) [22], complexity reduction of polymorphic sequences (CroPSs) [23], restriction site-associated DNA sequencing (RAD-seq) [24,25], and specific length amplified fragment sequencing (SLAF-seq) [26]. Specific-locus amplified fragment sequencing (SLAF-seq) technology, reported by Sun et al. (2013) [26], is an efficient strategy for the de novo SNP discovery and genotyping of large populations based on an enhanced reduced representation library (RRL) sequencing method. The advantages of SLAF-seq technology are: (i) Deep sequencing to ensure genotyping accuracy; (ii) a lower sequencing cost; (iii) pre-designed RRL scheme to optimize marker efficiency; (iv) and double barcode multiplexed sequencing system for large population and large numbers of loci. To date, this strategy has been applied to various species for SNP high-density genetic mapping, such as cucumber [27], *Agropyron gaertn* [28], and orchardgrass [29], due to its advantages of optimized marker efficiency, accurate genotyping, affordable price, and applicability for large populations. In the present research, an F_1 mapping population of 106 individuals was created from the cross between two tetraploid potato varieties, YSP-4 × MIN-021. We used the SLAF-seq approach to construct a high-density integrated SNP genetic linkage map of tetraploid potato, which will expedite map-based cloning efforts, QTL location for important traits, as well as marker-assisted selection breeding for tetraploid potato.

2. Materials and Methods

2.1. Plant Materials

The F_1 mapping population consisted of 106 individuals from a cross between two tetraploid potato varieties, YSP-4 (female) and MIN-021 (male). YSP-4 is a wild tetraploid potato material, which has a short growth period, moderate tuber numbers per plant, high commodity potato rate, and high starch content (ca. 18%). This material is also highly resistant to early blight and virus disease. MIN-021 is a tetraploid potato material, which has a short growth period, high yield, and high starch content (ca. 19%). All the materials were planted in the potato breeding base of Inner Mongolia Agricultural University. The field trial was arranged in randomized complete block design (RCBD) with three replications per plot. Each plot contained 20 plants, which were grown in 2 rows with a spacing of 30 cm within rows and 90 cm between rows, and the planting depth was about 12 cm. The experiment field had sandy soil with pH 7.8 to 8.2, good irrigation conditions with annual precipitation from 300 to 400 mm, and the geographic position is 111°42′ E, 45°57′ N, with an altitude of 1063 m.

2.2. DNA Extraction

At the potato squaring stage, the genomic DNA of all parents and 106 progenies was extracted from young fresh leaf tissue by the Plant Genomics DNA Kit (Tiangen, Beijing, China). Then, the quality of DNA was determined by electrophoresis on a 1% (w/v) agarose gel stained with ethidium bromide, and the concentration was quantified by an ND-1000 Spectrophotometer (Nano Drop, Wilmington, DE, USA) and adjusted to a concentration of 50 ng/μL.

2.3. SLAF Library Preparation and Sequencing

According to the genome size and GC (guanine-cysteine) content of the tested materials, the potato genome (http://solanaceae.plantbiology.msu.edu/pgsc_download.shtml) was selected as a reference genome to make predictions of the electronic enzyme, and finally determine the enzyme combination of Rsa I and Hae III to digest the genomic DNA of the 106 F_1 individuals and their two parents. The read length used for sequencing ranged from 264 to 394 bp. The SLAF labels (the length of fragments ranged from 314 to 364 bp) were selected for paired-end sequencing (125 bp per end) on an Illumina HiSeq 2500 sequencing platform, performed by the Beijing Biomarker Technologies Corporation (http://Biomarker.com.cn/). The SLAFs with a sequence depth of less than 10-fold were considered as low-depth SLAFs and filtered out. Several steps were defined to deal with SLAF-seq data: Samples were distinguished by barcodes and data grouping by sequence similarity; sequence error evaluation by control data; minor allele frequency (MAF) filtering and SLAF definition; correction of sequence errors; and definition and evaluation of genotypes. In addition, the quality score algorithm was developed to evaluate the quality of SNP discovery and genotyping, which can help researchers balance accuracy and cost during heterozygote detection using high-throughput sequencing technology. The Q30 (a quality score of 30; indicating 99.90% confidence) was used to evaluate the sequencing quality of reads, and examination of the base distribution was used to detect the GC content of the raw data for data quality control. The raw sequence reads were deposited in the NCBI-short read archive (SRA) database (accession: PRJNA597429).

2.4. SLAF Data Analysis and Development of SNP Markers

The approach of clustering among reads was used to develop and search for polymorphic SLAF labels from 106 F_1 individuals and their parents. All paired-end reads generated from SLAF-seq raw reads were compared according to their sequence similarity as detected by the BLAST-like alignment tool (BLAT) [30]. The F_1 individual sequence reads were aligned on the referenced potato genome using Burrows–Wheeler Aligner (BWA) software [31]. Identical reads from different samples were clustered, and the fragment with over 90% sequence identity was defined as an SLAF label. The SLAF labels with differences in high-depth fragments were also considered as SNP or indel markers. According to the differences among sequences or allele numbers, the SLAF labels were divided into three categories, including NoPoly (non-polymorphic), Poly (polymorphic), and Repeat (repetitive). After comparing the sequence differences on SLAFs from each sample, the polymorphic SLAF labels were screened for further analysis. Both Sequence Alignment/Map tools (SAMtools) [32] and Genome Analysis Toolkit (GATK) [33] were used to identify SNPs, and their intersection was identified as the candidate SNP dataset. Only biallelic SNPs were retained as the final SNPs. The SNP locus were confirmed from the polymorphic SLAF labels, with the screening criteria of MAF > 0.5.

2.5. Construction of High-Density Linkage Map

The HighMap software was used to construct a high-density genetic linkage map of tetraploid potato [34]. The single-linkage clustering algorithm was used to cluster the SNP markers, which were ordered into linkage groups. The high quality MLOD value among SLAF labels was calculated and used for linkage grouping. The genotyping errors were corrected using the module of error genotyping correction of HighMap sofware.

3. Results

3.1. SLAF Library Construction and SLAF Labels Development

The in silico restriction enzyme combination of *Rsa*I and *Hae*III was used for genome DNA digestion and the prediction of the potato reference genome. A total of 334,787 SLAF labels were predictably obtained, which were evenly distributed on the genome. The rice genome (*Oryza sativa*) was used as a control for the restriction enzyme digestion control trial, in order to indirectly monitor the progress of the potato SLAF library construction. Compared with the control, the ratio of paired-end mapping reads was 89.20%, and the digestion efficiency of the RsaI and HaeIII restriction enzymes was 90.91%, which indicated that the potato SLAF library was constructed normally and suitable for high-throughput sequencing.

After SLAF library construction and high-throughput sequencing, a total of 2269.98 million pair-end reads (556.71 Gb data) with a length of 100 bp were obtained. The Q30 ratio was 95.05%, and the average GC (guanine-cytosine) content was 35.51%. Of all the high-quality data, 48,849,737 reads were from the male parent MIN-21, 41,510,213 reads were from the female parent YSP-4, and the average 90,562,465 reads were from 106 offspring of the F_1 mapping population (Table 1). According to bioinformatics analysis, a total of 838,604 SLAF labels were developed, with an average depth of 26.14-fold and 15.36-fold for each SLAF of the parents and offspring, respectively. Of all the 838,604 high-quality SLAFs, 282,838 were polymorphic, of which 113,473 polymorphic SLAFs could be used for map construction.

Table 1. Basic statistic of the SLAF-seq data in tetraploid potato.

Sample	Total Reads	Total Base	Q30 Percentage (%)	GC Percentage (%)
MIN-21	48,849,737	11,944,501,020	95.18	40.28
YSP-4	41,510,213	10,126,475,890	95.66	40.18
offspring	20,562,465	5,043,758,955	95.04	39.49
Control (Rice)	9,873,113	2,422,915,844	94.69	39.49
Total	2,269,981,207	556,709,426,104	95.05	39.51

3.2. SNP Marker Detection

A total of 7638 SLAF labels were screened from 113,473 polymorphic SLAFs, which were successfully classified into four segregation patterns: Hk × hk, lm × ll, nn × np, and ef × eg (Table 2). The patterns, except aa × bb, were used for later genetic map construction which was suitable for the F_1 population, because the potato F_1 population was not obtained by a cross between two fully homozygous parents with genotype aa or bb. After filtering out the SNP markers with sequence depths no more than 4-fold, a total of 7329 SNP markers were detected from 7638 SLAFs for map construction.

Table 2. The number of SLAFs in different types of segregation patterns for map construction.

Type of Segregation Patterns	Number of SLAFs	Percentage (%)
Hk × hk	51	0.67
Lm × ll	3898	54.03
Nn × np	3638	47.63
ef × eg	51	0.67
Total	7638	100

3.3. Construction of the Genetic Linkage Map

After four quality control steps, the 7329 screened SNPs were used to calculate the modified logarithm of odds (MLOD) values between two markers [35]. The markers with an MLOD value of less than three were filtered, and the remaining markers were grouped into 12 linkage groups (LGs). The HighMap software was used to analyze the linear arrangements of all the grouped SNPs and the genetic distance between adjacent SNP markers within each LG. An integrated map as well as two separate linkage maps for the female and male parents were constructed, including 12 linkage groups.

In YSP-4, the maternal linkage map contained 1638 SNP markers, which covered a total length of 1383.86 cM, with an average marker distance of 0.83 cM. The number of markers in the linkage groups ranged from 43 to 341 markers, with an average of 137 markers. The length of LGs ranged from 32.82 to 282.89 cM, with an average size of 0.84 cM (Table A1). In MIN-021, the paternal linkage map consisted of 1402 SNP markers, and covered a total length of 1203.94 cM, with an average marker distance of 0.87 cM. The number of mapped markers in the LGs ranged from 542 to 243, with an average of 117 markers. The length of LGs ranged from 26.05 to 170.2 cM, with an average size of 100.33 cM (Table A2).

The integrated genetic map included 3001 SNP markers, which covered a total length of 1415.88 cM, and the average distance between adjacent markers was 0.47 cM. The number of markers in the linkage groups ranged from 43 to 341 markers, with an average of 137 markers. The length of LGs ranged from 45.02 to 282.89 cM, with an average size of 117.99 cM (Table 3, Figures 1 and A1). LG chr10 was not only the shortest but also the densest group, with 440 loci spanning 33.47 cM, which had an average marker density of 0.08 cM. LG chr2 was the longest group, with 225 loci spanning 205.09 cM. The largest gap on this map was 25.19 cM, located in LG chr7 (Table 3; Figure 1).

chr 1   chr 2   chr 3   chr 4   chr 5   chr 6   chr 7   chr 8   chr 9   chr 10   chr 11   chr 12

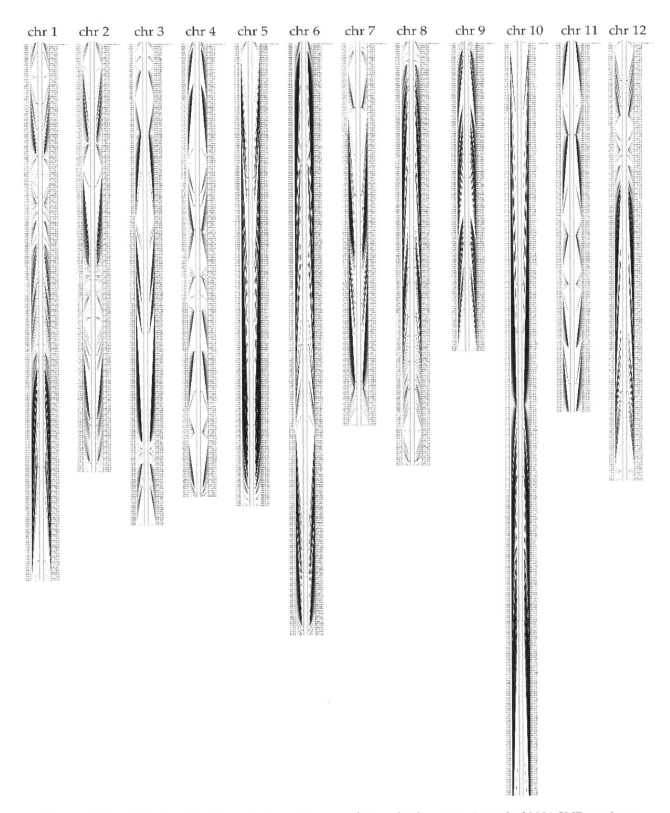

Figure 1. The high-density integrated genetic map of tetraploid potato. A total of 3001 SNP markers were distributed in 12 linkage groups, covering 1415.88 cM, with an average interval of 0.47 cM between markers.

Table 3. The integrated genetic map for tetraploid potato.

Linkage Group (ID)	Total Marker	Total DISTANCE (cM)	Average Distance (cM)	Max Gap (cM)	Gap ≥5 cM (%)
chr1	282	107.42	0.38	12.84	99.29
chr2	225	205.09	0.92	24.43	95.98
chr3	253	63.48	0.25	14.38	98.02
chr4	238	169.78	0.72	23.76	94.51
chr5	243	76.4	0.32	10.46	98.35
chr6	311	125.8	0.41	22.86	97.1
chr7	201	203.86	1.02	25.19	95
chr8	222	109.69	0.5	14.07	97.74
chr9	162	115.41	0.72	15.34	96.89
chr10	440	33.47	0.08	5.23	99.77
chr11	194	156.97	0.81	21.35	94.82
chr12	230	48.51	0.21	7.54	99.13
Total	3001	1415.88	0.47	25.19	97.22

The average depth of the SNP markers on the integrated map was 85.63-fold in the paternal parent MIN-021 and 65.10-fold in the maternal parent YSP-4, as well as 40.34-fold in the offspring of the F_1 population. Segregation distortion is occurs when the segregation ratio deviates from the expected Mendelian ratio, which is considered as a powerful driving force for organic evolution [36]. The Chi-square (χ^2) test ($\alpha = 0.05$) was used to analyze the goodness-of-fit to the expected segregation ratios for all the SNP markers. A total of 80 out of 3001 markers (2.7%) did not fit the expected segregation ratios at a level of $\alpha \leq 0.05$. The distorted SNP markers were mainly located on LG chr 3, chr 5, chr 7, chr 8, chr 11, and chr 12 (Table 4).

Table 4. The distorted SNP markers on integrated genetic map of tetraploid potato.

Linkage Group (ID)	The Number of Distorted SNP Markers
chr3	10
chr5	24
chr7	9
chr8	3
chr11	18
chr12	16
Total	80

3.4. Evaluation of the Genetic Map

The quality of this genetic map was evaluated by haplotype maps and heat maps, which directly revealed the recombination relationship among SNP markers in the 12 LGs. Haplotype maps were created to reflect the crossover events. The recombination events of the 12 LGs are shown on the haplotype maps (Figure A2). The haplotype maps from 12 LGs showed that all LGs had an extremely low double crossover rate, which indicated the genetic map had a high quality.

Heat maps were also constructed to evaluate the quality of the genetic map by using pair-wise recombination values for the 3001 SNP markers (Figure A3). It showed that most of the heat maps for 12 LGs performed well in visualization, which indicated that the markers were well-ordered, and the genetic distances of adjacent markers were accurate in each LG.

4. Discussion

4.1. The Development of SNP Markers Using SLAF-seq Technology

A genetic linkage map is the basis for QTL identification of important traits, map-based gene cloning, and molecular marker-assisted breeding of crops. Different types and numbers of polymorphic markers were used to construct genetic maps. For potato, most genetic linkage maps were mainly based on several conventional low-throughput molecular markers, including RFLP [6–10], AFLP [9,11], as well as RAPD, ISSR, and SSR markers [10]. However, it is time-consuming and costly to construct a high-density genetic map for potato using conventional molecular marker technologies. SNP markers are the most frequent polymorphisms and are suitable for high-throughput genotyping. In addition, many SNP markers are located within transcribed regions, which can generate more links between the genetic and physical maps. To date, high-density polymorphic SNP markers have been used in potato for large-scale genotyping and high-density genetic map construction [19–21].

SNP markers can be rapidly developed on a large scale by different high-throughput sequencing technologies and genotyping methods, such as genotyping by sequencing (GBS) [22], restriction site-associated sequencing (RAD-seq) [24,25], and SLAF-seq. The SLAF-seq technology, a combination of locus-specific amplification and high-throughput sequencing, provides a high-resolution strategy with a shorter period of time and lower cost for large-scale genotyping and can be generally applicable to various species and populations [26].

In the present research, we first used SLAF-seq technology in potato to develop SNP markers and construct the high-density genetic map. A total of 2269.98 million pair-end reads were obtained based on high-throughput sequencing. According to bioinformatics analysis, a total of 838,604 SLAF labels were generated, of which 282,838 were polymorphic. Finally, a total of 7329 polymorphic SNP markers were developed for high-density genetic map construction. The present study extends the utility of SLAF-seq technology to potato. The results showed that SLAF-seq was an effective tool to rapid develop large-scale SNP markers, which met the requirements for high-density genetic map construction of tetraploid potato.

4.2. Mapping Population and Strategies

In general, the F_2, backcross (BC), doubled haploid (DH), or recombinant inbred lines (RIL)population are used as an appropriate mapping population to construct genetic linkage maps [5,37,38]. However, for most autopolyploid species, it is very difficult to obtain a typical family-based population in potato because of its high heterozygosity. To date, most of the reported potato linkage maps have been established by applying a double pseudo-testcross strategy on an F1 population. The double pseudo-testcross strategy was first put forward by Grattapaglia and Sederoff (1994) [37] to construct the genetic linkage map for genetically heterozygous species of forest trees. An F_1 population was used as a mapping population by crossing between two irrelevant and highly heterozygous parents. The gene segregation patterns were assumed as a backcross. Afterwards, this strategy has been widely used to construct linkage maps for those heterozygous species, such as danshen [39], pineapple [40], rhodesgrass [41], and sweet potato [5].

In the present research, an F_1 segregation population from a cross YSP-4 × MIN-021 was created, of which 106 individuals were randomly selected and used for SNP genotyping and map construction based on the double pseudo-testcross strategy. In the pseudo-testcross, a total of 7638 polymorphic SLAF markers were classified into four segregation patterns, which were hk × hk, lm × ll, nn × np, and ef × eg. The 7329 SNP markers screened and confirmed from 7638 polymorphic SLAFs were then used to construct a genetic linkage map. In our study, among the 838,604 high-quality SLAFs, 282,838 were polymorphic, with a polymorphic rate of 33.7%. It indicates that there is considerable genetic difference between YSP-4 and MIN-021. Therefore, it is suitable to use them as mapping parents, and the F_1 population derived from the cross between them conforms to the requirements of the mapping population for high-density map construction.

4.3. The High-Density Genetic Map of Tetraploid Potato

Segregation distortion is a common phenomenon that has been observed in many studies [42–44]. It may generate from cytological attributes, genetic drift, gametophyte selection, or some biological reasons [45,46]. Segregation distortion could alter the estimation of recombination and cause a spurious linkage [47]. Therefore, distorted markers may affect the accuracy of genetic maps. In our study, only 2.7% SNP markers located on the integrated map were distorted markers, which indicated the high map accuracy.

To our knowledge, only one high-density SNP genetic linkage map for tetraploid potato was reported because of the high heterozygosity of autotetraploid potato [20]. In the present research, we first used the SLAF-seq method for genotyping and developing SNP markers, and constructed high-density genetic maps of tetraploid potato. The integrated map included 3001 SNP markers, and had a genetic length of 1415.88 cM, with an average distance between markers of 0.47 cM. Compared with the map obtained by Hackett et al. (2013) [20], the integrated map had more SNP markers (3001 vs. 1130), higher marker density (0.47 cM vs. 1.60 cM), and larger total length (1415.88 cM vs. 1087.5 cM). Thus, our map has better coverage of the potato genome and nearer marker density.

5. Conclusions

In the present study, the SLAF-seq technology was first successfully used for the development of large-scale SNP markers and the construction of high-density linkage maps in tetraploid potato. The integrated high genetic linkage map generated here has the best coverage of the potato genome and the nearest marker density reported for tetraploid potato until now. This work represents an important step forward in genomics and marker-assisted breeding of tetraploid potato. It also provides a foundation for QTL location and map-based gene cloning of important traits for potato, such as tuber yield, starch content, and protein content. In addition, the application of SLAF-seq strategy and the mapping population in our study will provide valuable references for other tetraploid plants.

Author Contributions: Z.Y. and X.Y. conceived and designed the research. Z.Y., X.Y. and M.Z. performed the experiments. X.Y. and M.Z. completed the writing of the article. D.Y., J.L. (Jingwei Li), G.W., and J.L. (Jiaqi Li) assisted in the completion of the experiments. All authors have read and agreed to the published version of the manuscript.

Appendix A

Table A1. The maternal genetic map for tetraploid potato.

Linkage Group (ID)	Total Marker	Total Distance (cM)	Average Distance (cM)	Max Gap (Cm)	Gap ≥ 5 cM (%)
chr1	43	54.06	1.29	26.6	95.24
chr2	70	199.96	2.9	66.56	86.96
chr3	138	58.29	0.43	48.73	99.27
chr4	107	149.13	1.41	30.15	93.4
chr5	167	45.02	0.27	21.7	98.8
chr6	235	172.13	0.74	31.9	95.73
chr7	153	282.89	1.86	88.65	94.74
chr8	153	116.85	0.77	15.34	96.71
chr9	80	91.37	1.16	55.92	98.73
chr10	341	32.82	0.1	7.32	99.85
chr11	57	123.36	2.2	31.91	91.07
chr12	94	57.8	0.62	16.63.	96.77
Total	1638	1383.68	0.84	88.65	95.62

Table A2. The paternal genetic map for tetraploid potato.

Linkage Group (ID)	Total Marker	Total Distance (cM)	Average Distance (cM)	Max Gap (cM)	Gap ≥ 5 cM (%)
chr1	243	117.29	0.48	12.87	99.17
chr2	156	167.27	1.08	28.45	96.77
chr3	125	51.51	0.42	15.34	96.77
chr4	132	151.04	1.15	51.29	95.42
chr5	76	83.47	1.11	10.46	93.33
chr6	83	77.56	0.95	37.57	96.34
chr7	52	101.5	1.99	71.52	96.08
chr8	71	88.68	1.27	70.27	98.57
chr9	82	108.18	1.34	16.63	93.83
chr10	102	52.42	0.52	10.46	98.02
chr11	138	178.97	1.31	41.74	96.35
chr12	142	26.05	0.18	3.92	100
Total	1402	1203.94	0.87	71.52	96.72

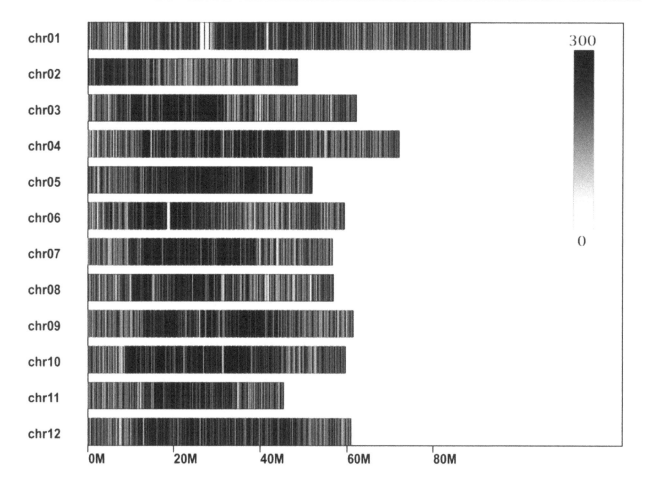

Figure A1. The SNP distribution on the potato genome. The *x*-axis represents the chromosome length and the *y*-axis indicates the chromosome code. Each band represents a chromosome, and the genome is divided according to the size of 1 M. The more SNPs in each band, the darker the color; the smaller the number of SNPs, the lighter the color. The darker areas in the figure are the areas where SNPs are concentrated.

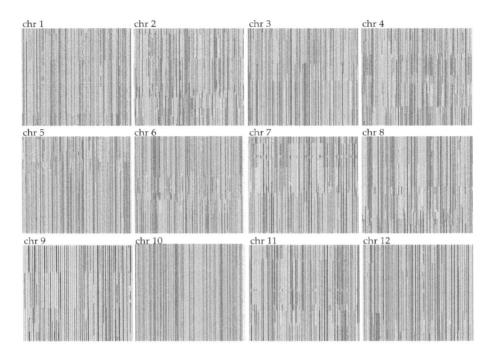

Figure A2. Haplotype maps for 12 linkage groups of the integrated genetic map for tetraploid potato. The haplotype maps consist of 12 maps from LG chr 1 to LG chr 12. Each two columns represent the genotype of an individual. Blank columns are used between two individuals. The first and second columns represent the paternal and maternal chromosome, respectively. Rows correspond to genetic markers. Green and blue boxes indicate one chromatid from parents, and gray boxes indicate missing data.

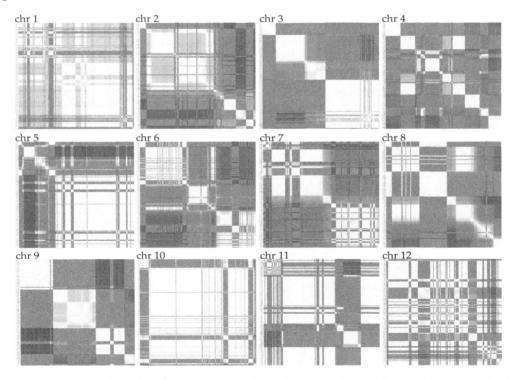

Figure A3. Heat maps for 12 linkage groups of the integrated genetic map for tetraploid potato. The heat maps consist of 12 maps from LG chr 1 to LG chr 12. Markers of each row and column are ranked according to the map order. Each small square represents the rate of recombination between two markers. Yellow color represents highly tight linkage; red color represents relatively weak linkage, the darker the red color, the less tight linkage; and blue color represents no linkage.

References

1. FAOSTAT. 2017. Available online: http://faostat.fao.org/site/339/default.aspx (accessed on 2 September 2019).
2. Hermsen, J.G.T. Nature, evolution and breeding of polyploids. *Iowa State J. Res.* **1984**, *58*, 411–420.
3. Sybenga, J. Chromosome pairing affinity and quadrivalent formation in polyploids: Do segmental allopolyploids exist? *Genome* **1996**, *39*, 1176–1184. [CrossRef]
4. Cervantes-Flores, J.C.; Yencho, G.C.; Kriegner, A.; Pecota, K.V.; Faulk, M.A.; Mwanga, R.O.M.; Sosinski, B.R. Development of a genetic linkage map and identification of homologous linkage groups in sweetpotato using multiple-dose AFLP markers. *Mol. Breed.* **2008**, *21*, 511–532. [CrossRef]
5. Zhao, N.; Yu, X.X.; Jie, Q.; Li, H.; Li, H.; Hu, J.; Zhai, H.; He, S.Z.; Liu, Q.C. A genetic linkage map based on AFLP and SSR markers and mapping of QTL for dry-matter content in sweetpotato. *Mol. Breed.* **2013**, *32*, 807–820. [CrossRef]
6. Gebhardt, C.; Ritter, E.; Barone, A.; Debener, T.; Walkemeier, B.; Schachtschabel, U.; Kaufmann, H.; Thompson, R.D.; Bonierbale, M.W.; Ganal, M.W. RFLP maps of potato and their alignment with the homoeologous tomato genome. *Theor. Appl. Genet.* **1991**, *83*, 49–57. [CrossRef] [PubMed]
7. Tanksley, S.D.; Ganal, M.W.; Prince, J.P.; De Vicente, M.C.; Bonierbale, M.W.; Broun, P.; Fulton, T.M.; Giovannoni, J.J.; Grandillo, S.; Martin, G.B. High density molecular linkage maps of the tomato and potato genomes. *Genetics* **1992**, *132*, 1141–1160. [PubMed]
8. Jacobs, J.M.E.; Van Eck, H.J.; Arens, P.; Verkerk-Bakker, B.; Te Lintel Hekkert, B.; Bastiaanssen, H.J.M.; El-Kharbotly, A.; Pereira, A.; Jacobsen, E.; Stiekema, W.J. A genetic map of potato (*Solanum tuberosum*) integrating molecular markers, including transposons, and classical markers. *Theor. Appl. Genet.* **1995**, *91*, 289–300. [CrossRef]
9. Caromel, B.; Mugniéry, D.; Lefebvre, V.; Andrzejewski, S.; Ellissèche, D.; Kerlan, M.C.; Rousselle, P.; Roussellebourgeois, F. Mapping QTLs for resistance against *Globodera pallida* (Stone) Pa2/3 in a diploid potato progeny originating from *Solanum spegazzinii*. *Theor. Appl. Genet.* **2003**, *106*, 1517–1523. [CrossRef]
10. Yamanaka, S.; Ikeda, S.; Imai, A.; Luan, Y.; Watanabe, J.A.; Watanabe, K.N. Construction of integrated genetic map between various existing DNA markers and newly developed P450-related PBA markers in diploid potato (*Solanum tuberosum*). *Breed. Sci.* **2005**, *55*, 223–230. [CrossRef]
11. Van Os, H.; Andrzejewski, S.; Bakker, E.; Barrena, I.; Glenn, J.B.; Caromel, B.; Ghareeb, B.; Isidore, E.; De Jong, W.; Van Koert, P. Construction of a 10,000-marker ultradense genetic recombination map of potato: Providing a framework for accelerated gene isolation and a genomewide physical map. *Genetics* **2006**, *173*, 1075–1087. [CrossRef]
12. Sa, K.J.; Park, J.Y.; Park, K.C.; Ju, K.L. Analysis of genetic mapping in a waxy/dent maize RIL population using SSR and SNP markers. *Genes. Genom.* **2012**, *34*, 157–164. [CrossRef]
13. Su, C.F.; Wang, W.; Gong, S.L.; Zuo, J.H.; Li, S.J.; Xu, S.Z. High density linkage map construction and mapping of yield trait QTLs in maize (*Zea mays*) using the genotyping-by-sequencing (GBS) technology. *Front. Plant Sci.* **2017**, *8*, 706–719. [CrossRef] [PubMed]
14. Zhai, H.J.; Feng, Z.Y.; Liu, X.Y.; Cheng, X.J.; Peng, H.R.; Yao, Y.Y.; Sun, Q.X.; Ni, Z.F. A genetic linkage map with 178 SSR and 1 901 SNP markers constructed using a RIL population in wheat (*Triticum aestivum* L.). *J. Integr. Agric.* **2015**, *14*, 1697–1705. [CrossRef]
15. Leon, T.B.D.; Linscombe, S.; Subudhi, P.K. Molecular dissection of seedling salinity tolerance in rice (*Oryza sativa* L.) using a high-density GBS-based SNP linkage map. *Rice* **2016**, *9*, 52–73. [CrossRef] [PubMed]
16. Yang, Q.; Yang, Z.J.; Tang, H.F.; Yu, Y.; Chen, Z.Y.; Wei, S.H.; Sun, Q.X.; Peng, Z.S. High-density genetic map construction and mapping of the homologous transformation sterility gene (*hts*) in wheat using *gbs* markers. *BMC Plant Biol.* **2018**, *18*, 301–309. [CrossRef] [PubMed]
17. Colasuonno, P.; Gadaleta, A.; Giancaspro, A.; Nigro, D.; Giove, S.; Incerti, O.; Mangini, G.; Signorile, A.; Simeone, R.; Blanco, A. Development of a high-density SNP-based linkage map and detection of yellow pigment content QTLs in durum wheat. *Mol. Breed.* **2014**, *34*, 1563–1578. [CrossRef]
18. Xu, X.; Pan, S.K.; Cheng, S.F.; Zhang, B.; Mu, D.S.; Ni, P.X.; Zhang, G.Y.; Yang, S.; Li, R.Q.; Wang, J. Genome sequence and analysis of the tuber crop potato. *Nature* **2011**, *475*, 189–195.
19. Felcher, K.J.; Coombs, J.J.; Massa, A.N.; Hansey, C.N.; Hamilton, J.P.; Veilleux, R.E.; Buell, C.R.; Douches, D.S. Integration of two diploid potato linkage maps with the potato genome sequence. *PLoS ONE* **2012**, *7*, e36347. [CrossRef]

20. Hackett, C.A.; Mclean, K.; Bryan, G.J. Linkage analysis and QTL mapping using SNP dosage data in a tetraploid potato mapping population. *PLoS ONE* **2013**, *8*, e63939. [CrossRef]

21. Endelman, J.B.; Jansky, S.H. Genetic mapping with an inbred line-derived F_2 population in potato. *Theor. Appl. Genet.* **2016**, *129*, 935–943. [CrossRef]

22. Elshire, R.J.; Glaubitz, J.C.; Sun, Q.; Poland, J.A.; Kawamoto, K.; Buckler, E.S.; Mitchell, S.E. A robus, simple genotyping by-sequencing (GBS) approach for high diversity species. *PLoS ONE* **2011**, *6*, e19379. [CrossRef] [PubMed]

23. Van Orsouw, N.J.; Hogers, R.C.; Janssen, A.; Yalcin, F.; Snoeijers, S.; Verstege, E.; Schneiders, H.; van der Poel, H.; Van Oeveren, J.; Verstegen, H.; et al. Complexity reduction of polymorphic sequences (CRoPS): A novel approach for large-scale polymorphism discovery in complex genomes. *PLoS ONE* **2007**, *2*, e1172. [CrossRef] [PubMed]

24. Miller, M.R.; Dunham, J.P.; Amores, A.; Cresko, W.A.; Johnson, E.A. Rapid and cost-effective polymorphism identification and genotyping using restriction site associated DNA (RAD) markers. *Genome Res.* **2007**, *17*, 240–248. [CrossRef] [PubMed]

25. Baird, N.A.; Etter, P.D.; Atwood, T.S.; Currey, M.C.; Shiver, A.L.; Lewis, Z.A.; Selker, E.U.; Cresko, W.A.; Johnson, E.A. Rapid SNP discovery and genetic mapping using sequenced RAD markers. *PLoS ONE* **2008**, *3*, e3376. [CrossRef]

26. Sun, X.W.; Liu, D.Y.; Zhang, X.F.; Li, W.B.; Liu, H.; Hong, W.G.; Jiang, C.B.; Guan, N.; Ma, C.X.; Zeng, H.P. SLAF-seq: An efficient method of large-scale de novo SNP discovery and genotyping using high-throughput sequencing. *PLoS ONE* **2013**, *8*, e58700. [CrossRef] [PubMed]

27. Xu, X.W.; Lu, L.; Zhu, B.Y.; Xu, Q.; Qi, X.H.; Chen, X.H. QTL mapping of cucumber fruit flesh thickness by SLAF-seq. *Sci. Rep.* **2015**, *5*, 15829. [CrossRef]

28. Zhang, Y.; Zhang, J.P.; Huang, L.; Gao, A.N.; Yang, X.M.; Liu, W.H.; Li, X.Q.; Li, L.H. A high-density genetic map for P genome of *Agropyron Gaertn.* based on specific-locus amplified fragment sequencing (SLAF-seq). *Planta* **2015**, *242*, 1335–1347. [CrossRef]

29. Zhao, X.X.; Huang, L.K.; Zhang, X.Q.; Wang, J.P.; Yan, D.F.; Li, J.; Tang, L.; Li, X.L.; Shi, T.W. Construction of high-density genetic linkage map and identification of flowering-time QTLs in orchardgrass using SSRs and SLAF-seq. *Sci. Rep.* **2016**, *6*, 29345. [CrossRef]

30. Ken, W.J. BLAT—The BLAST-like alignment tool. *Genome Res.* **2002**, *12*, 656–664.

31. Li, H.; Durbin, R. Fast and accurate short read alignment with Burrows-Wheeler transform. *Bioinformatics* **2009**, *25*, 1754–1760. [CrossRef]

32. Depristo, M.A.; Banks, E.; Poplin, R.E.; Garimella, K.V.; Maguire, J.R.; Hartl, C. A framework for variation discovery and genotyping using next-generation DNA sequencing data. *Nat. Genet.* **2011**, *43*, 491–498. [CrossRef] [PubMed]

33. Li, H.; Handsaker, B.; Wysoker, A.; Fennell, T.; Ruan, J.; Homer, N.; Marth, G.; Abecasis, G.; Durbin, R. The sequence Alignment/Map format and SAMtools. *Bioinformatics* **2009**, *25*, 2078–2079. [CrossRef] [PubMed]

34. Liu, D.Y.; Ma, C.X.; Hong, W.G.; Huang, L.; Liu, M.; Liu, H.; Zeng, H.P.; Deng, D.J.; Xin, H.G.; Song, J.; et al. Construction and analysis of high-density linkage map using high-throughput sequencing data. *PLoS ONE* **2014**, *9*, e98855. [CrossRef] [PubMed]

35. Vision, T.J.; Brown, D.G.; Shmoys, D.B.; Durrett, R.T.; Tanksley, S.D. Selective mapping: A strategy for optimizing the construction of high-density linkage maps. *Genetics* **2000**, *155*, 407–420.

36. Foisset, N.; Delourme, R.; Barret, P.; Hubert, N.; Landry, B.S.; Renard, M. Molecular-mapping analysis in *Brassica napus* using isozyme, RAPD and RFLF markers on a doubled-haploid progeny. *Theor. Appl. Genet.* **1996**, *93*, 1017–1025. [CrossRef]

37. Grattapaglia, D.; Sederoff, R. Genetic linkage maps of eucalyptus *grandis* and *eucalyptus urophylla* using a pseudo-testcross: Mapping strategy and RAPD markers. *Genetics* **1994**, *137*, 1121–1137.

38. Venkat, S.K.; Bommisetty, P.; Patil, M.S.; Reddy, L.; Chennareddy, A. The genetic linkage maps of anthurium species based on RAPD, ISSR and SRAP markers. *Sci. Hortic.* **2014**, *178*, 132–137. [CrossRef]

39. Liu, T.; Guo, L.L.; Pan, Y.L.; Zhao, Q.; Wang, J.H.; Song, Z.Q. Construction of the first high-density genetic linkage map of *salvia miltiorrhiza* using specific length amplified fragment (SLAF) sequencing. *Sci. Rep.* **2016**, *6*, 24070. [CrossRef]

40. Carlier, J.D.; Reis, A.; Duval, M.F.; Coppens d'Eeckenbrugge, G.; Leitao, J.M. Genetic maps of RAPD, AFLP and ISSR markers in *Ananas bracteatus* and *A. comosus* using the pseudo-testcross strategy. *Plant Breed.* **2010**, *123*, 186–192. [CrossRef]

41. Ubi, B.E.; Fujimori, M.; Mano, Y.; Komatsu, T. A genetic linkage map of rhodesgrass based on an F_1 pseudo-testcross population. *Plant Breed.* **2010**, *123*, 247–253. [CrossRef]

42. Cloutier, S.; Ragupathy, R.; Niu, Z.; Duguid, S. SSR-based linkage map of flax (*Linum usitatissimum* L.) and mapping of QTLs underlying fatty acid composition traits. *Mol. Breed.* **2011**, *28*, 437–451. [CrossRef]

43. Escudero, M.; Hahn, M.; Hipp, A.L. RAD-seq linkage mapping and patterns of segregation distortion in sedges: Meiosis as a driver of karyotypic evolution in organisms with holocentric chromosomes. *J. Evol. Biol.* **2018**, *31*, 833–843. [CrossRef] [PubMed]

44. Moreira, F.M.; Madini, A.; Marino, R.; Zulini, L.; Stefanini, M.; Velasco, R.; Kozma, P.; Grando, M.S. Genetic linkage maps of two interspecific grape crosses (*vitisspp.*) used to localize quantitative trait loci for downy mildew resistance. *Tree Genet. Genomes* **2011**, *7*, 153–167. [CrossRef]

45. Shappley, Z.W.; Jenkins, J.N.; Meredith, W.R.; Mccarty, J.C. An RFLP linkage map of upland cotton, *Gossypium hirsutum* L. *Theor. Appl. Genet.* **1998**, *97*, 756–761. [CrossRef]

46. Lacape, J.M.; Nguyen, T.B.; Thibivilliers, S.; Bojinov, B.; Courtois, B.; Cantrell, R.G.; Burr, B.; Hau, B. A combined RFLP-SSR-AFLP map of tetraploid cotton based on a *Gossypium hirsutum* x *Gossypium barbadense* backcross population. *Genome* **2003**, *46*, 612–626. [CrossRef] [PubMed]

47. Cloutier, S.; Cappadocia, M.; Landry, B.S. Analysis of RFLP mapping inaccuracy in *Brassica napus* L. *Theor. Appl. Genet.* **1997**, *95*, 83–91. [CrossRef]

8

Multi-Trait Regressor Stacking Increased Genomic Prediction Accuracy of Sorghum Grain Composition

Sirjan Sapkota [1,2,*], J. Lucas Boatwright [1,2], Kathleen Jordan [1], Richard Boyles [2,3] and Stephen Kresovich [1,2]

[1]   Advanced Plant Technology Program, Clemson University, Clemson, SC 29634, USA; jboatw2@clemson.edu (J.L.B.); kjorda7@clemson.edu (K.J.); skresov@clemson.edu (S.K.)

[2]   Department of Plant and Environmental Sciences, Clemson University, Clemson, SC 29634, USA; rboyles@clemson.edu

[3]   Pee Dee Research and Education Center, Clemson University, Florence, SC 29506, USA

*   Correspondence: ssapkot@clemson.edu

Abstract: Genomic prediction has enabled plant breeders to estimate breeding values of unobserved genotypes and environments. The use of genomic prediction will be extremely valuable for compositional traits for which phenotyping is labor-intensive and destructive for most accurate results. We studied the potential of Bayesian multi-output regressor stacking (BMORS) model in improving prediction performance over single trait single environment (STSE) models using a grain sorghum diversity panel (GSDP) and a biparental recombinant inbred lines (RILs) population. A total of five highly correlated grain composition traits—amylose, fat, gross energy, protein and starch, with genomic heritability ranging from 0.24 to 0.59 in the GSDP and 0.69 to 0.83 in the RILs were studied. Average prediction accuracies from the STSE model were within a range of 0.4 to 0.6 for all traits across both populations except amylose (0.25) in the GSDP. Prediction accuracy for BMORS increased by 41% and 32% on average over STSE in the GSDP and RILs, respectively. Prediction of whole environments by training with remaining environments in BMORS resulted in moderate to high prediction accuracy. Our results show regression stacking methods such as BMORS have potential to accurately predict unobserved individuals and environments, and implementation of such models can accelerate genetic gain.

Keywords: genomics; genomic selection; genomic prediction; marker-assisted selection; whole genome regression; grain quality; near infra-red spectroscopy; cereal crop; sorghum; multi-trait

1. Introduction

Cereal grains provide more than half of the total human caloric consumption globally and amount to over 80% in some of the poorest nations of the world [1]. Sorghum [*Sorghum bicolor* (L.) Moench], a drought-tolerant cereal crop, is a dietary staple for over half a billion people of semi-arid tropics which is inhabited by some of the most food insecure and malnourished populations [2]. In industrialized countries, such as the United States and Australia, grain sorghum is primarily grown for animal feed. But in recent years the uses of sorghum grain have expanded to baking, malting, brewing, and biofortification [3–5]. Therefore, genetic improvement of sorghum grain composition is crucial to mitigate the global malnutrition crisis, to increase efficiency of feed grains used in animal production, and to serve evolving niche markets for gluten-free grains.

In the last two decades, the use of genome-wide markers in prediction of genetic merit of individuals has revolutionized plant and animal breeding. Genomic prediction (GP) uses statistical models to estimate marker effects in a training population with phenotypic and genotypic data which is then used to predict breeding values of individuals solely from genetic markers [6,7]. Training population size, genetic relatedness between individuals in training and testing population, marker density, span of linkage disequilibrium and genetic architecture of traits are some of the factors that can affect the predictive ability of the models [8–10]. Genomic prediction models are routinely studied and applied by breeding programs around the world in several crops. Novel statistical methods that are capable of incorporating pedigree, genomic, and environmental covariates into statistical-genetic prediction models have emerged as a result of extensive computational research [11].

One of the main advantages of GP is that breeders can use phenotypic values from some lines in some environments to make predictions of new lines and environments. Genomic best linear unbiased prediction (GBLUP) proposed by VanRaden [12] is probably the most widely used genomic prediction model in both plant and animal breeding. Since then GBLUP model has been extended to include G × E interactions resulting in improved prediction accuracy of unobserved lines in environments [13,14]. Burgueño et al. [13] found an increase in prediction ability of unobserved wheat genotypes by about 20% in multi-environment GBLUP model compared to single environment model. Also an extension of the GBLUP model, Jarquín et al. [14] introduced a reaction norm model which introduces the main and interaction effects of markers and environmental covariates by using high-dimensional random variance-covariance structures of markers and environmental covariates. While most of the genomic prediction studies have been on individual traits, breeding programs use selection indices based on several traits to make breeding decisions. To address those challenges, expanded genomic prediction models that perform joint analysis of multiple traits have been studied using empirical and simulated data [15,16]. Subsequent improvement in prediction accuracy from multi-trait model over single-trait model depends on trait heritability and correlation between the traits involved [15,17].

Data generated in breeding programs span multiple environment and are recorded for multiple traits for each individual. While multi-environment models and multi-trait models are implemented separately, a single model to account for complexity of variance-covariance structure in a combined multi-trait multi-environment (MTME) model was lacking until Montesinos-López et al. [18] developed a Bayesian whole genome prediction model to incorporate and analyze multiple traits and multiple environments simultaneously. Montesinos-López et al. [18] also developed a computationally efficient Markov Chain Monte Carlo (MCMC) method that produces a full conditional distribution of the parameters leading to an exact Gibbs sampling for the posterior distribution. Another MTME model that employs a completely different method was proposed by Montesinos-López et al. [19]. This method, called the Bayesian multi-output regression stacking (BMORS), is a Bayesian version of multi-target regressor stacking (MTRS) originally proposed by Spyromitros-Xioufis et al. [20,21]. This method consists of training in two stages: first training multiple learning algorithms for the same dataset and then subsequently combining the predictions to obtain the final predictions.

Genomic prediction for grain quality traits has previously been reported in crops such as wheat [22–24], rye [25], maize [26], and soybean [27]. Hayes et al. [28] and Battenfield et al. [23] used near-infrared derived phenotypes in genomic prediction of protein content and end-use quality in wheat. Multi-trait genomic prediction models can simultaneously improve grain yield and protein content despite being negatively correlated [24,29]. In sorghum, grain macronutrients have shown to be inter-correlated among one another [30], which suggests the multi-trait models may increase predictive ability of individual grain quality traits. The ability to assess genetic merit of unobserved selection candidates across environments is promising for reducing evaluation cost and generation interval in the sorghum breeding pipeline where parental lines of commercial hybrids are currently selected on the basis of extensive progeny testing [31]. In order to extend capacities to performance index selection for multiple environments, we need to study and effectively implement MTME genomic prediction models in our breeding programs. In this study, we report the first implementation of genomic prediction for grain composition in sorghum, and the objective was to assess potential for improvement in prediction accuracy of multi-trait regressor stacking model over single trait model for five grain composition traits: amylose, fat, gross energy, protein and starch.

2. Results

2.1. Phenotypic Variation

A single calibration curve for near infra-red spectroscopy (NIRS) was used for the two populations studied. Table 1 outlines the summary statistics of NIRS predictions and phenotypic distribution and heritability of the grain composition traits. The cross validation accuracy (R^2) of the NIRS calibration curve was moderately high to high, except for fat which had a moderate R^2 value (0.41). We had a total of three environments (three years in one location) for the GSDP and four environments (two years in two locations) for the RILs. Traits were normally distributed except amylose in two 2014 environments in the RILs which had bimodal distribution (Figures S1 and S2). All traits showed significant variation in distribution across the environments, except starch in GSDP.

Table 1. Summary statistics of near infra-red spectroscopy (NIRS) calibration and phenotypic distribution. R^2 is the prediction accuracy and SECV is the standard error of cross validation for the NIRS calibration curve. Mean represents the phenotypic mean of the trait with its standard deviation (SD). h^2 is the estimate of genomic heritability.

Trait	NIRS		GSDP		RILs	
	R^2	SECV	Mean ± SD	h^2	Mean ± SD	h^2
Amylose	0.60	2.24	13.87 ± 2.98	0.24	11.49 ± 4.32	0.77
Fat	0.41	0.53	2.53 ± 0.57	0.54	3.07 ± 0.67	0.76
Gross energy	0.71	25.80	4108.33 ± 55.15	0.59	4124.56 ± 41.74	0.69
Protein	0.96	0.27	12.02 ± 1.45	0.39	11.43 ± 1.03	0.83
Starch	0.89	0.75	68.30 ± 2.44	0.44	68.37 ± 1.87	0.79

The genomic heritabilities of all traits except gross energy were significantly higher ($p < 0.05$) in the RILs than in the GSDP (Table 1). Trait heritabilities were high in the RILs, with protein and gross energy having the highest and lowest heritabilities, respectively. In the GSDP, genomic heritability was moderately high for fat and gross energy, moderate for protein and starch, and low for amylose. The poor genomic heritability (0.24) of amylose in the GSDP was expected because only a very small proportion (1%) of accessions have low amylose as a result of *waxy* gene (Mendelian trait).

Figure 1 shows correlation between the adjusted phenotypic means for trait and environment combination. Starch was negatively correlated ($p < 0.001$) with all other traits in both populations except for amylose in the RILs. Fat, protein and gross energy were significantly positively ($p < 0.001$) correlated to each other across environments in both populations. The strongest positive correlation was between gross energy and fat, whereas the strongest negative correlations were found between starch gross energy and starch protein. Moderate (0.4) to high (0.73) positive correlation was observed between years for all traits except for amylose (r = 0.08) between 2014 and 2017 in GSDP (Figure 1). We conducted a principal component analysis (PCA) of correlation matrix for the traits in each environment. In both populations, the first component separated amylose and starch from the other three traits, whereas, the second component separated amylose from starch and gross energy from protein and fat (Figure S3). The first component explained 78.8% and 75.9% of variation, and second component explained 6.3% and 9.8% of variation in the GSDP and RILs, respectively. The third principal component in the RILs separated proteins from fat and explained about 7.6% of the variation.

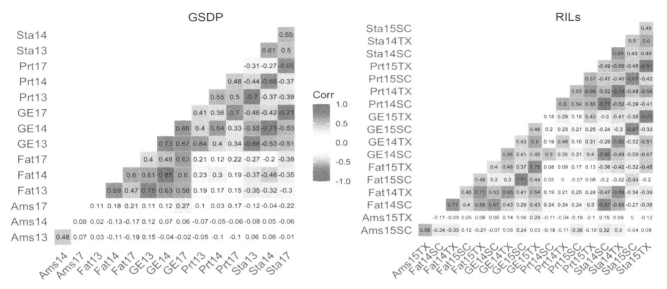

Figure 1. Correlation between traits across year and location combination for the two populations. Ams: amylose, GE: gross energy, Prt: protein, Sta: starch, SC: South Carolina, TX: Texas, and numbers in x and y-axes represent the year.

2.2. Prediction Performance in Single and Multiple Environment

We first implemented GBLUP prediction model for single-trait single-environment (STSE). Prediction accuracies of the STSE model varied across environments in both populations (Figure 2). The environments 2014 in the GSDP and TX2014 in the RILs had highest average prediction accuracy but were not always the best predicted environment for all traits. While poorly predicted for amylose, the environments 2017 in the GSDP and TX2015 in the RILs had higher prediction accuracy for starch compared to all or most environments. Despite variation across environments and populations, the average prediction accuracies from the STSE were within the range of 0.4 to 0.6 for all traits except amylose (0.25) in the GSDP. The average prediction accuracy of the STSE model in the GSDP was positively correlated (r = 0.86) with the genomic heritability of the traits. In the RILs, there was a positive correlation (r = 0.77) between average prediction accuracy and genomic heritability for amylose, fat and gross energy, but the traits (protein and starch) with the highest heritabilities had relatively lower average prediction accuracies.

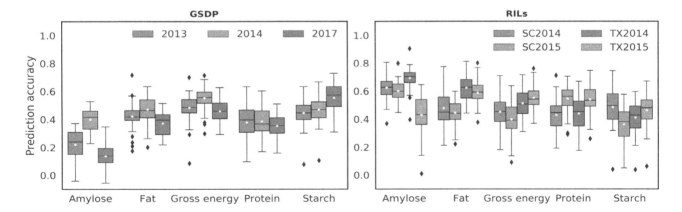

Figure 2. Prediction accuracy for single-trait single-environment (STSE) model. The y-axis shows prediction accuracy calculated as Pearson's correlation between observed values and predicted values of phenotypes. Legend represents the environment/years. SC: South Carolina, TX: Texas, GSDP: Grain sorghum diversity panel, RILs: recombinant inbred lines. Pale blue dots represent the mean of prediction accuracy.

We did not see substantial improvement in multi-environment (BME) model over the STSE prediction accuracies (Figure S4). In the GSDP, the multi-environment models resulted in a decline in average prediction accuracy compared to the STSE model for fat (21%), amylose (10%) and protein (4%), however, no significant change was observed for gross energy and starch (Table 2, Figure S5). The prediction accuracy in the RILs increased by an average of 3% in the BME compared to the STSE, however, the overall trend of prediction accuracy for traits and environments remained unchanged (Table 3, Figure S5). The environment SC2014 showed consistent increase in accuracy for BME over STSE model across all traits with about 10% increase for protein (Table 3). Amylose in TX2015 environment had the single greatest increase (12%) in average prediction accuracy in the BME among all trait-environment combinations for the RILs (Table 3).

Table 2. Percent change in mean prediction accuracy (r) over the single trait single environment (STSE) model in the diversity panel (GSDP). BME: Bayesian multi-environment, and BMORS: Bayesian multi-output regressor stacking. Values were rounded to the nearest whole number.

Trait	2013		2014		2017	
	BME	BMORS	BME	BMORS	BME	BMORS
Amylose	−11	66	−5	11	−13	−13
Fat	−24	47	−12	47	−27	58
Gross energy	3	54	−2	40	1	57
Protein	−3	56	−1	55	−8	52
Starch	4	37	−2	38	1	17
Average	−6	52	−4	38	−9	34

Table 3. Percent change in mean prediction accuracy (r) over the single trait single environment (STSE) model in the recombinant inbred lines (RILs). BME: Bayesian multi-environment, BMORS: Bayesian multi-output regressor stacking, SC: South Carolina, and TX: Texas. Values were rounded to the nearest whole number.

Trait	SC2014		SC2015		TX2014		TX2015	
	BME	BMORS	BME	BMORS	BME	BMORS	BME	BMORS
Amylose	2	28	0	28	−1	25	12	43
Fat	5	33	1	15	2	18	2	20
Gross energy	7	28	−3	27	1	18	3	17
Protein	10	51	1	23	5	60	−4	33
Starch	5	36	−4	40	7	54	4	37
Average	6	35	−1	27	3	35	3	30

2.3. Bayesian Multi-Output Regression Stacking

We tested two different prediction schemes in the BMORS prediction model using the two functions *BMORS()* and *BMORS_Env()* as described in Montesinos-López et al. [19]. While the *BMORS()* function was used for a five-fold CV as described in the methods section, the *BMORS_Env()* was used to assess the prediction performance of whole environments while using the remaining environments as training. So in *BMORS_Env*, an environment was left out during training and correlation between the predicted values (obtained from training with remaining environments) and observed values for the test environment was measured as prediction accuracy for that environment in BMORS_Env model.

2.3.1. Five-Fold CV

The prediction accuracy from five-fold CV in BMORS increased by 41% and 32% on average over the STSE model in GSDP and RILs, respectively. Figure 3 shows the prediction accuracy of BMORS for each trait and environment combination across the two populations. While the percent change in accuracy varied across environments, the BMORS model nonetheless had higher average prediction accuracy than the STSE and BME models for all traits (Figure S4). The increase in average accuracy in BMORS over STSE ranged from 11% (amylose, 2014) to 66% (amylose, 2013) in the GSDP with exception of amylose in 2017 (13% decrease), and 15% (fat, SC2015) to 60% (protein, TX2014) in the RILs (Tables 2 and 3). The increase in average prediction accuracy was higher (35%) for both locations in 2014 for the RILs, whereas, the year 2013 in the GSDP increased the most. Among the traits, protein (54%) had the greatest average increase in prediction accuracy in the GSDP, whereas in the RILs, protein and starch (42%) both showed the greatest increase.

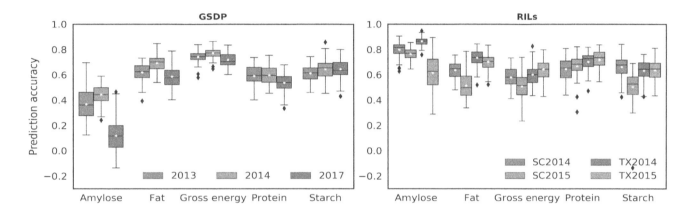

Figure 3. Prediction accuracy of Bayesian multi-output regressor stacking (BMORS) model using five-fold cross validation. Legend represents the years/environment. The y-axis shows prediction accuracy calculated as Pearson's correlation between observed values and predicted values of phenotypes. SC: South Carolina, TX: Texas, GSDP: Grain sorghum diversity panel, and RILs: Recombinant inbred lines. Pale blue dots show mean of the prediction accuracy.

2.3.2. Prediction of Whole Environment

Predicting a whole environment using the BMORS model usually yielded higher accuracy than the mean prediction accuracy from the STSE or BME model where only portions of each environment was tested instead of whole environment as in *BMORS_Env* model (Figures 2 and 4, Figure S5, Table 4). This shows that BMORS model can be reliably used in predicting unobserved environment with the same accuracy as from STSE or BME models from training within the environments. The distribution of prediction accuracy across trait and environment combination were, however, similar to the results from the STSE model. In the GSDP, little variation in prediction accuracies was observed across environments for gross energy, starch and protein, whereas, amylose and fat showed greater variability in prediction accuracy between environments. In the RILs, prediction accuracy for all traits except protein had high variability across the environments (Table 4).

In order to assess predictability by location or year in the RILs, we tested one location or year by training the BMORS model using the other location or year, respectively (Table 4). The Texas location had higher accuracy of prediction for fat (+0.11) and gross energy (+0.1) compared to South Carolina, but rest of the traits had similar prediction accuracy (difference < 0.02). Prediction accuracy of whole years varied across traits, amylose (+0.09) and fat(+0.04) were higher in 2014, protein was higher (+0.05) in 2015, and starch and gross energy were similar.

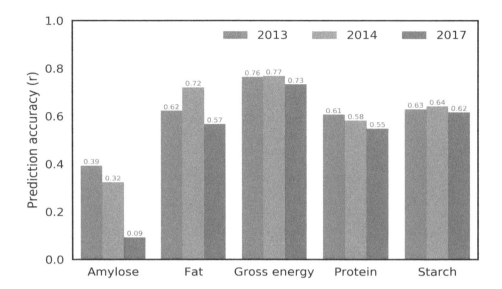

Figure 4. Prediction accuracy of whole environment predicted using the Bayesian multi-output regressor stacking (BMORS_Env) in the diversity panel (GSDP). The y-axis shows prediction accuracy calculated as Pearson's correlation between observed values and predicted values of phenotypes. Values on top of the bar represent the height of the bar.

Table 4. Prediction accuracy of the test environments predicted using the Bayesian multi-output regressor stacking (BMORS_Env) in the recombinant inbred lines (RILs). SC: South Carolina, TX, Texas. Prediction accuracy was calculated as Pearson's correlation between observed values and predicted values of phenotypes.

Trait	Year × Location				Location		Year	
	SC2014	**SC2015**	**TX2014**	**TX2015**	**SC**	**TX**	**2014**	**2015**
Amylose	0.79	0.80	0.88	0.60	0.76	0.74	0.74	0.65
Fat	0.69	0.49	0.78	0.74	0.60	0.71	0.64	0.60
Gross energy	0.56	0.49	0.62	0.66	0.48	0.58	0.56	0.56
Protein	0.65	0.66	0.66	0.70	0.59	0.58	0.61	0.66
Starch	0.64	0.52	0.68	0.60	0.55	0.56	0.56	0.55

3. Discussion

Phenotyping for grain compositional traits is—(1) challenging and labor-intensive, (2) destructive for most accurate results, and (3) only performed after plants reach physiological maturity and are harvested. The use of genomic prediction for compositional traits will be extremely valuable because it increases selection intensity and decreases generational interval by overcoming the phenotyping challenges. Moreover, these traits are complex and quantitatively inherited so will benefit from genomic prediction's ability to account for many small effect QTLs in estimating breeding values.

3.1. Trait Architecture and Prediction Accuracy

While the accuracy of NIRS calibration for traits in this study ranged from moderate to high, there was prediction error associated with NIRS prediction. However, it is unclear if and what effects NIRS prediction error had on genomic prediction. No direct correlation was observed between the genomic prediction accuracy and NIRS statistics for the traits studied. The trait with the lowest NIRS R^2, fat, was predicted as well as or better than starch, protein and gross energy, which had NIRS $R^2 > 0.7$. Despite varying strength of correlations between traits across the two populations studied, the nature of relationship was similar for a given pair of traits, which is also in agreement with previous studies [30,32,33]. The strong negative relationship of starch and amylose to protein, fat and gross energy was further elucidated by the PCA analysis of phenotypic correlation matrix (Figure S3). Since starch, protein and fat were measured on a percent dry matter basis, the strong correlation between them is expected.

Genetic relatedness and trait architecture are known to affect the accuracy of genomic prediction [8,34]. The genetic relatedness between individuals and heritability of the traits were higher in the RILs than in GSDP (Figure S6, Table 1). Those factors could be contributing to higher average prediction accuracy in the RILs. However, the average prediction accuracies for gross energy and starch were comparable between GSDP and RILs (Figure S4). Prediction accuracy in the GSDP could have been boosted by greater genetic diversity despite lower genetic relatedness [35]. Heffner et al. [22] observed a prediction accuracy of 0.5–0.6 for wheat flour protein in two biparental populations. Guo et al. [26] reported prediction accuracies of 0.44 and 0.8 for protein and amylose in rice diversity panel. Similar results were observed in our STSE models for protein content (Figure 2). Whereas, lower prediction accuracy of amylose in our diversity panel is probably due to the lack of sufficient low-amylose lines with the *waxy* gene [30]. While genomic prediction study for starch, fat and gross energy has not been reported in sorghum, these traits are nutritionally one of the most important traits for any cereal grain. The moderate to high prediction accuracy observed suggests implementation of genomic selection can improve genetic gain for these grain quality traits.

3.2. Multi-Trait Regressor Stacking

One of the daunting tasks of genomic prediction is estimating the effects of unobserved individuals and environments. As multiple traits are analyzed across several environments, the ability to combine information from multiple traits and environments can be crucial in increasing accuracy of prediction [13,15,16]. When the correlations among traits are high, prediction accuracies of complex traits can be increased by using multivariate model that takes this correlation into account [15,18]. We fit a Bayesian multi-environment (BME) model (2) that takes the genotype × environment effects into consideration. In the GSDP, where environments were three years at the same location, the BME model showed a slight decline (7%) in average prediction accuracy which was mostly due to the two traits, amylose and fat (Table 2). The RILs showed slight increase (2–3%) in prediction accuracy of traits when averaged over the environments, but there was variability across the environments (Table 3).

We implemented two functions [*BMORS()* and *BMORS_Env()*] which are not only used to evaluate prediction accuracy but are also computationally efficient [19]. The BMORS model (3) performs two-stage training by stacking the multi-environment models from all the traits. The five-fold cross validation conducted for BMORS was similar to the CV1 strategy of Montesinos-López et al. [18]. The use of multi-trait models has been consistently shown to increase prediction accuracy over single-trait models across different crops and traits [15–17,36]. The multi-target regressor stacking increased average prediction accuracy by 41% and 32% in the GSDP and RILs, respectively, as compared to the STSE prediction accuracy. Average prediction accuracy of all traits improved in BMORS over STSE and BME across both the populations (Figure S4). Consistent improvement in accuracy of BMORS is a result of the ability to use not only correlation between traits but also between environment in the model training [18,19]. The ability to accurately estimate genetic merit of lines in unobserved environments is of tremendous value in plant breeding. Our results show potential of *BMORS_Env()* function for predicting the whole environment. Testing a whole environment by training BMORS model using all other environments resulted in higher prediction accuracy for that trait-environment combination than using STSE or BME model. Prediction accuracy of all environments were 0.5 or higher with exception of amylose in GSDP, the reason for which we have discussed above (Figure 4, Table 4).

3.3. Application for Breeding

Grain quality traits such as starch and protein content have been under selection since the inception of phenotypic selection in modern breeding practices. More recently, total energy supplement of grain has gained attention for increasing feed efficiency in animal production, and a need exists for increasing total calories for human nutrition in the wake of global malnutrition crisis. Despite high correlations among these traits, the genetic variation underlying starch, protein and fat can be decoupled. Boyles et al. [30] showed major and minor effect QTLs underlying the three traits are distributed across the genome and are segregating in biparental populations. However, in practice, selection would be conducted simultaneously for these traits using a selection index rather than for individual traits. Velazco et al. [31] observed an increase in predictive ability by using a multi-trait model for grain yield and stay green in sorghum, and argue that such an exercise would allow for using selection index for implementation of genomic selection for correlated traits. Increased prediction accuracy, improved selection index, and estimation of precise genetic, environmental and residual co-variances makes multi-trait multi-environment models preferable over univariate models [18]. The multi-trait regression stacking model we tested shows large scale improvement in model prediction and can be used in tandem with Bayesian multi-trait multi-environment (BMTME) model for parameter estimation and assessing prediction accuracy. The ability to estimate genetic effects and breeding values of unobserved environments will be of great advantage to predict performance in diverse environments and for implementation of selection theory.

4. Materials and Methods

4.1. Plant Material

4.1.1. Grain Sorghum Diversity Panel

A grain sorghum diversity panel (GSDP) of 389 diverse sorghum accessions was planted in randomized complete block design with two replications in 2013, 2014, and 2017 field seasons at the Clemson University Pee Dee Research and Education Center in Florence, SC. The GSDP included a total of 332 accessions from the original United States sorghum association panel (SAP) developed by Casa et al. [37]. The details on experimental field design and agronomic practices are described in Boyles et al. [38] and Sapkota et al. [35]. Briefly, the experiments were planted in a two row plots each 6.1 m long, separated by row spacing of 0.762 m with an approximate planting density of 130,000 plants ha^{-1}. Fields were irrigated only when signs of drought stress was seen across the field.

4.1.2. Recombinant Inbred Population

A biparental population of 191 recombinant inbred lines (RILs) segregating for grain quality traits was studied along with the GSDP. The parents of the RIL population were BTx642, a yellow-pericarp drought tolerant line, and BTxARG-1, a white pericarp waxy endosperm (low amylose) line. The population was planted in two replicated plots in randomized complete block design across two years (2014 and 2015) in Blackville, SC and College Station, TX. Field design and agronomic practices have previously been described in detail in Boyles et al. [30].

4.2. Phenotyping

The primary panicle of three plants selected from each plot were harvested at physiological maturity. The plants from beginning and end of the row were excluded to account for border effect. Panicles were air dried to a constant moisture (10–12%) and threshed. A 25 g subsample of cleaned and homogenized grain ground to 1-mm particle size with a CT 193 Cyclotec Sample Mill (FOSS North America) was used in near-infrared spectroscopy (NIRS) for compositional analysis.

Grain composition traits such as total fat, gross energy, crude protein, and starch content can be measured using NIRS. Previous studies have shown high NIRS predictability of the traits used in feed analysis [39,40]. We used a DA 7250™ NIR analyzer (Perten Instruments). The ground sample was packed in a gradually rotating Teflon dish positioned under the instrument's light source and predicted phenotypic values was generated based on calibration curve for spectral measurements. The calibration curve was built using wet chemistry values from a subset of samples. The wet chemistry was performed by Dairyland Laboratories, Inc. (Arcadia, WI, USA) and the Quality Assurance Laboratory at Murphy-Brown, LLC (Warsaw, NC, USA). The details on the prediction curves and wet chemistry can be found in Boyles et al. [30].

4.3. Genotypic Data

Genotyping-by-sequencing (GBS) was used for genetic characterization of the GSDP and RILs populations [30,38,41]. Sequenced reads were aligned to the BTx623 v3.1 reference assembly (phytozome) using Burrows-Wheeler aligner [42]. SNP calling, imputation and filtering was done using TASSEL 5.0 pipeline [43]. The TASSEL plugin FILLIN for GSDP and FSFHap for RILs population were used to impute for missing genotypes. Following imputation SNPs with minor allele frequency (MAF) < 0.01, and sites missing in more than 10% and 30% of the genotypes in GSDP and RILs, respectively, were filtered. The number of genotypes studied for each population represent those with at least 70% of SNP sites. The genotype matrix with 224,007 SNPs from GSDP and 56,142 SNPs from RILs population was used for whole genome regression.

4.4. Statistical Analysis

The statistical software environment 'R' was used for model building and analysis [44]. The phenotypic values of the traits were adjusted for random effects of replications within environment using 'lme4' package in R [45]. Principal component analysis was done using the R package 'factoextra' [46]. Marker-based estimates of narrow sense (genomic) heritabilities were calculated using the SNP genotype matrix and phenotypic values using the R package 'heritability' [47]. A matrix with dummy variables '1' and '0' representing combinations of environmental variables (replication and year for GSDP, and replication, year and location for RILs) was used as co-variate in heritability estimation.

4.4.1. Single-Trait Single-Environment (STSE) Model:

The following genomic best linear unbiased prediction (GBLUP) model was used to assess prediction performance of an individual trait from a single environment:

$$y_j = \mu + g_j + e_j, \tag{1}$$

where y_j is a vector of adjusted phenotypic mean of the jth line ($j = 1, 2, ..., J$). μ is the overall mean which is assigned a flat prior, g_j is a vector of random genomic effect of the jth line, with $g = (g_1, ..., g_j)^T \sim N(0, G\sigma_1^2)$, σ_1^2 is a genomic variance, G is the genomic relationship matrix in the order $J \times J$ and is calculated [12] as $G = \frac{ZZ^T}{2\sum p_j q_j}$, where q_j and p_j denote major and minor allele frequency of jth line respectively, and Z is the design matrix for markers of order $J \times p$ (p is total number of markers). Further, e_j is residual error assigned the normal distribution $e \sim N(0, I\sigma_e^2)$ where I is identity matrix and σ_e^2 is the residual variance with a scaled-inverse Chi-square density.

4.4.2. Bayesian Multi-Environment (BME) GBLUP Model

Considering genotype $\times$ environment interaction can contribute to substantial amount of phenotypic variance in complex traits, we fit the following univariate linear mixed model to account for environmental effects in prediction performance:

$$y_{ij} = E_i + g_j + gE_{ij} + e_{ij}, \tag{2}$$

where y_j is a vector of adjusted phenotypic mean of the jth line in the ith environment ($i = 1, 2, ..., I$, $j = 1, 2, ..., J$). E_i represents the effect of ith environment and g_j represents the genomic effect of the jth line as described in Equation (1). The term gE_{ij} represents random interaction between the genomic effect of jth line and the ith environment with $gE = (gE_{11}, ..., gE_{IJ})^T \sim N(0, \sigma_2^2 I_I \otimes G)$, where σ_2^2 is an interaction variance, and e_{ij} is a random residual associated with the jth line in the ith environment distributed as $N(0, \sigma_e^2)$ where σ_e^2 is the residual variance.

4.4.3. Bayesian Multi-Output Regressor Stacking (BMORS)

BMORS is the Bayesian version of multi-trait (or multi-target) regressor stacking method [48]. The multi-target regressor stacking (MTRS) was proposed by Spyromitros-Xioufis et al. [20,21] based on multi-labeled classification approach of Godbole and Sarawagi [49]. In BMORS or MTRS, the training is done in two stages. First, L univariate models are implemented using the multi-environment GBLUP model given in Equation (2), then instead of using these models for prediction, MTRS performs the second stage of training using a second set of L meta-models for each of the L traits. The following model is used to implement each meta-model:

$$y_{ij} = \beta_1 \hat{Z}_{1ij} + \beta_2 \hat{Z}_{2ij} + ... + \beta_L \hat{Z}_{Lij} + e_{ij}, \tag{3}$$

where the covariates $\hat{Z}_{1ij}, \hat{Z}_{2ij}, ..., \hat{Z}_{Lij}$ represent the scaled prediction from the first stage training with the GBLUP model for L traits, and $\beta_1, ..., \beta_L$ are the regression coefficients for each covariate in the model. The scaling of each prediction was performed by subtracting its mean (μ_{lij}) and dividing by its corresponding standard deviation (σ_{lij}), that is, $\hat{Z}_{lij} = (\hat{y}_{lij} - \mu_{lij})\sigma_{lij}^{-1}$, for each $l = 1, ..., L$. The scaled predictions of its response variables yielded by the first-stage models as predictor information by the BMORS model. Simply put, the multi-trait regression stacking model is based on the idea that a second stage model is able to correct the predictions of a first-stage model using information about the predictions of other first-stage models [20,21].

4.4.4. Performance of Prediction Model:

All prediction models were fit using Bayesian approach in statistical program 'R'. The STSE model (1) was fit using the R package 'BGLR' [50], BME model (2) and BMORS model (3) were fit using the R package 'BMTME' [19]. A minimum of 20,000 iterations with 10,000 burn-in steps was used for each Bayesian run.

The evaluation of prediction performance of models was done using a five-fold cross validation (CV), which means 80% of the samples were used as training set and testing was done on the remaining 20% for each cross-validation fold. The individuals were randomly assigned into five mutually exclusive folds. Four folds were used to train prediction models and to predict the genomic estimated breeding values (GEBVs) of the individuals in fifth fold (validation/test set). The accuracy of prediction for each fold was calculated as Pearson's correlation coefficient (r) between predicted values and adjusted phenotypic means for the individuals in validation set. Each cross validation run, therefore, resulted in five estimates of prediction accuracy. The same set of individuals were assigned to training and validation across different traits and models tested by using *set.seed()* function in R. In order to avoid bias due to sampling, we performed 10 different cross-validation runs to calculate the mean and dispersion of the prediction accuracies.

5. Conclusions

Phenotyping of grain compositional traits using near-infrared spectroscopy is labor-intensive, generally destructive, and time limiting. Therefore, the use of genomic selection for these traits will be extremely valuable. This study establishes the potential to improve genomics-assisted selection of grain composition traits by using multi-trait multi-environment model. The phenotypic measurements obtained from NIRS prediction were amenable to genomic selection as shown by moderate to high prediction accuracy for single trait prediction. While multi-environment model alone did not lead to much improvement over single environment model, stacking of regression from multiple traits showed substantial improvement in prediction accuracy. The prediction accuracy increased by 32% and 41% in the RILs and GSDP, respectively, when using the Bayesian multi-output regressor stacking (BMORS) model compared to a single trait single environment model. The ability to predict line performance in an unobserved environment is of great importance to breeding programs, and results show high accuracy for predicting whole environments using BMORS.

Supplementary Materials
s1. The supplementary file contains six figures: Figure S1. Phenotypic distribution of grain composition traits in the RILs. In the x-axes, SC: South Carolina, TX: Texas, numbers represent years. Values are percentage dry basis for protein, fat and starch; gross energy is in KCal/lb; and amylose is in percent of starch. Figure S2. Phenotypic distribution of grain composition traits in the GSDP. Numbers in x-axes represent years. Values are percentage dry basis for protein, fat and starch; gross energy is in Cal/g; and amylose is in percent of starch. Figure S3. PCA analysis of correlation matrix between traits. a. GSDP, and b. RILs. Ams: amylose, GE: gross energy, Prt: protein, Sta: starch, SC: South Carolina, TX: Texas. The numbers in the text represent years of the environment. Figure S4. Overall prediction accuracy of traits across all the environment for the three prediction methods in the two populations. The y-axis shows prediction accuracy calculated as Pearson's correlation between observed values and predicted values of phenotypes. Legend represents the environment/years. SC: South Carolina, TX: Texas, GSDP: Grain sorghum diversity panel, RILs: recombinant inbred lines. Figure S5. Prediction accuracy using five-fold CV in Bayesian multi-environment (BME) model. a. GSDP, and b. RILs. Legend represents the environment/years. SC: South Carolina, TX: Texas. Pale blue dots represent the mean of prediction accuracy. Figure S6. Heatmap for genomic relationship matrix calculated using vanRaden (2008). a. GSDP, b. RILs. Trees show hierarchical clustering using Euclidean distance.

Author Contributions: S.S. conceptualized the study, performed data analysis, and wrote the manuscript; R.B. helped with experimental design and field phenotyping; J.L.B. helped in computation and data analysis; K.J. helped in near infra-red phenotyping; S.K. helped with conception of the study, acquisition of fund, and management of the study. All authors have read and agreed to the published version of the manuscript.

Acknowledgments: The authors would like to thank William L. Rooney and Brian K. Pfeiffer for their contributions to phenotyping of the recombinant inbred population at College Station, TX. Our appreciation goes to the Wade Stackhouse Fellowship, and Robert and Lois Coker Endowment for their support during the study.

Abbreviations

The following abbreviations are used in this manuscript:

GSDP	Grain sorghum diversity panel
RIL	Recombinant inbred line
NIRS	Near infra-red spectroscopy
GP	Genomic prediction
GBLUP	Genomic best linear unbiased prediction
STSE	Single trait single environment
BME	Bayesian multi-environment
BMORS	Bayesian multi-output regressor stacking
MTME	Multi-trait multi-environment
CV	Cross validation
SC	South Carolina
TX	Texas
QTL	Quantitative trail loci
SNP	Single nucleotide polymorphism

References

1. Awika, J.M. Major cereal grains production and use around the world. In *Advances in Cereal Science: Implications to Food Processing and Health Promotion*; ACS Publications: Washington, DC, USA, 2011; pp. 1–13.

2. Mace, E.S.; Tai, S.; Gilding, E.K.; Li, Y.; Prentis, P.J.; Bian, L.; Cruickshank, A. Whole-genome sequencing reveals untapped genetic potential in Africa's indigenous cereal crop sorghum. *Nat. Commun.* **2013**, *4*, 2320. [CrossRef]

3. Taylor, J.R.; Schober, T.J.; Bean, S.R. Novel food and non-food uses for sorghum and millets. *J. Cereal Sci.* **2006**, *44*, 252–271. [CrossRef]

4. Taylor, J. Food product development using sorghum and millets: opportunities and challenges. *Qual. Assur. Saf. Crop. Foods* **2012**, *4*, 151. [CrossRef]

5. Zhu, F. Structure, physicochemical properties, modifications, and uses of sorghum starch. *Compr. Rev. Food Sci. Food Saf.* **2014**, *13*, 597–610. [CrossRef]

6. Meuwissen, T.H.E.; Hayes, B.J.; Goddard, M.E. Prediction of Total Genetic Value Using Genome-Wide Dense Marker Maps. *Genetics* **2001**, *157*, 1819–1829.

7. Bernardo, R.; Yu, J. Prospects for genomewide selection for quantitative traits in maize. *Crop. Sci.* **2007**, *47*, 1082–1090. [CrossRef]

8. Habier, D.; Fernando, R.L.; Dekkers, J.C. The impact of genetic relationship information on genome-assisted breeding values. *Genetics* **2007**, *177*, 2389–2397. [CrossRef] [PubMed]

9. Zhong, S.; Dekkers, J.C.; Fernando, R.L.; Jannink, J.L. Factors affecting accuracy from genomic selection in populations derived from multiple inbred lines: A barley case study. *Genetics* **2009**, *182*, 355–364. [CrossRef] [PubMed]

10. Combs, E.; Bernardo, R. Accuracy of genomewide selection for different traits with constant population size, heritability, and number of markers. *Plant Genome* **2013**, *6*, 1–7. [CrossRef]

11. Crossa, J.; Pérez-Rodríguez, P.; Cuevas, J.; Montesinos-López, O.; Jarquín, D.; de los Campos, G.; Dreisigacker, S. Genomic selection in plant breeding: Methods, models, and perspectives. *Trends Plant Sci.* **2017**, *22*, 961–975. [CrossRef]

12. VanRaden, P.M. Efficient methods to compute genomic predictions. *J. Dairy Sci.* **2008**, *91*, 4414–4423. [CrossRef] [PubMed]

13. Burgueño, J.; de los Campos, G.; Weigel, K.; Crossa, J. Genomic prediction of breeding values when modeling genotype×environment interaction using pedigree and dense molecular markers. *Crop. Sci.* **2012**, *52*, 707–719. [CrossRef]

14. Jarquín, D.; Crossa, J.; Lacaze, X.; Du Cheyron, P.; Daucourt, J.; Lorgeou, J.; Burgueño, J. A reaction norm model for genomic selection using high-dimensional genomic and environmental data. *Theor. Appl. Genet.* **2014**, *127*, 595–607. [CrossRef] [PubMed]

15. Jia, Y.; Jannink, J.L. Multiple-trait genomic selection methods increase genetic value prediction accuracy. *Genetics* **2012**, *192*, 1513–1522. [CrossRef]

16. Guo, G.; Zhao, F.; Wang, Y.; Zhang, Y.; Du, L.; Su, G. Comparison of single-trait and multiple-trait genomic prediction models. *BMC Genet.* **2014**, *15*, 30. [CrossRef]

17. Lado, B.; Vázquez, D.; Quincke, M.; Silva, P.; Aguilar, I.; Gutiérrez, L. Resource allocation optimization with multi-trait genomic prediction for bread wheat (*Triticum aestivum* L.) baking quality. *Theor. Appl. Genet.* **2018**, *131*, 2719–2731. [CrossRef]

18. Montesinos-López, O.A.; Montesinos-López, A.; Crossa, J.; Toledo, F.H.; Pérez-Hernández, O.; Eskridge, K.M.; Rutkoski, J. A genomic Bayesian multi-trait and multi-environment model. *G3 Genes Genomes Genet.* **2016**, *6*, 2725–2744. [CrossRef]

19. Montesinos-López, O.A.; Montesinos-López, A.; Luna-Vázquez, F.J.; Toledo, F.H.; Pérez-Rodríguez, P.; Lillemo, M.; Crossa, J. An R package for Bayesian analysis of multi-environment and multi-trait multi-environment data for genome-based prediction. *G3 Genes Genomes Genet.* **2019**, *9*, 1355–1369. [CrossRef]

20. Spyromitros-Xioufis, E.; Tsoumakas, G.; Groves, W.; Vlahavas, I. Multi-target regression via input space expansion: Treating targets as inputs. *Mach. Learn.* **2016**, *104*, 55–98.

21. Spyromitros-Xioufis, E.; Tsoumakas, G.; Groves, W.; Vlahavas, I. Multi-label classification methods for multi-target regression. *arXiv* **2012**, arXiv:12116581.

22. Heffner, E.L.; Jannink, J.L.; Iwata, H.; Souza, E.; Sorrells, M.E. Genomic selection accuracy for grain quality traits in biparental wheat populations. *Crop. Sci.* **2011**, *51*, 2597–2606. [CrossRef]
23. Battenfield, S.D.; Guzmán, C.; Gaynor, R.C.; Singh, R.P.; Peña, R.J.; Dreisigacker, S.; Poland, J.A. Genomic selection for processing and end-use quality traits in the CIMMYT spring bread wheat breeding program. *Plant Genome* **2016**, *9*, 1–12. [CrossRef] [PubMed]
24. Haile, J.K.; N'Diaye, A.; Clarke, F.; Clarke, J.; Knox, R.; Rutkoski, J.; Pozniak, C.J. Genomic selection for grain yield and quality traits in durum wheat. *Mol. Breed.* **2018**, *38*, 75. [CrossRef]
25. Schulthess, A.W.; Wang, Y.; Miedaner, T.; Wilde, P.; Reif, J.C.; Zhao, Y. Multiple-trait-and selection indices-genomic predictions for grain yield and protein content in rye for feeding purposes. *Theor. Appl. Genet.* **2016**, *129*, 273–287. [CrossRef] [PubMed]
26. Guo, Z.; Tucker, D.M.; Basten, C.J.; Gandhi, H.; Ersoz, E.; Guo, B.; Gay, G. The impact of population structure on genomic prediction in stratified populations. *Theor. Appl. Genet.* **2014**, *127*, 749–762. [CrossRef] [PubMed]
27. Duhnen, A.; Gras, A.; Teyssèdre, S.; Romestant, M.; Claustres, B.; Daydé, J.; Mangin, B. Genomic selection for yield and seed protein content in Soybean: A study of breeding program data and assessment of prediction accuracy. *Crop. Sci.* **2017**, *57*, 1325–1337. [CrossRef]
28. Hayes, B.; Panozzo, J.; Walker, C.; Choy, A.; Kant, S.; Wong, D.; Spangenberg, G.C. Accelerating wheat breeding for end-use quality with multi-trait genomic predictions incorporating near infrared and nuclear magnetic resonance-derived phenotypes. *Theor. Appl. Genet.* **2017**, *130*, 2505–2519. [CrossRef]
29. Rapp, M.; Lein, V.; Lacoudre, F.; Lafferty, J.; Müller, E.; Vida, G.; Leiser, W.L. Simultaneous improvement of grain yield and protein content in durum wheat by different phenotypic indices and genomic selection. *Theor. Appl. Genet.* **2018**, *131*, 1315–1329. [CrossRef]
30. Boyles, R.E.; Pfeiffer, B.K.; Cooper, E.A.; Rauh, B.L.; Zielinski, K.J.; Myers, M.T.; Kresovich, S. Genetic dissection of sorghum grain quality traits using diverse and segregating populations. *Theor. Appl. Genet.* **2017**, *130*, 697–716. [CrossRef]
31. Velazco, J.G.; Jordan, D.R.; Mace, E.S.; Hunt, C.H.; Malosetti, M.; Van Eeuwijk, F.A. Genomic prediction of grain yield and drought-adaptation capacity in sorghum is enhanced by multi-trait analysis. *Front. Plant Sci.* **2019**, *10*, 997. [CrossRef]
32. Murray, S.C.; Sharma, A.; Rooney, W.L.; Klein, P.E.; Mullet, J.E.; Mitchell, S.E.; Kresovich, S. Genetic improvement of sorghum as a biofuel feedstock: I. QTL for stem sugar and grain nonstructural carbohydrates. *Crop. Sci.* **2008**, *48*, 2165–2179. [CrossRef]
33. Sukumaran, S.; Xiang, W.; Bean, S.R.; Pedersen, J.F.; Kresovich, S.; Tuinstra, M.R.; Yu, J. Association mapping for grain quality in a diverse sorghum collection. *Plant Genome* **2012**, *5*, 126–135. [CrossRef]
34. Jannink, J.L.; Lorenz, A.J.; Iwata, H. Genomic selection in plant breeding: From theory to practice. *Briefings Funct. Genom.* **2010**, *9*, 166–177. [CrossRef] [PubMed]
35. Sapkota, S.; Boyles, R.; Cooper, E.; Brenton, Z.; Myers, M.; Kresovich, S. Impact of sorghum racial structure and diversity on genomic prediction of grain yield components. *Crop. Sci.* **2020**, *60*, 132–148. [CrossRef]
36. Bhatta, M.; Gutierrez, L.; Cammarota, L.; Cardozo, F.; Germán, S.; Gómez-Guerrero, B.; Castro, A.J. Multi-trait Genomic Prediction Model Increased the Predictive Ability for Agronomic and Malting Quality Traits in Barley (*Hordeum vulgare* L.). *G3 Genes Genomes Genet.* **2020**, *10*, 1113–1124. [CrossRef] [PubMed]
37. Casa, A.M.; Pressoir, G.; Brown, P.J.; Mitchell, S.E.; Rooney, W.L.; Tuinstra, M.R.; Kresovich, S. Community resources and strategies for association mapping in sorghum. *Crop. Sci.* **2008**, *48*, 30–40. [CrossRef]
38. Boyles, R.E.; Cooper, E.A.; Myers, M.T.; Brenton, Z.; Rauh, B.L.; Morris, G.P.; Kresovich, S. Genome-wide association studies of grain yield components in diverse sorghum germplasm. *Plant Genome* **2016**, *9*, 1–17. [CrossRef]
39. Kays, S.E.; Barton, F.E. Rapid prediction of gross energy and utilizable energy in cereal food products using near-infrared reflectance spectroscopy. *J. Agric. Food Chem.* **2002**, *50*, 1284–1289. [CrossRef]
40. De Alencar Figueiredo, L.F.; Sine, B.; Chantereau, J.; Mestres, C.; Fliedel, G.; Rami, J.F.; Courtois, B. Variability of grain quality in sorghum: Association with polymorphism in Sh2, Bt2, SssI, Ae1, Wx and O2. *Theor. Appl. Genet.* **2010**, *121*, 1171–1185. [CrossRef]
41. Morris, G.P.; Ramu, P.; Deshpande, S.P.; Hash, C.T.; Shah, T.; Upadhyaya, H.D.; Riera-Lizarazu, O.; Brown, P.J.; Acharya, C.B.; Mitchell, S.E.; et al. Population genomic and genome-wide association studies of agroclimatic traits in sorghum. *Proc. Natl. Acad. Sci. USA* **2013**, *110*, 453–458. [CrossRef]

42. Li, H.; Durbin, R. Fast and accurate long-read alignment with Burrows–Wheeler transform. *Bioinformatics* **2010**, *26*, 589–595. [CrossRef]

43. Glaubitz, J.C.; Casstevens, T.M.; Lu, F.; Harriman, J.; Elshire, R.J.; Sun, Q.; Buckler, E.S. TASSEL-GBS: A high capacity genotyping by sequencing analysis pipeline. *PLoS ONE* **2014**, *9*, e90346. [CrossRef] [PubMed]

44. R Core Team. R: A Language and Environment for Statistical Computing. 2019. Available online: https://www.R-project.org/ (accessed on 26 April 2019).

45. Bates, D.; Mächler, M.; Bolker, B.; Walker, S. Fitting Linear Mixed-Effects Models Using lme4. *J. Stat. Softw.* **2015**, *67*, 1–48. [CrossRef]

46. Kassambara, A.; Mundt, F. Factoextra: extract and visualize the results of multivariate data analyses. *R Package Version* **2017**, *1*, 337–354.

47. Kruijer, W.; Boer, M.P.; Malosetti, M.; Flood, P.J.; Engel, B.; Kooke, R.; van Eeuwijk, F.A. Marker-based estimation of heritability in immortal populations. *Genetics* **2015**, *199*, 379–398. [CrossRef] [PubMed]

48. Montesinos-López, O.A.; Montesinos-López, A.; Crossa, J.; Cuevas, J.; Montesinos-López, J.C.; Gutiérrez, Z.S.; Singh, R. A Bayesian genomic multi-output regressor stacking model for predicting multi-trait multi-environment plant breeding data. *G3 Genes Genomes Genet.* **2019**, *9*, 3381–3393. [CrossRef] [PubMed]

49. Godbole, S.; Sarawagi, S. Discriminative methods for multi-labeled classification. In Proceedings of the Pacific-Asia Conference on Knowledge Discovery and Data Mining, Sydney, Australia, 26–28 May 2004; Springer: Berlin/Heidelberg, Germany, 2004; p. 22–30.

50. Pérez, P.; de Los Campos, G. Genome-wide regression and prediction with the BGLR statistical package. *Genetics* **2014**, *198*, 483–495. [CrossRef]

9

Single-Molecule Long-Read Sequencing of Avocado Generates Microsatellite Markers for Analyzing the Genetic Diversity in Avocado Germplasm

Yu Ge, Xiaoping Zang, Lin Tan, Jiashui Wang, Yuanzheng Liu, Yanxia Li, Nan Wang, Di Chen, Rulin Zhan * and Weihong Ma *

Haikou Experimental Station, Chinese Academy of Tropical Agricultural Sciences, Haikou 570102, China
* Correspondence: zhanrulin@catas.cn (R.Z.); zjwhma@catas.cn (W.M.)

Abstract: Avocado (*Persea americana* Mill.) is an important fruit crop commercially grown in tropical and subtropical regions. Despite the importance of avocado, there is relatively little available genomic information regarding this fruit species. In this study, we functionally annotated the full-length avocado transcriptome sequence based on single-molecule real-time sequencing technology, and predicted the coding sequences (CDSs), transcription factors (TFs), and long non-coding RNA (lncRNA) sequences. Moreover, 76,777 simple sequence repeat (SSR) loci detected among the 42,096 SSR-containing transcript sequences were used to develop 149,733 expressed sequence tag (EST)-SSR markers. A subset of 100 EST-SSR markers was randomly chosen for an analysis that detected 15 polymorphicEST-SSR markers, with an average polymorphism information content of 0.45. These 15markers were able to clearly and effectively characterize46 avocado accessions based on geographical origin. In summary, our study is the first to generate a full-length transcriptome sequence and develop and analyze a set of EST-SSR markers in avocado. The application of third-generation sequencing techniques for developing SSR markers is a potentially powerful tool for genetic studies.

Keywords: *Persea americana*; SMRT sequencing; simple sequence repeat; genetic relationship

1. Introduction

Avocado (*Persea americana* Mill.) belonging to the family Lauraceae of the order Laurales is native to Mexico and Central and South America, and is one of the most economically important subtropical/tropical fruit crops worldwide [1]. Taxonomic treatments differ considerably in terms of the circumscription and defining of infraspecific avocado entities [2–5]. Additionally, researchers have long considered that geographical isolation has likely resulted in the following three ecological races of avocado: Mexican (*P. americana* var. *drymifolia*), Guatemalan (*P. americana* var. *guatemalensis*), and West Indian (*P. Americana* var. *americana*) [1]. The Mexican race adapted to a Mediterranean climate, whereas the Guatemalan race originated in a tropical highland climate, and the West Indian race adapted to humid tropical lowland conditions [1].

Avocado is rich in lipids, sugars, proteins, minerals, vitamins, and other active ingredients [6–8]. Moreover, avocado production has increased worldwide [1]. One factor contributing to the increases in production and consumption is the expansion of avocado products into new global markets where avocado was previously unknown or scarce, includingChina, which is an emerging market for the production and consumption of avocado [1,9]. After avocado was first introduced and cultivatedin China in the late 1950s, selective breeding by some national scientific research bodies and other state farms have resulted in the development of more than 10 superior avocado accessions [9,10]. Additionally, natural crosses among avocado accessions have generated new hybrids on state and private farms, andsome nativeaccessions are increasingly produced in somewhat remote areas with distinct local environmental conditions [9,10]. Avocado is broadly grown and exploited in some provinces in southern China, including Hainan, Guangxi, Yunnan, and Taiwan [9,10]. The climatic conditions in these provinces are subtropical to tropical, which are ideal conditions for the cultivation of avocado [9,10].

The avocado germplasm should be precisely characterized to maximize its utility to breeders worldwide [1]. Specifically, a molecular characterization is required for analyses of the genetic relationships among avocado germplasm. Over the past two decades, studies involving various types of molecular markers have examined the genetic relationships among avocado germplasm [11–20]. Of the many available DNA markers, simple sequence repeats (SSRs) are commonly used for investigating plant genetics and breeding because they are widely distributed and abundant in plant genomes. They are also genetically codominant, highly reproducible, multi-allelic, and perfectly suitable for high-throughput genotyping [21–25].

Expressed sequence tag (EST)-derived markers in the genomic coding regions have an advantage over genomic DNA-derived markers, and can be efficiently amplified to reveal conserved sequences among related species [26]. There has recently been increasing interest in developing EST-SSR markers viahigh-throughput transcriptome sequencing. Thus, there has been rapid progress in the development of EST-SSR markers based on transcriptome data produced with second-generation sequencing technology for *Lilium brownii* var. *viridulum* Baker [27], *crataegus Pinnatifida* Bunge [28], *Acer miaotaiense* P. C. Tsoong [29], and *Rosa hybrida* hort. ex Lavalle [30]. Among the third-generation sequencing platforms, PacBio RS II, which is regarded as the first commercialized third-generation sequencer, is based on single-molecule real-time (SMRT) technology [31]. The PacBio RS II system can produce much longer reads than second-generation sequencing platforms, and has been applied to effectively capture full-length transcriptsequences for EST-derived marker development [32]. However, there are few reports regarding the application ofEST-SSR markers developed with SMRT technology for crop breeding.

Single-molecule real-time technology has the following threemain advantages over second-generation sequencing options: it generates longer reads, it has higher consensus accuracy, and it is less biased [33]. A previous study revealed that SMRT technology can precisely ascertain alternative polyadenylation sites and full-length splice isoforms, and also detect a higher isoform density than that for the reference genome [34]. The application of SMRT technology for nearly 3 years has helped to elucidate the complexity of the transcriptome and molecular mechanism underlying the metabolite synthesisin safflower [31], *Zanthoxylum bungeanum* Maxim. [32], *Trifolium pratense* L. [34], *Saccharum officinarum* L. [35], *Panicum virgatum* L. [36], *Medicago sativa* L. [37], *Zanthoxylum planispinum* Sieb. [38], *Cynodon dactylon* L. Pers. [39], *Camellia sinensis* L. O. Ktze. [40], and *Cassia obtusifolia* L. [41].

In the previous study, we had generated the first full-length transcriptome sequence of avocadobased on SMRT technology andthe short-reads obtained in this previous study involving second-generation transcriptome sequencing were used to correct the transcripts that were obtained with SMRT technology [42]. In this study, we functionally annotated sequences andcompleted SSR mining experiments from SMRT technology in avocado mesocarp. We also predicted the coding sequences (CDSs), transcription factors (TFs), and long non-coding RNA (lncRNA) sequences. Furthermore, we identified a set of EST-SSR markers, and assessed their utility for determining the genetic diversity among 46 selected avocado accessions from various locations in southern China. The generated data enabled the broad and distinct visualization of the genetic diversity in the analyzed avocado germplasm. The results of this study represent useful genetic and transcriptome information to support future research on avocado.

2. Materials and Methods

2.1. Sample Collection, DNA Extraction, and RNA Extraction

For transcriptome analyses, avocado fruits (cultivar 'Hass') were harvested from April to September 2018 from six 10-year-old trees (grafted onto Zutano clonal rootstock) growing at the Chinese Academy of Tropical Agricultural Sciences (CATAS; Danzhou, Hainan, China; latitude 19°31' N, longitude 109°34' E, and 20 m above sea level). Each biological replicate comprised samples from two trees. Specifically, fruits that developed during the main flowering season (i.e., February 2018) were marked, after which samples were collected at five time-points (75, 110, 145, 180, and 215 days after full bloom) until the fruits reached physiological maturity (i.e., able to ripen after harvest). The fruits were randomly collected for each biological replicate during each developmental stage. Fruits were quickly brought to the laboratory, after which the mesocarp (pulp) was separated from the seedand then immediately frozen at −80 °C for subsequent transcriptome analyses. Total RNA was extracted with a Plant RNA Kit (OMEGA Bio-Tek, Norcross, GA, USA).

For kompetitive allele-specific PCR (KASP) genotyping and EST-SSR detection, seven commercial cultivars and 39 native accessions were selected. These native accessions were obtained from the CATAS (Danzhou, Hainan, China; latitude 19°31' N, longitude 109°34' E, and 20 m above sea level), Daling State Farm (DLSF; Baisha, Hainan, China; latitude 19°14' N, longitude 109°14' E, and 60 m above sea level), Mengmao State Farm (MMSF; Ruili, Yunnan, China; latitude 24°00' N, longitude 97°50' E, and 240 m above sea level), and Guangxi Vocational and Technical College (GVTC; Nanning, Guangxi, China; latitude 22°29' N, longitude 108°11' E, and 79 m above sea level). Details regarding the avocado germplasm are provided in Table S1. Genomic DNA was extracted from fresh leaves as described by Ge [43].

2.2. PacBiocDNA Library Construction and Sequencing

Poly-T oligo-attached magnetic beads were used to purify the mRNA from the total RNA extracted from 15 mesocarp (pulp) samples collected at each analyzed developmental stage. The mRNA from all five developmental stages was combined to serve as the template to synthesize cDNA with the SMARTer PCR cDNA Synthesis Kit (Clontech, Mountain View, CA, USA). After a PCR amplification, quality control check, and purification, full-length cDNA fragments were acquired according to the BluePippin Size Selection System protocol, ultimately resulting in the construction of a cDNA library (1–6 kb). Selected full-length cDNA sequences were ligated to the SMRT bell hairpin loop. The concentration of the cDNA library was then determined with the Qubit 2.0 fluorometer, whereas the quality of the cDNA library was assessed with the 2100 Bioanalyzer (Agilent). Finally, one SMRT cell was sequenced with the PacBio RSII system (Pacific Biosciences, Menlo Park, CA, USA).

2.3. IlluminacDNA Library Construction and Sequencing

Oligo-(dT) magnetic beads were used to purify the mRNA from the total RNA extracted from 15 mesocarp (pulp) samples from five developmental stages. Three replicates were analyzed for each developmental stage. Samples from each developmental stage underwent an RNA-sequencing analysis, with three biological replicates per sample. The fragmentation step was completed with divalent cations in heated 5× NEBNext First Strand Synthesis Reaction Buffer. First-strand cDNA was synthesized with a series of random hexamer primers and reverse transcriptase, after which the second-strand cDNA was generated with DNA polymerase I and RNase H. The cDNA libraries were constructed by ligating the cDNA fragments to sequencing adapters and amplifying the fragments by PCR. The libraries were then sequenced with the Illumina HiSeq 2000 platform (Nanxin Bioinformatics Technology Co., Ltd., Guangzhou, China).

2.4. Quality Filtering and Correction of PacBio Long-Reads

Raw reads were processed into error-corrected reads of insert (ROIs) using an isoform sequencing pipeline, with minimum full pass = 0.00 and minimum predicted accuracy = 0.80. Next, full-length, non-chimeric transcripts were detected by searching for the poly-A tail signal and the 5′ and 3′ cDNA primer sequences in the ROIs. Iterative clustering for error correction was used to obtain high-quality consensus isoforms, which were then polished with QuiverVersion 1.0. The low-quality full-length transcript isoforms were corrected based on Illumina short-reads with the default setting of the Proovread program. High-quality and corrected low-quality transcript isoforms were confirmed as nonredundant with the CD-HIT software.

2.5. Functional Annotation

Genes were functionally annotated based on a BLASTX search (E-value threshold of 10^{-5}) of the following databases: Clusters of Orthologous Groups of proteins (KOG/COG) (available online: http://www.ncbi.nlm.nih.gov/KOG/; available online: http://www.ncbi.nlm.nih.gov/COG/), Non-supervised Orthologous Groups (eggNOG) (available online: http://eggnogdb.embl.de/#/app/home), Swiss-Prot (a manually annotated and reviewed protein sequence database, available online: http://www.uniprot.org/), Pfam (assigned with the HMMER3.0 package, available online: https://pfam.xfam.org/), and NCBI nonredundant protein sequence (Nr) (availableonline: http://www.ncbi.nlm.nih.gov/). Additionally, the KEGG Automatic Annotation Server [44] was used to assign these genes to Kyoto Encyclopedia of Genes and Genomes (KEGG) metabolic pathways (available online: http://www.genome.jp/kegg/). The unigenes were annotated with gene ontology (GO) terms (available online: http://www.geneontology.org/) with the Blast2GO (version 2.5) program [45] based on the BLASTX matches in the Pfam and Nr databases (E-value threshold of 10^{-6}).

2.6. Mining of EST-SSR Markers

The MISA (version 1.0) program, with the following default settings, was used to locate SSRs: a minimum of five repeats; a minimum motif length of 5 for tri- and hexanucleotides, 6 for dinucleotides, and 10 for single nucleotides.

2.7. Analyses of Detected Coding Sequences, Transcription Factors, and Long Non-Coding RNA Features

The open reading frames (ORFs) detected with the TransDecoder (version 3.0.0) program were designated as putative CDSs if they satisfied the following criteria: (1) An ORF was detected in a transcript sequence; (2) the log-likelihood score was >0, and was similar to what was calculated with the GeneID software; (3) the score was higher when the ORF was in the first reading frame than when the ORF was in the other five reading frames; (4) if a candidate ORF was within another candidate ORF, the longer one was reported. However, a single transcript could be associated with multiple ORFs (because of operons and chimeras); and (5) the putative encoded peptide matched a Pfam domain.

Transcription factor gene families were identified based on categorically defined TF families and criteria from the KO, KOG, GO, Swiss-Prot, Pfam, Nr, and Nt databases. Specifically, the default parameters of the iTAK (version 1.2) program were used. The methods used to identify and classify TFs were previously described by Perez-Rodriguez [46].

The following four computational tools were combined to sort non-protein-coding RNA candidates from putative protein-coding RNAs among the transcripts: the Coding Potential Calculator (CPC), Coding-Non-Coding Index (CNCI), Coding Potential Assessment Tool (CPAT), and Pfam database. Transcripts longer than 200 nt, with more than two exons, were selected as lncRNA candidates and were further screened with CPC/CNCI/CPAT/Pfam, which distinguished the protein-coding genes from the non-coding genes.

2.8. Assignment of the Native Avocado Accessions with an Unknown Race

To validate the origins of the 33 native accessions with an unknown race, six primers for race-specific single nucleotide polymorphism (SNP) loci were used for KASP genotyping listed in Table S2 [47]. The primer mix, which was prepared and used as described by KBioscience (http://www.kbioscience.co.uk), comprised 46 μL dH$_2$O, 30 μL common primer (100 μM), and 12 μL each tailed primer (100 μM). The SNPs were amplified by PCR in a thermal cycler with a 5-μL solution consisting of 1× KASP Master mix, 10 ng genomic DNA, and the SNP-specific KASP assay mix. The following PCR amplification conditions were the same as those used for each SNP assay: 94 °C for 15 min; 10 touchdown cycles of 94 °C for 20 s, and 58–61 °C for 60 s (decreasing by 0.8 °C per cycle); 35 cycles of 94 °C for 20 s and 57 °C for 60 s. The resulting data were analyzed with the Roche LightCycler 480 (version 1.50.39) program.

2.9. Identification of EST-SSR Markers

To screen the EST-SSR loci, primers based on the sequences flanking the selected microsatellite loci were designed with the Primer3 program; the PCR products ranged from 100 to 300 bp. All assigned marker names included Pa-eSSR to indicate their association with *P. Americana* and EST-SSRs. A subset of 100 EST-SSR primer pairs was randomly selected for validation by a PCR amplification with the same conditions as those described by Ge [43]. The PCR products were analyzed with the 96-capillary 3730xl DNA Analyzer (Applied Biosystems, Foster City, CA, USA). The detection system included 8.9 μL HIDI (Applied Biosystems), 0.1 μL LIZ (Applied Biosystems), and 1 μL PCR products (1:10 dilution). A lack of amplification was considered indicative of a null allele.

2.10. Data Analysis

The number of observed alleles (Na), effective number of alleles (Ne), observed heterozygosity (Ho), expected heterozygosity (He), and polymorphism information content (PIC) of each EST-SSR was assessed with the POPGEN (version 1.32) program [48]. A cluster analysis was performed with PowerMarker (version 3.25) [49]. The cophenetic correlation coefficient was computed for the dendrogram after the construction of a cophenetic matrix to measure the goodness of fit between the original similarity matrix and the dendrogram. Bootstrap support values were obtained from 1000 replicates. A neighbor-joining tree was constructed based on shared alleles, and visualized with the MEGA6.0 software [50].

3. Results

3.1. General Properties and Functional Annotations Based on Public Databases of Single-Molecule Long-Reads

Figure 1 presents the length distribution of 651,260 reads of insert in avocado mesocarp, and the classification of the reads of insert in avocado mesocarpis listed in Figure 2. The SMRT and Illumina HiSeq 2000 sequencing data were deposited in the GenBank database (accession numbersPRJNA551932 and PRJNA541745, respectively). Gene annotations according to a BLASTX algorithm indicated that the 71,627 avocado transcripts significantly matched sequences in the COG, GO, KEGG, KOG, Pfam, Swiss-Prot, eggNOG, and Nr databases, respectively (Table S3). The species with the most matches for the transcripts were *Nelumbo nucifera* Gaertn. (41.18% of transcripts), *Vitis vinifera* L. (10.76% of transcripts), *Elaeis guineensis* Jacq. (8.88% of transcripts), and *Phoenix dactylifera* L. (6.90% of transcripts). The homology with the other species was relatively low (1.14%–2.54% of transcripts; Figure 3). To further predict and classify the functions of the annotated transcripts, we analyzed their matching GO terms, eggNOG classifications, and KEGG pathway assignments. A total of 45,134 transcripts were assigned to 51 subcategories of the three main GO functional categories as follows: 106,390 transcripts for biological processes, 45,931 transcripts for cellular components, and 69,120 for molecular functions (Figure 4a, Table S4). Next, 70,205 transcripts were functionally classified into 25eggNOG categories (Figure 4b, Table S5). Among the 26 categories, the most heavily represented group was posttranslational modification, protein turnover, chaperones (6410 transcripts, 8.94%), followed by signal transduction mechanisms (4189 transcripts, 5.84%) and transcription (3868 transcripts, 5.39%). Only 20 and 6 transcripts belonged to the cell motility and nuclear structure categories, respectively. Finally, 33,310 transcripts were assigned to 129 KEGG pathways (Table S6). The most represented pathways were related to carbon metabolism (1678 transcripts), protein processing in endoplasmic reticulum (1649 transcripts), and biosynthesis of amino acids (1503 transcripts).

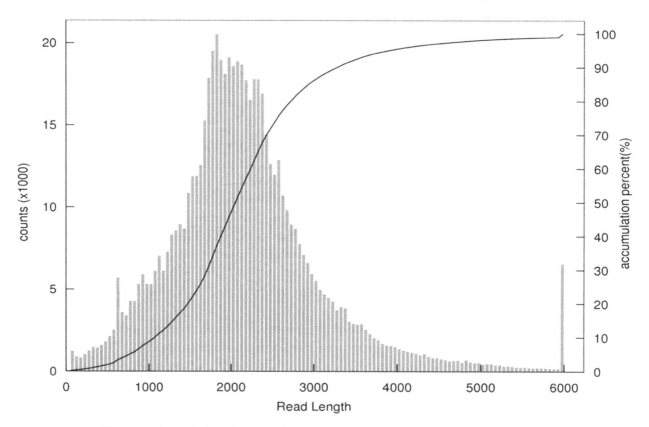

Figure 1. Length distribution of 651,260 reads of insert in avocadomesocarp.

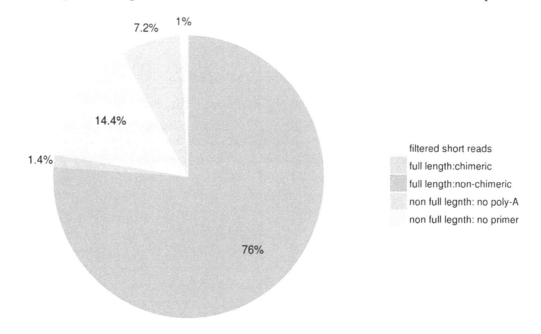

Figure 2. Classification of reads of insert in avocadomesocarp.

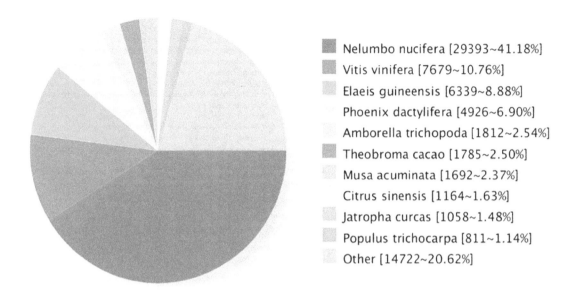

Figure 3. Species most closely related to avocado based on the NCBI nonredundant protein sequence database.

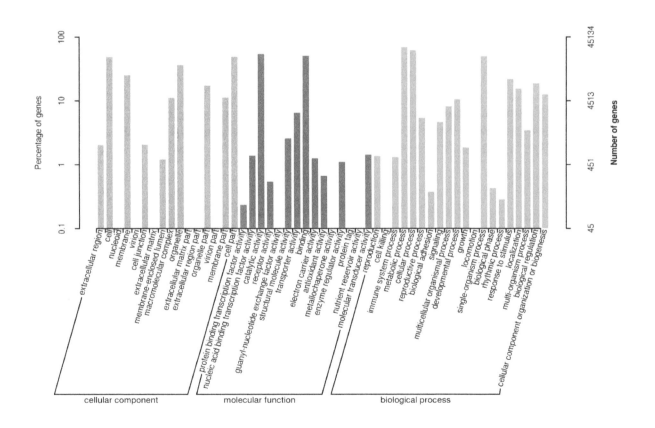

(a)

Figure 4. *Cont.*

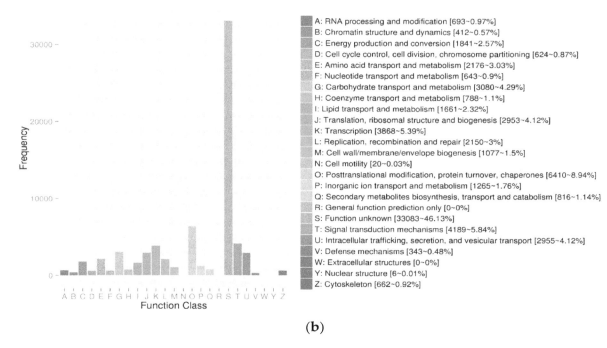

Figure 4. Functional classification of transcripts. The predicted functions were based on Gene Ontology (**a**) and Non-supervised Orthologous Groups (**b**) databases.

3.2. *Predictions of ORFs, TFs, and lncRNAs*

A total of 73,946 ORFs were predicted, 61,523 of which were complete CDSs. The number and length distribution of proteins encoded by the CDS regions are presented in Figure 5 and Additional file 1. A total of 7969 putative avocado TFs distributed in 203 families were identified (Table S7). The most abundant TF categories included RLK-Pelle_DLSV (241) and C3H (240). Additionally, the CPC, CNCI, CPAT, and Pfam database were combined to distinguish lncRNA candidates from putative protein-coding RNAs among the unannotated transcripts. Analyses with the CPC, CNCI, CPAT, and Pfam database revealed 7869, 6444, 16,464, and 15,579 transcripts longer than 200 nt with more than two exons as lncRNA candidates. A total of 3596 lncRNA transcripts were predicted (Figure 6).

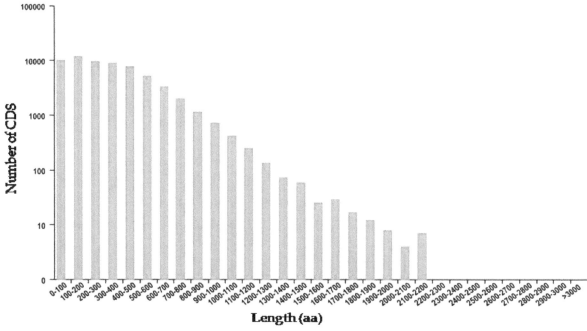

Figure 5. Distribution of 61,523 complete coding sequences for the avocado open reading frames.

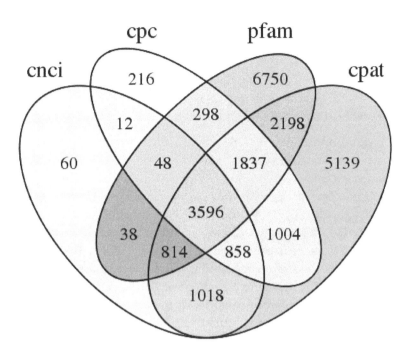

Figure 6. The number of long non-coding RNA transcripts predicted in avocado based on the Coding Potential Calculator, Coding-Non-Coding Index, Coding Potential Assessment Tool, and Pfam database.

3.3. Frequency and Distribution of Various Types of EST-SSR Loci

The 75,946 transcript sequences comprising 170,959,769 bp detected in this study included 42,096 sequences containing 76,777 SSR loci (Table 1). Of these SSR-containing transcript sequences, 19,825 harbored more than one SSR locus. Mononucleotide motifs were the most abundant (44,800, 58.35%), followed by di- (18,903; 24.62%), tri- (11,724, 15.27%), tetra- (788, 0.01%), hexa- (321, 0.00%), and pentanucleotide (241, 0.00%) motif repeats (Table 2).

Table 1. Details regarding the simple sequence repeats (SSRs) identified from single-molecule real-time (SMRT) sequencing in avocado mesocarp.

Source	Number
Total number of sequences examined	75,956
Total size of examined sequences (bp)	170,959,769
Total number of identified SSRs	76,777
Number of SSR containing sequences	42,096
Number of sequences containing more than 1 SSR	19,825
Number of SSRs present in compound formation	12,675

Table 2. Details regarding the number of repeating units at avocado expressed sequence tag-simple sequence repeat (EST-SSR) loci.

SSR Motif Length	Repeat Unit Number								
	5	6	7	8	9	10	>10	Total	%
Mono-	-	-	-	-	-	8951	35,849	44,800	58.35
Di-	-	4017	2773	2489	2013	1547	6064	18,903	24.62
Tri-	6129	2838	1237	720	403	193	204	11,724	15.27
Tetra-	541	175	41	21	8	-	2	788	0.01
Penta-	172	67	1	1	-	-	-	241	0.00
Hexa-	228	72	15	3	2	-	1	321	0.00
Total	7070	7169	4067	3234	2426	10,691	42,120	76,777	
%	9.21	9.34	5.30	4.21	3.16	13.92	45.14		

There were 5–1343 SSRs per locus. Moreover, SSRs with more than 10 repeats were the most abundant, followed by those with 10, 6, and 5 random repeats. Among the 139 different repeat types, $(A/T)_n$ was the most common (56.63%). The six other main motif types were $(AG/CT)_n$ (19.14%), $(AAG/CTT)_n$ (5.97%), $(AT/AT)_n$ (3.35%), $(AGC/CTG)_n$ (2.18%), and $(AC/GT)_n$ (2.04%) (Table S8).

3.4. Development of Polymorphic EST-SSR Markers, Analysis of Genetic Diversity, and KASP genotyping

Using Primer3, we developed 149,733 EST-SSR markers from the 49,911 SSR loci (Table S9). To verify the amplification of the EST-SSR markers, a subset of 100 EST-SSR markers was randomly chosen and tested with seven accessions from various regions in southern China (Table S10). The primers for 30 of the tested markers generated amplification products, whereas 37 primer pairs amplified nonpolymorphic products and 33 did not produce clear amplicons. The 30 polymorphic EST-SSR markers, which included 15 di-, 5 tri-, 5 tetra-, 2 penta-, and 3hexanucleotidemotif-based markers, were further verified with 46 avocado accessions. Finally, 15 polymorphic EST-SSR markers, with missing allele frequencies <10% for all 46 avocado accessions, were selected for subsequent analyses of genetic diversity (Table S11). A total of 71 alleles in the 46 avocado accessions carried the 15 polymorphic EST-SSR markers. Eight of these alleles were considered to be accession-specific and the other 63 alleles were generally found in multiple accessions (Table S11). The eight accession-specific alleles were from the following accessions: Renong No. 4, Renong, No. 5; Renong No. 6, Guiyan No. 8, Daling No. 5, Daling No. 6, RL chang, and RL yuan.

The 15polymorphic EST-SSRs were applied to evaluate diversity parameters (Table 3). The Na amplified per SSR locus varied from 2 to 10, with a mean of 4.73. The Ne varied from 1.04 to 4.39, with an average of 2.31, and Ho ranged from 0.04 to 0.93, with an average of 0.49. The He ranged from 0.04 to 0.77, with an average of 0.50, and PIC values ranged from 0.04 to 0.74, with an average of 0.45.

Table 3. Diversity parameters associated with 15 polymorphic EST-SSRs analyzed in 46 avocado accessions.

Marker Name	Transcript ID	Na [1]	Ne [2]	Ho [3]	He [4]	PIC [5]
Pa-eSSR-17	F01_cb7709_c10/f1p0/2063	8	3.02	0.61	0.67	0.62
Pa-eSSR-18	F01_cb7876_c2/f1p0/2226	10	3.09	0.61	0.68	0.65
Pa-eSSR-19	F01_cb1803_c26/f1p0/2838	6	2.04	0.63	0.51	0.46
Pa-eSSR-20	F01_cb10663_c1/f1p0/2458	3	1.87	0.50	0.46	0.40
Pa-eSSR-21	F01_cb15691_c2/f1p0/2049	3	2.41	0.50	0.58	0.50
Pa-eSSR-22	F01_cb3034_c12/f2p0/2705	5	2.85	0.67	0.65	0.60
Pa-eSSR-23	F01_cb12182_c0/f6p2/1774	3	1.47	0.28	0.32	0.29
Pa-eSSR-24	F01_cb13109_c0/f3p0/1635	5	2.80	0.48	0.64	0.58
Pa-eSSR-25	F01_cb1901_c3/f1p1/2722	2	1.04	0.04	0.04	0.04
Pa-eSSR-26	F01_cb7204_c7/f10p1/2700	3	2.65	0.93	0.62	0.55
Pa-eSSR-27	F01_cb10594_c1/f1p0/4058	3	1.40	0.33	0.29	0.27
Pa-eSSR-28	F01_cb9432_c36/f1p2/1811	5	1.56	0.43	0.36	0.33
Pa-eSSR-29	F01_cb15387_c0/f3p0/1548	8	4.39	0.49	0.77	0.74
Pa-eSSR-30	F01_cb12814_c24/f1p0/3423	4	2.67	0.53	0.62	0.55
Pa-eSSR-31	F01_cb10835_c0/f4p0/2019	3	1.33	0.28	0.25	0.22
	Total	71				
	Mean	4.73	2.31	0.49	0.50	0.45

[1] Number of observed alleles; [2] effective number of alleles; [3] observed heterozygosity; [4] expected heterozygosity; [5] polymorphism information content.

Six race-specificKASP markers were used to determine the race of 33 avocado accessions with an unknown race. The KASP genotyping results demonstrated that all 33 avocado accessions were Guatemalan × West Indian hybridsbased on the corresponding genotype of each racial avocado (Table S2).

3.5. Analyses of Genetic Relationships Based on Polymorphic EST-SSRs from SMRT Sequencing Data

A cluster analysis grouped the 46 accessions into two major sections (Figure 7). The dendrogram revealed a clear separation between the native avocado accessions from Hainan province and those from Guangxi and Yunnan provinces. In cluster I, 19 Guatemalan × West Indian hybrids were clustered into two sub-sections. Sub-cluster I-I consisted of 13native Guatemalan × West Indian hybrids from Guangxi province. Sub-cluster I-II contained two native Guatemalan × West Indian hybrids from Yunnan province.Cluster II comprised 27 Guatemalan × West Indian hybrids from Hainan province. Among these hybrids, 15 and 6were obtained from the CATAS and DLSF, respectively.

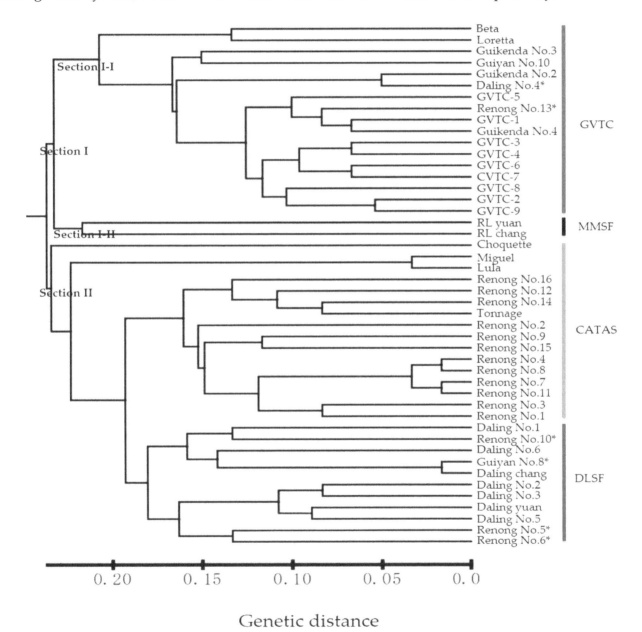

Genetic distance

Figure 7. Neighbor-joining consensus tree of 1000 bootstrap replicates revealing the phylogenetic relationships among the 46 analyzed avocado accessions based on the shared alleles for the 15 EST-SSR markers. GVTC, native avocado accessions from Guangxi Vocational and Technical College; MMSF, native avocado accessions from Mengmao State Farm; CATAS, native avocado accessions from the Chinese Academy of Tropical Agricultural Sciences; and DLSF, native avocado accessions from Daling State Farm. The native avocado accessionslabeled withan asterisk originated from other regions.

Figure 8 presents the distribution of the 46 avocado accessions for the first two principal coordinates of a principal coordinate analysis (PCoA). On the basis of the first coordinate, which accounted for 21.71% of the total variation, the accessions were generally distributed in two groups. The native avocado accessions from Hainan and Yunnan provinceswere basically grouped separately from the native avocado accessions from Guangxi province. The second coordinate accounted for 10.06% of the total variation.Finally, we observed that the native avocado accessions were generally grouped according to their geographical origins.

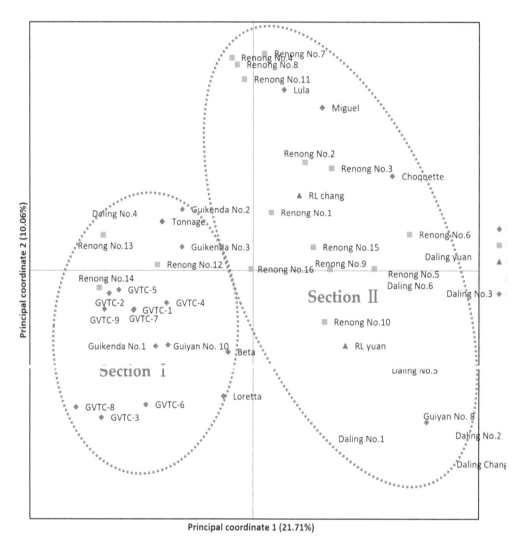

Figure 8. Principal coordinate analysis of 46 avocado accessions based on the 15 EST-SSR markers. POP1, avocado accessions fromFlorida, USA; POP2, native avocado accessions from theChinese Academy of Tropical Agricultural Sciences; POP3, native avocado accessions from Mengmao State Farm; POP4, native avocado accessions from Daling State Farm; and POP5, native avocado accessionsfrom Guangxi Vocational and Technical College.

4. Discussion

Transcriptome sequencing is a useful technique for obtaining a large number of transcripts for organisms lacking a reference sequence, at least partly because it is inexpensive and can be completed rapidly [51–53]. To date, several short-read next-generation sequencing (NGS) transcriptome databases have been developed for avocado mesocarp samples [54,55] and avocado mixed tissue samples [18,56]. However, both the number and length of the transcript sequences derived from these short-read NGS studies have hamperedtheirapplication ingenetics and molecular biology research [41]. One of the advances in sequencing technology has been the development of the long-read SMRT sequencing technique, which enables researchers to obtain a substantial number of full-length sequences from a cDNA library [32]. In the current study, we applied the PacBio SMRT system to generate and analyze the full-length transcriptome of avocado mixed mesocarp samples collected at various developmental stages. The 25.79 Gb SMRT data produced in this study provide the first comprehensive insights into the avocado mesocarp, which is the most economically valuable organ of this fruit species, and might serve as the genetic basis for future research on avocado. Interestingly, the full-length transcriptome sequence described herein is also the first such sequence for a plant species from the family Lauraceae.

In this study, 93.82% (71,627 of 76,345) of the nonredundant transcripts were annotated based on similarities with sequences in public databases. Thus, a greater proportion of transcripts were annotated in this study than in previous investigations involving NGS data for various avocado races (49.00%) [18] and for avocado mesocarp samples (57.50%) [55]. We determined that the mean length of the avocado nonredundant transcripts was2330 bp, implying that our sequences were long enough to represent full-length transcripts. Additionally, this mean length was in between the mean lengths obtained for other species, including *Z. bungeanum* (3414 bp) [32], *T. pretense* (2789 bp) [34], *M. sativa* (1706 bp) [37], *Z. planispinum* (1781 bp) [38], *C. sinensis* (1781 bp) [40], and *Arabidopsis pumila* (2194 bp) [57]. Moreover, the 76,345 nonredundant transcripts derived from the 25.79 Gb clean PacBio SMRT data produced in this study may facilitate future research on the physiology, biochemistry, and molecular genetics of avocado and related species.

A previous study indicated that lncRNAs may be important for the gene regulation in eukaryotic cells, especially during some key biological processes [58]. However, the number of lncRNAs encoded in genomes as well as their characteristics remains largely unknown [59]. Predicting and functionally annotating lncRNAs is challenging, but valuable because they are not orthologous and there is a lack of homologous sequences between closely related species [38]. Unfortunately, very few of the lncRNA functions have been elucidated [60,61]. Hence, the lncRNA information for one species is not suitable for predicting the lncRNAs in another species. In this study, 3596 avocado transcript sequences (accounting for 4.71% of the total number of nonredundant transcripts) were putatively predicted aslncRNAs. This almost completely uncharacterized gene pool may include genes associated with agronomically relevant traits related to the most economically valuable organ (mesocarp).

The accurate identification of avocado germplasm races is needed to ensure that germplasm collections are optimally used by plant breeders and farmers worldwide [1]. The traditional assignment of avocado races based on morphological traits is imprecise because of environmental effects and a limited number of applicable characteristics [17]. Molecular-based characterizations are more consistent and valid for assigning avocado genotypes. We previously confirmed the universality of six race-specific KASP markers [47]. These markers were used in the current study to identify avocado accessions with an unknown race, with implications for the application of available avocado germplasms for breeding and resource conservation. Interestingly, the KASP genotyping results revealed that all of the native avocado accessions included in this study are Guatemalan × West Indian hybrids. The reason for this observation might be related to theintroduction of avocado cultivars and the climates of the sample collection regions. First, the major avocado cultivars grown commercially are typically hybrids of three races (i.e., mainly Guatemalan × West Indian and Guatemalan × Mexican hybrids) [1]. Since the late 1950s, Guatemalan × West Indian and Guatemalan × Mexican hybrids have been brought into China from other countries for cultivation in Southern China [9]. Second, the native avocado accessions included in the present studyare mainly from three geographical regions, namely Nanning located in the central and southern region of Guangxi province, Danzhou and Baisha located in the central and western region of Hainan province, and Ruili located in the western region ofYunnan province. These locations are characterized by a warm and humid oceanic climatewith a relatively low altitude in the central and southern region of Guangxi province and the central and western region of Hainan province. Although Ruili is located in the western region ofYunnan province and far from the ocean, it still has a subtropical monsoon climate. The climates of these three regions resemble that of the areas in which theWest Indian races originated, and are favorablefor the growth of Guatemalan ×

West Indian hybrids. Therefore, Guatemalan × West Indian hybridsmay have graduallybecome the dominant native avocado accessions because of artificial selection or via naturally occurring crosses.

The 100 EST-SSR markers randomly selected for validation in the present study had an amplification rate of 67%, and 30 were determined to be polymorphic. This polymorphism level is generally consistent with that of our previous study [18]. In subsequent analyses of the genetic diversity of these polymorphic EST-SSR markers among 46 avocado accessions, 15 markers produced 4.73 alleles per locus, which was fewer than the 6.13 alleles per locus of Ge [18], the 11.40 alleles per SSR locus of Gross-German and Viruel [17], the 18.8 alleles per SSR locus of Schnell [16], and the 9.75 alleles per SSR locus of Alcaraz and Hormaza [62]. Additionally, a PIC value > 0.5 is generally considered to represent a high polymorphism rate [63]. In this study, 7 of 15 polymorphic EST-SSRs had a PIC value < 0.5. This result may have been because the 46 avocado accessions in this study are genotypically the same (Guatemalan × West Indian hybrids), with relatively low genetic diversity.

In this study, a cluster analysis and a PCoA grouped the native avocado accessions according to where they originated. Additionally, some of the native avocadoaccessions derived from different regions was included in the same sub-cluster. For example, Renong No. 13 from Hainan province clustered with the native accessions from Guangxi province. One factor leading to this promiscuous clustering is the fact that avocado germplasm resources have been exchanged among researchers and breeders since the late 1980s. The CATAS, which is a national scientific research unit, was commissioned to popularize superior avocado accessions among breeders at adjacent state farms or at other national scientific research units. Some superior native accessions from the CATAS may be the male or female parent of other native accessions from various state farms orother national scientific research units, which is consistent with our study results. Furthermore, a cluster analysis grouped two native avocado accessions from Yunnan province with the native avocado accessions from Guangxi province. In contrast, our PCoA indicated that these two native avocado accessions from Yunnan province belong to the same groupas the native avocado accessions from Hainan province. We speculate that the relatively few native avocado accessions from Yunnan province (i.e., two) may have led to these contradictory results based on two statistical analyses. At many avocado plantations in Yunnan province, the local avocado accessions have been replaced by"Hass," which is the most economically valuable avocado cultivar, ultimately making it difficult to collect local avocado accessions. Thus, maximizing the economic benefits of cultivating specific avocado cultivars, while ensuring avocado genetic resources are conserved will need to be addressed.

5. Conclusions

We annotated SMRT sequencing data based on the COG, GO, KEGG, KOG, Pfam, Swiss-Prot, eggNOG, and Nr databases. Among 71,627 transcripts, 45,134, 52,125, and 33,310 were annotated according to GO, eggNOG, and KEGG classifications, respectively. We detected 76,777 SSR loci in 42,096 transcript sequences and used them to develop 149,733 EST-SSR markers. From a randomly selected subset comprising 100 EST-SSR markers, we finally identified 15 polymorphic EST-SSR markers on 71 alleles, which had 2–10 of these markers per locus. A cluster analysis and a PCoA separated the 46 avocado accessions according to their geographical origins. These 15 newly developed EST-SSR markers may be useful for future analyses of avocado accessions and may contribute to the improved management of avocado resources for germplasm conservation and breeding programs.

Supplementary Materials
Table S1. Sources of the 46 avocado accessions evaluated in this study. Table S2. KASP primer information and KASP genotyping results. Table S3. Gene annotations of the 71,627 avocado transcripts. Table S4. Characteristics of the GO annotation of avocado transcripts. Table S5. Characteristics of eggNOG classifications of avocado transcripts. Table S6. Characteristics of KEGG pathways ofavocado transcripts. Table S7. Transcription factors identified in the avocado transcripts. Table S8. Frequencies of different repeat motifs in EST-SSRs from avocado. Table S9. Characteristics ofavocado EST-SSR markers in this study. Table S10. Summary of 100 EST-SSR markers used for amplification. Table S11. Summary of 15 EST-SSRs in 46 avocado accessions.Additional file 1. Coding sequences predicted with TransDecoder.

Author Contributions: Y.G., R.Z., and W.M. conceived and designed the experiments; J.W., Y.L. (Yuanzheng Liu), and N.W. performed the experiments; L.T. and D.C. analyzed the data; Y.L. (Yanxia Li) helped complete the experiments; X.Z. contributed materials; and Y.G. wrote the manuscript.

Acknowledgments: We gratefully acknowledge Pingzhen Lin from the Haikou Experimental Station of the Chinese Academy of Tropical Agricultural Sciences for supporting the collection of avocado resources. We thank Yajima for editing the English text of a draft of this manuscript.

References

1. Schaffer, B.; Wolstenholme, B.N.; Whiley, A.W. *The Avocado: Botany, Production and Uses*, 2nd ed.; CPI Group (UK) Ltd.: Croydon, UK, 2012.

2. Kopp, L.E. A taxonomic revision of the genus *Persea* in the western hemisphere (*Persea-Lauraceae*). *Mem. N. Y. Bot. Gard.* **1966**, *14*, 1–120.

3. Williams, L.O. The avocado, a synopsis of the genus *Persea*, subg. *Persea Econ. Bot.* **1977**, *31*, 315–320. [CrossRef]

4. Schaffer, B.; Wolstenholme, B.N. *The Avocado: Botany, Production and Uses*; CAB International: Wallingford, UK, 2002.

5. Van der Werff, H. A synopsis of *Persea* (Lauraceae) in Central America. *Novon* **2002**, *12*, 575–586. [CrossRef]

6. Dreher, M.L.; Davenport, A.J. Hass avocado composition and potential health effects. *Crit. Rev. Food Sci.* **2013**, *53*, 738–750. [CrossRef]

7. Galvão, M.D.S.; Narain, N.; Nigam, N. Influence of different cultivars on oil quality and chemical characteristics of avocado fruit. *Food Sci. Technol.* **2014**, *34*, 539–546. [CrossRef]

8. Ge, Y.; Si, X.Y.; Cao, J.Q.; Zhou, Z.X.; Wang, W.L.; Ma, W.H. Morphological characteristics, nutritional quality, and bioactive constituents in fruits of two avocado (*Perseaamericana*) varieties from hainan province, China. *J. Agric. Sci.* **2017**, *9*, 8–17. [CrossRef]

9. Ge, Y.; Si, X.Y.; Lin, X.E.; Wang, J.S.; Zang, X.P.; Ma, W.H. Advances in avocado (*Perseaamericana* Mill.). *South China Fruit* **2017**, *46*, 63–70. [CrossRef]

10. Zhang, L.; Zhang, D.S.; Liu, K.D. Environmental analysis and countermeasures for industrial development of Hainan avocado. *Chin. J. Agric. Resour. Reg. Plan.* **2015**, *36*, 78–84.

11. Fiedler, J.; Bufler, G.; Bangerth, F. Genetic relationships of avocado (*Perseaamericana* Mill.) using RAPD markers. *Euphytica* **1998**, *101*, 249–255. [CrossRef]

12. Mhameed, S.; Sharon, D.; Kaufman, D.; Lahav, E.; Hillel, J.; Degani, C.; Lavi, U. Genetic relationships within avocado (*Perseaamericana* Mill.) cultivars and between *Persea* species. *Theor. Appl. Genet.* **1997**, *94*, 279–286. [CrossRef]

13. Furnier, G.R.; Cummings, M.P.; Clegg, M.T. Evolution of the avocados as revealed by DNA restriction site variation. *J. Hered.* **1990**, *81*, 183–188. [CrossRef]

14. Davis, J.; Henderson, D.; Kobayashi, M. Genealogical relationships among cultivated avocado as revealed through RFLP analysis. *J. Hered.* **1998**, *89*, 319–323. [CrossRef]

15. Ashworth, V.E.T.M.; Clegg, M.T. Microsatellite markers in avocado (*Perseaamericana* Mill.). genealogical relationships among cultivated avocado genotypes. *J. Hered.* **2003**, *94*, 407–415. [CrossRef]

16. Schnell, R.J.; Brown, J.S.; Olano, C.T.; Power, E.J.; Krol, C.A.; Kuhn, D.N.; Motamayor, J.C. Evaluation of avocado germplasm using microsatellite markers. *J. Am. Soc. Hortic. Sci.* **2003**, *128*, 881–889. [CrossRef]

17. Gross-German, E.; Viruel, M.A. Molecular characterization of avocado germplasm with a new set of SSR and EST-SSR markers: Genetic diversity, population structure, and identification of race-specific markers in a group of cultivated genotypes. *Tree Genet. Genomes* **2013**, *9*, 539–555. [CrossRef]

18. Ge, Y.; Tan, L.; Wu, B.; Wang, T.; Zhang, T.; Chen, H.; Zou, M.; Ma, F.; Xu, Z.; Zhan, R. Transcriptome sequencing of different avocado ecotypes: De novo transcriptome assembly, annotation, identification and validation of EST-SSR markers. *Forests* **2019**, *10*, 411. [CrossRef]

19. Chen, H.; Morrel, P.L.; Ashwoth, V.E.T.M.; De la Cruz, M.; Clegg, M.T. Nucleotide diversity and linkage disequilibrium in wild avocado (*Perseaamericana* Mill.). *J. Hered.* **2008**, *99*, 382–389. [CrossRef]

20. Chen, H.; Morrel, P.L.; Ashwoth, V.E.T.M.; De la Cruz, M.; Clegg, M.T. Tracing the geographic origins of mayor avocado cultivars. *J. Hered.* **2009**, *100*, 56–65. [CrossRef]

21. Hou, M.Y.; Mu, G.J.; Zhang, Y.J.; Cui, S.L.; Yang, X.L.; Liu, L.F. Evaluation of total flavonoid content and analysis of related EST-SSR in Chinese peanut germplasm. *Crop Breed. Appl. Biotechnol.* **2017**, *17*, 221–227. [CrossRef]

22. Azevedo, A.O.N.; Azevedo, C.D.O.; Santos, P.H.A.D.; Ramos, H.C.C.; Boechat, M.S.B.; Arêdes, F.A.S.; Ramos, S.R.R.; Mirizola, L.A.; Perera, L.; Aragão, W.M.; et al. Selection of legitimate dwarf coconut hybrid seedlings using DNA fingerprinting. *Crop Breed. Appl. Biotechnol.* **2018**, *18*, 409–416. [CrossRef]

23. Ahmad, A.; Wang, J.D.; Pan, Y.B.; Sharif, R.; Gao, S.J. Development and use of simple sequence repeats (SSRs) markers for sugarcane breeding and genetic studies. *Agronomy* **2018**, *8*, 260. [CrossRef]

24. Ferreira, F.; Scapim, C.A.; Maldonado, C.; Mora, F. SSR-based genetic analysis of sweet corn inbred lines using artificial neural networks. *Crop Breed. Appl. Biotechnol.* **2018**, *18*, 309–313. [CrossRef]

25. Ge, Y.; Hu, F.C.; Tan, L.; Wu, B.; Wang, T.; Zhang, T.; Ma, F.N.; Cao, J.Q.; Xu, Z.N.; Zhan, R.L. Molecular diversity in a germplasm collection of avocado accessions from the tropical and subtropical regions of China. *Crop Breed. Appl. Biotechnol.* **2019**, *19*, 153–160. [CrossRef]

26. Wu, J.; Cai, C.F.; Cheng, F.Y.; Cui, H.L.; Zhou, H. Characterization and development of EST-SSR markers in tree peony using transcriptome sequences. *Mol. Breed.* **2014**, *34*, 1853–1866. [CrossRef]

27. Biswas, M.K.; Nath, U.K.; Howlader, J.; Bagchi, M.; Natarajan, S.; Kayum, M.A.; Kim, H.T.; Park, J.I.; Kang, J.G.; Nou, I.S. Exploration and exploitation of novel SSR markers for candidate transcription factor genes in Lilium species. *Genes* **2018**, *9*, 97. [CrossRef]

28. Ma, S.L.Y.; Dong, W.S.; Lyu, T.; Lyu, Y.M. An RNA sequencing transcriptome analysis and development of EST-SSR markers in Chinese hawthorn through Illumina sequencing. *Forests* **2019**, *10*, 82. [CrossRef]

29. Li, X.; Li, M.; Hou, L.; Zhang, Z.Y.; Pang, X.M.; Li, Y.Y. De novo transcriptome assembly and population genetic analyses for an endangered Chinese endemic *Acer miaotaiense* (Aceraceae). *Genes* **2018**, *9*, 378. [CrossRef]

30. Qi, W.C.; Chen, X.; Fang, P.H.; Shi, S.C.; Li, J.J.; Liu, X.T.; Cao, X.Q.; Zhao, N.; Hao, H.Y.; Li, Y.J.; et al. Genomic and transcriptomic sequencing of *Rosa hybrida* provides microsatellite markers for breeding, flower trait improvement and taxonomy studies. *BMC Plant Biol.* **2018**, *18*, 119. [CrossRef]

31. Chen, J.; Tang, X.H.; Ren, C.X.; Wei, B.; Wu, Y.Y.; Wu, Q.H.; Pei, J. Full-length transcriptome sequences and the identification of putative genes for flavonoid biosynthesis in safflower. *BMC Genom.* **2018**, *19*, 548. [CrossRef]

32. Tian, J.Y.; Feng, S.J.; Liu, Y.L.; Zhao, L.L.; Tian, L.; Hu, Y.; Yang, T.X.; Wei, A.Z. Single-molecule long-read sequencing of *Zanthoxylumbungeanum* maxim. transcriptome: Identification of aroma-related genes. *Forests* **2018**, *9*, 765. [CrossRef]

33. Roberts, R.J.; Carneiro, M.O.; Schatz, M.C. The advantages of SMRT sequencing. *Genome Biol.* **2013**, *14*, 405–409. [CrossRef]

34. Chao, Y.H.; Yuan, J.B.; Li, S.F.; Jia, S.Q.; Han, L.B.; Xu, L.X. Analysis of transcripts and splice isoforms in red clover (*Trifoliumpratense* L.) by single-molecule long-read sequencing. *BMC Plant Biol.* **2018**, *18*, 300. [CrossRef]

35. Hoang, N.V.; Furtado, A.; Mason, P.J.; Marquardt, A.; Kasirajan, L.; Thirugnanasambandam, P.P.; Botha, F.C.; Henry, R.J. A survey of the complex transcriptome from the highly polyploid sugarcane genome using full-length isoform sequencing and de novo assembly from short read sequencing. *BMC Genom.* **2017**, *18*, 395. [CrossRef]

36. Zuo, C.M.; Blow, M.; Sreedasyam, A.; Kuo, R.C.; Ramamoorthy, G.K.; Torres-Jerez, I.; Li, G.F.; Wang, M.; Dilworth, D.; Barry, K.; et al. Revealing the transcriptomic complexity of switchgrass by PacBio long-read sequencing. *Biotechnol. Biofuels* **2018**, *11*, 170. [CrossRef]

37. Chao, Y.H.; Yuan, J.B.; Guo, T.; Xu, L.X.; Mu, Z.Y.; Han, L.B. Analysis of transcripts and splice isoforms in *Medicago sativa* L. by single-molecule long-read sequencing. *Plant Mol. Biol.* **2019**, *99*, 219–235. [CrossRef]

38. Kim, J.A.; Roy, N.S.; Lee, I.H.; Choi, A.Y.; Choi, B.S.; Yu, Y.S.; Park, N.I.; Park, K.C.; Kim, S.; Yang, H.S.; et al. Genome-wide transcriptome profiling of the medicinal plant *Zanthoxylumplanispinum* using a single-molecule direct RNA sequencing approach. *Genomics* **2019**, *111*, 973–979. [CrossRef]

39. Zhang, B.; Liu, J.X.; Wang, X.S.; Wei, Z.W. Full-length RNA sequencing reveals unique transcriptome composition in bermudagrass. *Plant Physiol. Biochem.* **2018**, *132*, 95–103. [CrossRef]

40. Xu, Q.S.; Zhu, J.Y.; Zhao, S.Q.; Hou, Y.; Li, F.D.; Tai, Y.L.; Wan, X.C.; Wei, C.L. Transcriptome profiling using single-molecule direct RNA sequencing approach for in-depth understanding of genes in secondary metabolism pathways of *Camellia sinensis*. *Front. Plant Sci.* **2017**, *8*, 1205. [CrossRef]

41. Deng, Y.; Zheng, H.; Yan, Z.C.; Liao, D.Y.; Li, C.L.; Zhou, J.Y.; Liao, H. Full-length transcriptome survey and

expression analysis of *Cassia obtusifolia* to discover putative genes related to aurantio-obtusin biosynthesis, seed formation and development, and stress response. *Int. J. Mol. Sci.* **2018**, *19*, 2476. [CrossRef]

42. Ge, Y.; Cheng, Z.H.; Si, X.Y.; Ma, W.H.; Tan, L.; Zang, X.P.; Wu, B.; Xu, Z.N.; Wang, N.; Zhou, Z.X.; et al. Transcriptome profiling provides insight into the genes in carotenoid biosynthesis during the mesocarp and seed developmental stages of avocado (*Persea Americana*). *Int. J. Mol. Sci.* **2019**, *20*, 4117. [CrossRef]

43. Ge, Y.; Ramchiary, N.; Wang, T.; Liang, C.; Wang, N.; Wang, Z.; Choi, S.R.; Lim, Y.P.; Piao, Z.Y. Development and linkage mapping of unigene-derived microsatellite markers in *Brassica rapa* L. *Breed. Sci.* **2011**, *61*, 160–167. [CrossRef]

44. Kanehisa, M.; Araki, M.; Goto, S.; Hattori, M.; Hirakawa, M.; Itoh, M.; Katayama, T.; Kawashima, S.; Okuda, S.; Tokimatsu, T.; et al. KEGG for linking genomes to life and the environment. *Nucleic Acids Res.* **2008**, *36*, 480–484. [CrossRef]

45. Götz, S.; García-Gómez, J.M.; Terol, J.; Williams, T.D.; Nagaraj, S.H.; Nueda, M.J.; Robles, M.; Talon, M.; Dopazo, J.; Conesa, A. High-throughput functional annotation and data mining with the Blast2GO suite. *Nucleic Acids Res.* **2008**, *36*, 3420–3435. [CrossRef]

46. Perez-Rodriguez, P.; Riano-Pachon, D.M.; Correa, L.G.; Rensing, S.A.; Kersten, B.; Mueller-Roeber, B. PlnTFDB: Updated content and new features of the plant transcription factor database. *Nucleic Acids Res.* **2010**, *38*, 822–827. [CrossRef]

47. Ge, Y.; Zhang, T.; Wu, B.; Tan, L.; Ma, F.N.; Zou, M.H.; Chen, H.H.; Pei, J.L.; Liu, Y.Z.; Chen, Z.H.; et al. Genome-wide assessment of avocado germplasm determined from specific length amplified fragment sequencing and transcriptomes: Population structure, genetic diversity, identification, and application of race-specific markers. *Genes* **2019**, *10*, 215. [CrossRef]

48. Krawczak, M.; Nikolaus, S.; von Eberstein, H.; Croucher, P.J.; El Mokhtari, N.E.; Schreiber, S. PopGen: Population based recruitment of patients and controls for the analysis of complex genotype-phenotype relationships. *Community Genet.* **2006**, *9*, 55–61. [CrossRef]

49. Liu, K.; Muse, S.V. PowerMarker: An integrated analysis environment for genetic marker analysis. *Bioinformatics* **2005**, *21*, 2128–2129. [CrossRef]

50. Tamura, K.; Stecher, G.; Peterson, D.; Filipski, A.; Kumar, S. MEGA6: Molecular evolutionary genetics analysis version 6.0. *Mol. Biol. Evol.* **2013**, *30*, 2725–2729. [CrossRef]

51. Du, M.; Li, N.; Niu, B.; Liu, Y.; You, D.; Jiang, D.; Ruan, C.Q.; Qin, Z.Q.; Song, T.W.; Wang, W.T. De novo transcriptome analysis of *Bagariusyarrelli* (Siluriformes: Sisoridae) and the search for potential SSR markers using RNA-Seq. *PLoS ONE* **2018**, *13*, e0190343. [CrossRef]

52. Liu, F.M.; Hong, Z.; Yang, Z.J.; Zhang, N.N.; Liu, X.J.; Xu, D.P. De Novo transcriptomeanalysis of *Dalbergiaodorifera* T. Chen (Fabaceae) and transferability of SSR markers developed from the transcriptome. *Forests* **2019**, *10*, 98. [CrossRef]

53. Li, W.; Zhang, C.P.; Jiang, X.Q.; Liu, Q.C.; Liu, Q.H.; Wang, K.L. De Novotranscriptomicanalysis and development of EST–SSRs for *Styrax japonicas*. *Forests* **2018**, *9*, 748. [CrossRef]

54. Kilaru, A.; Cao, X.; Dabbs, P.B.; Sung, H.J.; Rahman, M.M.; Thrower, N.; Zynda, G.; Podicheti, R.; Ibarra-Laclette, E.; Herrera-Estrella, L.; et al. Oil biosynthesis in a basal angiosperm: Transcriptome analysis of *Persea Americana* mesocarp. *BMC Plant Biol.* **2015**, *15*, 203. [CrossRef]

55. Vergara-Pulgar, C.; Rothkegel, K.; González-Agüero, M.; Pedreschi, R.; Campos-Vargas, R.; Defilippi, B.G.; Meneses, C. De novo assembly of *Perseaamericana* cv. 'Hass' transcriptome during fruit development. *BMC Genom.* **2019**, *20*, 108. [CrossRef]

56. Ibarra-Laclette, E.; Méndez-Bravo, A.; Pérez-Torres, C.A.; Albert, V.A.; Mockaitis, K.; Kilaru, A.; López-Gómez, R.; Cervantes-Luevano, J.I.; Herrera-Estrell, L. Deep sequencing of the Mexican avocado transcriptome, an ancient angiosperm with a high content of fatty acids. *BMC Genom.* **2015**, *16*, 599. [CrossRef]

57. Yang, L.F.; Jin, Y.H.; Huang, W.; Sun, Q.; Liu, F.; Huang, X.Z. Full-length transcriptome sequences of ephemeral plant *Arabidopsis pumila* provides insight into gene expression dynamics during continuous salt stress. *BMC Genom.* **2018**, *19*, 717. [CrossRef]

58. Liu, J.; Wang, H.; Chua, N.H. Long noncoding RNA transcriptome of plants. *Plant Biotechnol. J.* **2015**, *13*, 319–328. [CrossRef]

59. Yandell, M.; Ence, D. A beginner's guide to eukaryotic genome annotation. *Nat. Rev. Genet.* **2012**, *13*, 329–342. [CrossRef]

60. Liu, J.; Jung, C.; Xu, J.; Wang, H.; Deng, S.; Bernad, L.; Arenas-Huertero, C.; Chua, N.H. Genome-wide analysis uncovers regulation of long intergenic noncoding RNAs in *Arabidopsis*. *Plant Cell* **2012**, *24*, 4333–4345. [CrossRef]

61. Ochogavía, A.; Galla, G.; Seijo, J.G.; González, A.M.; Bellucci, M.; Pupilli, F.; Barcaccia, G.; Albertini, E.; Pessino, S. Structure, target-specifificity and expression of *PN_LNC_N13*, a long non-coding RNA differentially expressed in apomictic and sexual *Paspalumnotatum*. *Plant Mol. Biol.* **2018**, *96*, 53–67. [CrossRef]

62. Alcaraz, M.L.; Hormaza, J.I. Molecular characterization and genetic diversity in an avocado collection of cultivars and local Spanish genotypes using SSRs. *Heredity* **2007**, *144*, 244–253. [CrossRef]

63. Botstein, D.; White, R.L.; Skolnick, M.; Davis, R.W. Construction of a genetic linkage map in man using restriction fragment length polymorphisms. *Am. J. Hum. Genet.* **1980**, *32*, 314–331.

10

Cost-Effective and Time-Efficient Molecular Assisted Selection for Ppv Resistance in Apricot based on *ParPMC2* Allele-Specific PCR

Ángela Polo-Oltra [1], Carlos Romero [2], Inmaculada López [1], María Luisa Badenes [1]

[1] Citriculture and Plant Production Center, Instituto Valenciano de Investigaciones Agrarias (IVIA), CV-315, km 10.7, 46113 Moncada, Valencia, Spain; polo_ang@externos.gva.es (Á.P.-O.); lopez_inmcap@gva.es (I.L.); badenes_mlu@gva.es (M.L.B.)

[2] Instituto de Biología Molecular y Celular de Plantas (IBMCP), Consejo Superior de Investigaciones Científicas (CSIC)—Universidad Politécnica de Valencia (UPV), Ingeniero Fausto Elio s/n, 46022 Valencia, Spain; cromero@ibmcp.upv.es

* Correspondence: garcia_zur@gva.es

Abstract: Plum pox virus (PPV) is the most important limiting factor for apricot (*Prunus armeniaca* L.) production worldwide, and development of resistant cultivars has been proven to be the best solution in the long-term. However, just like in other woody species, apricot breeding is highly time and space demanding, and this is particularly true for PPV resistance phenotyping. Therefore, marker-assisted selection (MAS) may be very helpful to speed up breeding programs. Tightly linked *ParPMC1* and *ParPMC2*, meprin and TRAF-C homology (MATH)-domain-containing genes have been proposed as host susceptibility genes required for PPV infection. Contribution of additional genes to PPV resistance cannot be discarded, but all available studies undoubtedly show a strong correlation between *ParPMC2*-resistant alleles (*ParPMC2res*) and PPV resistance. The *ParPMC2res* allele was shown to carry a 5-bp deletion (*ParPMC2*-del) within the second exon that has been characterized as a molecular marker suitable for MAS (PMC2). Based on this finding, we propose here a method for PPV resistance selection in apricot by combining high-throughput DNA extraction of 384 samples in 2 working days and the allele-specific genotyping of PMC2 on agarose gel. Moreover, the PMC2 genotype has been determined by PCR or by using whole-genome sequences (WGS) in 175 apricot accessions. These results were complemented with phenotypic and/or genotypic data available in the literature to reach a total of 325 apricot accessions. As a whole, we conclude that this is a time-efficient, cost-effective and straightforward method for PPV resistance screening that can be highly useful for apricot breeding programs.

Keywords: apricot; MAS; breeding; MATH; PPV resistance; agarose; *ParPMC*; *ParPMC2*-del

1. Introduction

Most cultivated apricots belong to the *Prunus armeniaca* L. species, a member of the Rosaceae family, *Prunus* genus and section Armeniaca (Lam.) Koch [1]. World apricot production reached 3.84 million tonnes in 2018, with Turkey, Uzbekistan and Iran as the main producers (http://www.fao.org/faostat/). This means an increase of about 45% since 1998 mainly due to Asian countries. By contrast, European production in this period has just increased slightly while the cultivated area declined up to 19%. Despite its wide geographical spread, apricot has very specific ecological requirements. Consequently, each region usually grows locally adapted cultivars. For this reason, significant breeding efforts have been undertaken since the first apricot breeding program started in 1925 at the Nikita Botanical Garden in Yalta (Crimea, Ukraine) [2]. However, apricot breeding based on biparental controlled crosses and subsequent selection of the best new allelic combinations is hardly limited by the capacity to evaluate trees in the field [3]. On one side, fruit trees show high space requirements to be grown. On the other, their juvenile phase is quite long and reliable pomological phenotyping requires several cropping seasons, which means that at least ten years are needed to release a new variety. Therefore, the implementation of marker-assisted selection (MAS) has a great potential to improve breeding efficiency in fruit trees, including apricot.

Sharka disease, caused by *Plum pox* virus (PPV), is currently the most important viral disease affecting stone fruit trees (*Prunus* spp.) [4]. To date, nine PPV strains (D, M, C, EA, W, Rec, T, CR and An) are identified [5]. However, PPV genetic diversity may be even bigger, as observed by Chirkov et al. [6], who recently described the new Tat isolates affecting sour cherry (*Prunus cerasus*). PPV-D and M are the most widespread and economically important strains [5,7]. A clear host preference is observed: PPV-D/plum/apricot and PPV-M/peach. However, underlying genetic determinants are still unknown [8].

Particularly in apricot, PPV-D has severely hindered production in the last three decades, especially in endemic areas. In this context, development of PPV-resistant varieties is the main objective of apricot breeding programs. However, resistant sources are scarce. Just a handful of North American PPV-resistant cultivars have been identified to date, and they are commonly used as donors in all apricot resistance breeding programs currently in progress [9]. Several independent works aimed at dissecting the genetic control of PPV resistance in apricot have identified the major dominant *PPVres* locus in the upper part of linkage group 1 [10–17]. According to the pedigree and fine mapping data, a single common ancestor carrying *PPVres* has been suggested for all PPV-resistant cultivars [16,18–20]. Moreover, other minor loci contributing to PPV resistance have been suggested [13–16], but their role has not yet been well defined. More recently, transcriptomic and genomic analyses of *PPVres* locus have pointed out *ParPMC1* and *ParPMC2*, two members of a cluster of meprin and TRAF-C homology domain (MATHd)-containing genes, as host susceptibility paralogous genes required for PPV infection [21]. The *ParPMC2* allele linked in coupling with PPV resistance (*ParPMC2res*) accumulates 15 variants, including a 5 nt deletion (*ParPMC2*-del) that results in a premature stop codon. Moreover, cultivars carrying the *ParPMC2res* allele show that *ParPMC2* and especially *ParPMC1* genes are downregulated. As a result, this *ParPMC2res* was proposed to be a pseudogene that confers PPV resistance by silencing functional homologs, the non-mutated *ParPMC2* allele and/or *ParPMC1*. Another plausible scenario involves epigenetic modifications to explain *ParPMC* silencing in the resistant cultivars [22].

In spite of evidence supporting linkage with the *PPVres* locus, some genotype-phenotype incongruencies (GPIs) have been detected in biparental populations segregating for PPV resistance [17,23,24]. In other words, some phenotypically susceptible individuals carrying *ParPMC2res* were classified as genetically resistant. Possible causes underlying these discrepancies, including other loci contributing to PPV resistance, are still unresolved. However, the potential benefit of using a *ParPMC2* allele-specific marker (PMC2) for MAS is still very high since sharka resistance phenotyping is a major bottleneck in apricot breeding programs. The most reliable method for apricot PPV resistance phenotyping is based on a biological test that uses GF-305 peach rootstocks as woody indicators and graft-inoculation with PPV [25]. This procedure is time-consuming and requires visual inspection during two to four growing seasons in several replicates per genotype followed by ELISA [26] and RT-PCR tests [27]. It should be noted that the plant to be tested must be of a significant size in order to have enough buds for grafting replicates, so it takes a couple of years from the time of crossing. As a result of a genetic mapping approach, Soriano et al. [18] reported the first successful MAS application for PPV resistance using 3 SSRs within the *PPVres* locus resolved by capillary electrophoresis. Afterwards, these SSRs were combined with a single sequence length polymorphism marker (ZP002) interrogating the *ParPMC2*-del resolved by capillary or acrylamide electrophoresis [24] and by high resolution melting [28]. However, specialized DNA testing services are needed to adopt these MAS approaches, and together with the economic costs, this could be a challenge [29].

Here, we report a method combining high-throughput DNA extraction of 384 samples in 2 days and PMC2 genotyping by allele-specific PCR amplification and agarose gel electrophoresis. This method is proven to be an easily implemented tool for MAS of PPV-resistant seedlings in almost any apricot breeding program. Therefore, bioassays for PPV resistance evaluation will be needed to confirm the phenotype in selected materials. Moreover, PMC2 genotype has been determined and/or revised for 325 worldwide cultivated apricot accessions providing useful information for breeders to select parental genotypes.

2. Materials and Methods

2.1. High-Throughput DNA Isolation in 96-Well Plate

The genomic DNA extraction protocol was optimized from the original Doyle and Doyle method [30] to manage 384 samples per isolation using 8-well 1.2-mL strip tubes (VWR International). For each accession, 2 leaf discs were collected and placed into a tube with 3 glass beads (VWR International). The strips were frozen in liquid N2 and stored at −20 °C before DNA isolation. Frozen tissue was ground for 1 min with a frequency of 26/s using a Qiagen TissueLyser 85210 (Qiagen, Hilden, Germany). Then, 340 µL of preheated CTAB isolation buffer (with 0.2% 2-mercaptoethanol) was added to the ground tissue and incubated at 65 °C for 40 min, shaking gently every 10 min. After a short spin, 340 µL of chloroform-isoamyl alcohol (24:1) was added and mixed inverting the plates. Tubes were centrifuged for 10 min at 3000 rpm and 4 °C. The clean aqueous phase was transferred to new strip tubes, and 1.5 vol of 100% ethanol and 15 mM ammonium acetate were added and mixed gently. After overnight incubation at −20 °C, tubes were centrifuged for 10 min at 3000 rpm at 4 °C. The supernatant was discarded inverting the tubes, and 300 µL of 70% ethanol was added. After centrifugation for 10 min at 3000 rpm at 4 °C, the supernatant was discarded and finally 75 µL of TE was added. DNA at 1:10 dilution was used for PCR. Some random DNA samples from each plate were subjected to quality control. DNA integrity was checked on an agarose gel, and quantification was performed using a Nanodrop ND-1000 spectrophotometer (Nanodrop Technologies, Wilmington, DE, USA).

2.2. PMC2 Genotype by Allele-Specific PCR Assay

PMC2 marker genotyping was performed using the allele-specific forward primer (PMC2-F-alleleR: 5′-GTCATTTTCATTGATGTCATTCA-3′ or PMC2-F-alleleS: 5′-GTCATTTTCATTGATGTCATTCA -3′) and one common reverse primer (PMC2-R: 5′- GTCATTTTCATTGATGTCATTCA -3′), as described by Zuriaga et al. [21]. PCRs were performed in a final volume of 20 µl containing 1 × DreamTaq buffer, 0.2 mM of each dNTP, 5 µM of each primer, 1 U of DreamTaq DNA polymerase (Thermo Fisher) and 2 µL of DNA extraction (diluted 1:10). Cycling conditions were as follows: an initial denaturing of 95 °C for 5 min; 35 cycles of 95 °C for 30 s, 55 °C for 45 s and 72 °C for 45 s; and a final extension of 72 °C for 10 min. PCR products were electrophoresed in 1% (w/v) agarose gels.

Available DNA samples from 120 apricot cultivars and accessions were PCR screened in this work. Part of this collection is currently kept at the collection of the Instituto Valenciano de Investigaciones Agrarias (IVIA) in Valencia (Spain), while other samples were provided by the Departamento de Mejora y Patología Vegetal del CEBAS-CSIC in Murcia (Spain), the University of St. Istvan (Budapest, Hungary) or by SharCo project (FP7-KBBE-2007-1) partners.

2.3. WGS Mapping and PMC2 Screening

WGSs of 73 cultivars were used in this study. Twenty-four of these WGSs and the 454 sequenced BAC clones belonging to the "Goldrich" *PPVres* locus R-haplotype were already screened in our previous works [20–31]. The other 49 WGSs were downloaded from the SRA repository (https://www.ncbi.nlm.nih.gov/sra). All raw reads were processed using the "run_trimmomatic_qual_trimming.pl" script from the Trinity software [32]. After removing the low-quality regions as well as vector and adaptor contaminants, cleaned reads were aligned to the peach genome v.2.0.a1 [33] using Bowtie2 v.2.2.4 software [34]. The presence/absence of the *ParPMC2*-del was visually inspected using IGV v.2.4.16 [35].

3. Results and Discussion

3.1. High-Throughput DNA Extraction and ParPMC2-del Genotyping for MAS

MAS offers great advantages over traditional seedling selection based just on phenotypic evaluations in fruit breeding [36]. DNA tests in segregating populations can improve the cost efficiency and/or the genetic gain for each seedling selection cycle [29], allowing to identify a few seedlings from among many thousands that have the genetic potential for desired performance levels [37]. As a result, agronomical evaluation in field trials is restricted to the promising selected materials. Implementation of MAS is especially valuable for traits that are difficult and/or expensive to phenotype as PPV resistance. As previously explained, the most reliable PPV resistance phenotyping is based on a biological test that uses graft-inoculated GF-305 peach seedlings [25] (Figure 1A). This protocol requires several replicates per genotype and visual symptoms inspection during 2–4 growing seasons, which entails the main bottleneck in apricot breeding programs. For instance, following this method at the IVIA's greenhouse and cold chamber facilities, we can phenotype no more than 3000 plants per year, which equals 500 seedlings (i.e., 6 replicates are needed for each seedling).

In this work, we present a new strategy to speed up while reducing costs of the current application of MAS for PPV resistance in apricot [18,24,28]. Here, we combine a high-throughput DNA extraction protocol that does not need sophisticated robotic systems and can be implemented in any regular laboratory, with PMC2 allele-specific PCR amplification using previously described primers [21] and agarose electrophoresis (Figure 1B). Both forwards primers differ at the 3′-end, allowing to easily discriminate the presence/absence of the 5-bp *ParPMC2*-del (Figure 2). With this DNA extraction method, one person can easily process up to 384 samples (four 96-well sample plates) in 2 working days, enabling high throughput sample preparation. This is 4 times more samples than a standard CTAB method using individual tubes, while the cost of reagents and consumables is similar in both cases (around 0.29–0.30 € per sample) (Table S2). DNA obtained has enough quantity and quality to ensure subsequent regular PCRs. A 1:10 dilution of the DNA obtained was directly used for PCR amplification, without any additional purification step. In contrast, commercial kits are much more expensive in terms of reagents and consumables with costs around 4€ per sample. Then, using this DNA, 3 different methods could be applied for PPV MAS in apricot: the fluorescent labelling of PCR fragments that are resolved using capillary electrophoresis [18], the high-resolution melting (HRM) approach [28], and the use of standard PCR resolved by agarose gel electrophoresis [21]. It should be noted that the first two methods require the use of special equipment that could not be available for some laboratories and that also make the protocol more expensive. For instance, just the capillary electrophoresis costs around 1.5–2€ per sample (PCR not included) and the fluorescently labelled primers needed for PCR (136€ 10 nm) are much more expensive than the non-labelled ones (4€ 20 nm). On the other hand, commercial kits for HRM are not very expensive (around 1€ per sample) but requires the use of real-time PCR machines specially calibrated for this type of experiments and the analysis software. As a resume, although prices differ between laboratories or countries, our rough estimate of the cost points to first and second approaches as 13 and 8 times more expensive, respectively, in terms of reagents and consumables than the protocol proposed in this work (Table S2).

Practical advantages of PMC2 genotyping over classical phenotyping may be illustrated by the following example (Figure 1). The estimated time needed for evaluating 1000 samples at the IVIA's facilities using bioassays is about 16 months (500 samples/8 months), taking into account that plants should be big enough to be ready-to-graft (approximately 2 years old). In contrast, just about 4 weeks are needed to conduct PMC2 genotyping just after seed germination. This estimated time was calculated assuming a 40-h workweek. As 1000 samples could be distributed into 10.4 96-well plates, ideally the DNA extraction would need 5.2 days (4 plates each 2 days), the 2 allele-specific PCRs would need 7.8 days (3 h each plate) and the agarose electrophoresis would last 2.6 days (2 PCR 96-well plates and 2 h per gel). In total, we would need 15.6 working days to genotype 1000 samples. This improvement removes the phenotyping bottleneck since all seedlings obtained from a particular cross can be PCR screened that same year. Hence, this quick and high-throughput method for DNA testing is expected to have an important effect on the cost efficiency of MAS, as suggested by Edge-Garza et al. [37].

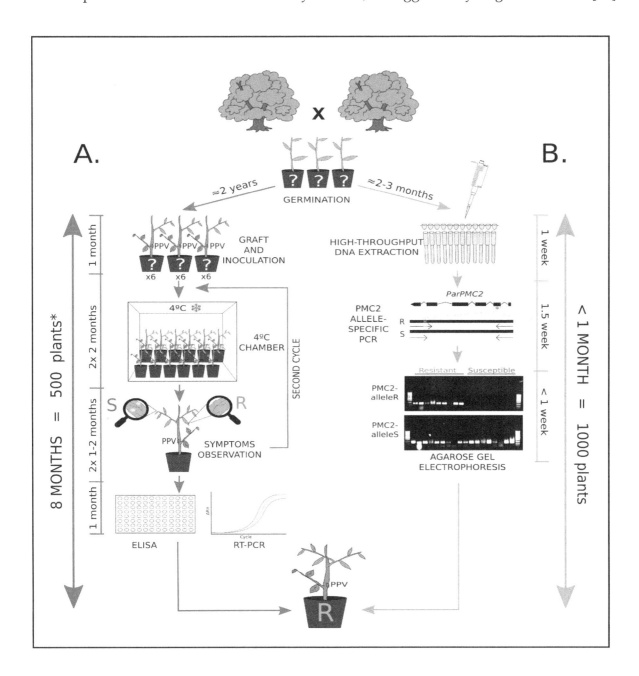

Figure 1. Comparison between traditional Plum pox virus (PPV) resistance phenotyping (**A**) and high-throughput marker-assisted selection (MAS) based on PMC2 allele-specific PCR (**B**). (*) Estimated duration based on Instituto Valenciano de Investigaciones Agrarias (IVIA) facilities.

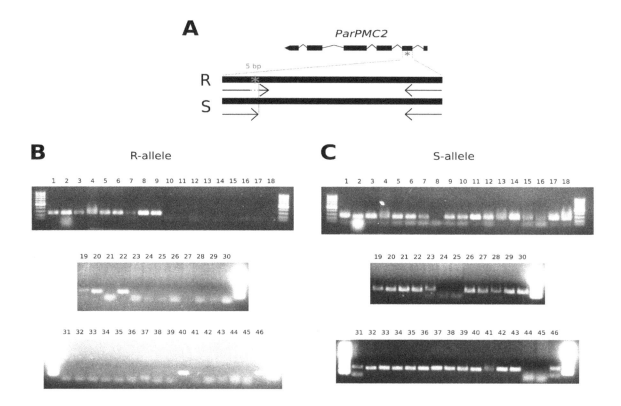

Figure 2. PMC2 genotyping by allele-specific PCR using forward primers differing at the 3′-end (**A**): R-allele (**B**) and S-allele (**C**) amplifications in 1% agarose gel electrophoresis for 46 apricot accessions (1: Goldrich, 2: Harlayne, 3: Henderson, 4: Lito, 5: Orange Red, 6: Pandora, 7: SEO, 8: Stella, 9: Veecot, 10: Bebeco, 11: Bergeron, 12: Canino, 13: Currot, 14: Ginesta, 15: Katy, 16: Mitger, 17: Palau, 18: Tyrinthos, 19: Piera, 20: Selene, 21: Colorao, 22: Moixent, 23: Perla, 24: Dama Vermella, 25: Maravilla, 26: Ninfa, 27: Palabras, 28: Sublime, 29: Dorada, 30: Castlebrite, 31: Martinet, 32: Corbató, 33: Gandía, 34: Cristalí, 35: Manri, 36: Gavatxet, 37: Pisana, 38: Xirivello, 39: Velazquez, 40: Mirlo Rojo, 41: Rojo Carlet, 42: Bulida, 43: ASP, 44: Silvercot, 45: Bora and 46: Roxana).

3.2. ParPMC2-del Highly Correlates with PPV Resistance in Apricot Germplasm

One of the main pillars of plant breeding relies on skilful parental selection to create new genetic variation by controlled crossing. Usually, breeders just connect the concept of DNA-informed breeding with the use of molecular markers for seedling selection, but it also can be very helpful for parental selection [36]. This is the case in apricot breeding for PPV resistance. Two decades ago, Martínez-Gómez et al. [9] reviewed phenotypic information regarding apricot cultivar behaviour against PPV. Similarly, here, we compile the PMC2 genotype of a wide set of apricot accessions to facilitate parental selection tasks incorporating also their resistance phenotype, pedigree and origin data from the literature when available. The PPV strain used for phenotyping was also included because differences in severity of the induced symptoms have been observed [10,16]. As a result, after screening 120 accessions by PCR and other 49 by WGS and reviewing the available literature, PMC2 genotype was determined in a total of 325 apricot cultivars or accessions that represent a wide range of geographic origins (Figure 3). A significant part of the materials come from European countries directly involved in PPV resistance research during the last decades, such as Italy (20.9%), Spain (15.7%) or France (14.8%) [38–42]. Regarding viral strain, PPV-M was more frequently used for phenotyping except for PPV-D in Spain and PPV-T in Turkey (Figure 3), in agreement with the prevalence of these two strains in every country [5,43].

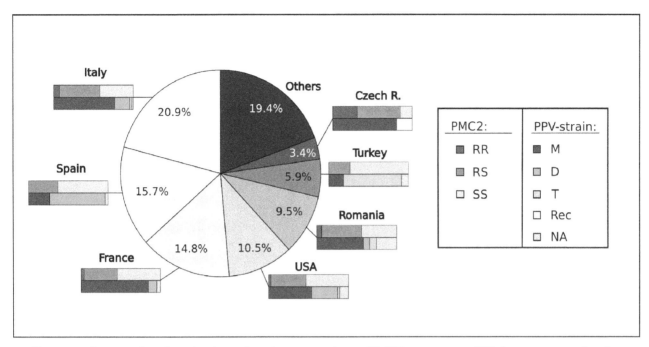

Figure 3. Geographic distribution of apricot accessions: PMC2 genotypes (RR: homozygous for the resistant allele; SS: homozygous for the susceptible allele; and RS: heterozygous) and PPV strain used for phenotyping are also indicated.

In total, 110 accessions were considered phenotypically resistant (Table 1), 108 were susceptible (Table 2) and 11 showed uncertain phenotype against the same or different PPV strains (Table 3). *ParPMC2*-del highly correlates with PPV resistance, as evidenced by its presence in 92.8% of the resistant accessions (Table 1) and its absence in 92.6% of the susceptible accessions (Table 2). Only 16 out of 219 (7.3%) accessions phenotypically classified as resistant or susceptible showed genotype-phenotype incongruences (GPIs). GPIs were previously reported mainly when using segregating populations [18,23,24,28,44], but clarifying reasons underlying GPIs was found difficult, as quite different factors may be involved. These factors include complex phenotyping protocols, loci other than *PPVres* contributing to PPV resistance, environmental conditions and/or gene–environment interactions. Additionally, putative misclassifications could also explain some genotypic discrepancies observed in this work.

For instance, Sunglo, the resistant donor parent of Goldrich, has been phenotyped as resistant by several authors using PPV-M [15,45,46] and PPV-D [47] and genotypically showed the SSR-resistant alleles targeting the *PPVres* locus [18]. However, WGS data (SRR2153157) supposedly corresponding to this accession do not have the *ParPMC2*-del. Something similar occurs with Mirlo Naranja, classified as resistant [48], that was found to carry one copy of the *ParPMC2*-del by PCR in this work but not in that of Passaro [49]. Detailed accession documentation may be helpful to resolve these discrepancies, but 13 of the 16 identified GPIs have no pedigree data available. This information would be very valuable to increase the efficiency of apricot breeding programs and germplasm management.

Table 1. Apricot PPV-resistant accessions genotyped for PMC2.

Name	Country [a]	Origin	Pedigree	PPV Resistance Phenotype [b]	PPV Strain Used	First Phenotype Ref	PMC2 Genotype [c]	PMC2 Genotype Ref
A4316	IT			R	M	[15]	RS	WGS
A4804	IT			R	M	[15]	RS	WGS
Adriana (= Le-3241)	CR	Horticulture Faculty, Lednice	Vestar × SEO [50]	R	M	[51]	RR	[24]
					Rec	[52]		
Alfred (= NY345)	USA	Geneva, NY State Expt Sta, by Robert C. Lamb	OP seedling of selection from (Doty × Geneva)	R	M	[53]	RS	WGS
Andswee	IR			R	M	[15]	RS	WGS
Anegat	FR	INRA, CEP Innovation		R	M/D	[54]	RS	[49]
Bergarouge (= Avirine A2914)	FR	INRA	Bergeron × Orange Red [55]	R	D	[23]	RS	[49]
					M	[56]		
Bergeval (= Aviclo, A3950)	FR	INRA		R	M	[56]	RS	[49]
BO03615011	IT		Goldrich × Harlayne [28]	R	M*	[49]	RS	[49]
BO03615025	IT		Goldrich × Harlayne [28]	R	M*	[49]	RR	[49]
BO03615034	IT		Goldrich × Harlayne [28]	R	M*	[28]	RR	[28]
BO03615049	IT		Goldrich × Harlayne [28]	R	M*	[28]	RR	[28]
BO03615053	IT		Goldrich × Harlayne [28]	R	M*	[28]	RS	[28]
BO03615070	IT		Goldrich × Harlayne [28]	R	M*	[49]	RR	[49]
BO04624031	IT		Portici × Goldrich [28]	R	M*	[28]	RS	[28]
BO04624039	IT		Portici × Goldrich [28]	R	M*	[49]	SS	[49]
BO05636034	IT		Kyoto × Priscilla [28]	R	M*	[28]	RS	[28]
BO06609012	IT		Silvercot × Bora [28]	R	M*	[49]	RS	[49]
BO06609013	IT		Silvercot × Bora [28]	R	M*	[49]	RS	[49]
BO06609024	IT		Silvercot × Bora [28]	R	M*	[49]	RS	[49]
BO06609033	IT		Silvercot × Bora [28]	R	M*	[49]	RS	[49]
BO06609036	IT		Silvercot × Bora [28]	R	M*	[49]	RS	[49]
BO06609037	IT		Silvercot × Bora [28]	R	M*	[49]	RS	[49]
BO06609039	IT		Silvercot × Bora [28]	R	M*	[49]	RS	[49]
BO06609045	IT		Silvercot × Bora [28]	R	M*	[49]	RS	[49]

Table 1. *Cont.*

Name	Country [a]	Origin	Pedigree	PPV Resistance Phenotype [b]	PPV Strain Used	First Phenotype Ref	PMC2 Genotype [c]	PMC2 Genotype Ref
BO06609048	IT		Silvercot × Bora [28]	R	M*	[28]	RS	[28]
BO06609055	IT		Silvercot × Bora [28]	R	M*	[28]	RS	[28]
BO06609060	IT		Silvercot × Bora [28]	R	M*	[49]	RS	[49]
BO06609068	IT		Silvercot × Bora [28]	R	M*	[49]	RS	[49]
BO06609074	IT		Silvercot × Bora [28]	R	M*	[49]	RS	[49]
BO06609079	IT		Silvercot × Bora [28]	R	M*	[49]	RS	[49]
BO06609083	IT		Silvercot × Bora [28]	R	M*	[49]	RS	[49]
BO06609087	IT		Silvercot × Bora [28]	R	M*	[49]	RS	[49]
BO06609099	IT		Silvercot × Bora [28]	R	M*	[49]	RS	[49]
BO06609104	IT		Silvercot × Bora [28]	R	M*	[49]	RS	[49]
BO06609113	IT		Silvercot × Bora [28]	R	M*	[49]	RS	[49]
BO06609129	IT		Silvercot × Bora [28]	R	M*	[49]	RS	[49]
BO06609133	IT		Silvercot × Bora [28]	R	M*	[49]	RS	[49]
BO06609136	IT		Silvercot × Bora [28]	R	M*	[49]	RS	[49]
BO96621002	IT		Goldrich × Lito [28]	R	M	[57]	RR	[28]
BO96621030	IT		Goldrich × Lito [28]	R	M	[57]	RS	[28]
Bora (BO90610010)	IT	University of Bologna and Milan, by D. Bassi	Early Blush × PA 7005-2 [58]	R	M/D	[58]	RS	[21,28]
Candela (= LE-2927)	CR	Horticulture Faculty, Lednice	Hungarian Best × SEO [59]	R	M	[60]	RR	[49]
Cebir	TU			R	T	[61]	RS	[61]
Congat	FR	INRA, CEP Innovation		R	-	[62]	RS	[49]
Early Blush (= RUTBHART, NJA53, Aurora46)	US	Rutgers Horticultural Research Farm, New Brunswick, N.J.	RR17–62 × NJA-13 [63]	R	D / M / T	[64] / [65] / [61]	RS	PCR; [21,28,61]
Farlis	FR	Marie-France BOIS, France (IPS)		R	M*	[28,49]	RS	[28]
Farmingdale (=NY346)	USA	Geneva, NY State Expt Sta, by Robert C. Lamb	OP seedling of selection from (Doty × Geneva) [66]	R	M	[53]	RS	[28]

Table 1. *Cont.*

Name	Country[a]	Origin	Pedigree	PPV Resistance Phenotype[b]	PPV Strain Used	First Phenotype Ref	PMC2 Genotype[c]	PMC2 Genotype Ref
Flavor cot (=Bayoto)	USA	Washington State University Research, by Tom Toyama		R	M	[57]	RS	[28]
Flopria	FR	PSB Producción Vegetal S.L.		R	M*	[28]	RS	PCR; [28]
GG9310	SP	IVIA, Moncada, Valencia	Goldrich × Ginesta [18]	R	D	IVIA	RS	PCR
GG9318	SP	IVIA, Moncada, Valencia	Goldrich × Ginesta [18]	R	D	IVIA	RS	PCR
GG937	SP	IVIA, Moncada, Valencia	Goldrich × Ginesta [18]	R	D	IVIA	RS	PCR
GG941	SP	IVIA, Moncada, Valencia	Goldrich × Ginesta [18]	R	D	IVIA	RS	PCR
GG979	SP	IVIA, Moncada, Valencia	Goldrich × Ginesta [18]	R	D	IVIA	RS	PCR
GG9869	SP	IVIA, Moncada, Valencia	Goldrich × Ginesta [18]	R	D	IVIA	RS	PCR
Gilgat	FR	INRA / CEP INNOVATION		R	M*	[28]; [49]	RS	[28]
GP9817	SP	IVIA, Moncada, Valencia	Goldrich × Palau [18]	R	D	IVIA	RS	PCR
Dama Rosa (GG9871)	SP	IVIA, Moncada, Valencia	Goldrich × Ginesta [18]	R	D	IVIA	RS	PCR; [49]
Dama Taronja (GK988)	SP	IVIA, Moncada, Valencia	Goldrich × Katy [18]	R	D	IVIA	RS	PCR; [49]
Dulcinea	IT	Pisa University	Moniqui OP [67]	R	D	[64]	SS	PCR; [49]
Fracasso	IT			Tolerant	T	[61]	SS	[61]
Harlayne	C	Agr. Canada, Res. Station, Harrow, Ontario, by REC Layne	V51092 ((Reliable × OP) × OP) × Sun Glo [66]	R	D / M	[68] / [45]	RS	PCR; [20,21,24, 28,61]
Harval (=HW437)	C	Agr. Canada, Res. Station, Harrow, Ontario, by REC Layne	Veecot × HW435 (Rouge du Roussillon × NJA2 (Morden604 OP)) [66]	R	M	[69]	RS	[28]
Henderson	USA	Geneva, NY, by GW Henderson	Unknown [66]	R	M / D	[46] / [70]	RS	PCR; [21]
Kaniş (=M2252)	TU			R	T	[61]	SS	[61]
Karum	TU			R	T	[61]	RS	[61]
Lady cot (=HYB 3-3)	FR	COT International		R	M*	[28]	RS	[28]
Laycot	C		V51092 ((Reliable o.p.) o.p.) × NJA1 [71]	R	M	[15]	RR	WGS

Table 1. *Cont.*

Name	Country [a]	Origin	Pedigree	PPV Resistance Phenotype [b]	PPV Strain Used	First Phenotype Ref	PMC2 Genotype [c]	PMC2 Genotype Ref
LE-2904	CR	Horticulture Faculty, Lednice	Velkopavlovická × SEO [19]	R	M	[72]	RS	[49]
LE-3205	CR	Horticulture Faculty, Lednice		R	M*	[49]	RR	[49]
Le-3246	CR	Horticulture Faculty, Lednice	Vestar × SEO [51]	R	M	[51]	RS	[24]
LE-3662	CR	Horticulture Faculty, Lednice		R	M	[72]	RR	[49]
Lifos	TU			R	T	[61]	RS	[61]
Lillycot	FR	SDR Fruit Llc (US)	Unknown [73]	R	M*	[28]	RS	[28]
Lito	GR		SEO × Tirynthos [18]	R	M / D	[74] / IVIA	RS	PCR; [24,28]
Mediabel (=Mediabell)	FR	Newcot and IPS		R	M*	[28]	RS	[28]
Mirlo Naranja (= Mirlo anaranjado)	SP	CEBAS-CSIC, Murcia	Rojo Pasión × Búlida Precoz [48]	R	D	[48]	RS / SS	PCR / [49]
Mirlo Blanco	SP	CEBAS-CSIC, Murcia	Rojo Pasión × Búlida Precoz [48]	R	D	[48]	RS	[28]
Mirlo Rojo	SP	CEBAS-CSIC, Murcia	Rojo Pasión × Búlida Precoz [48]	R	D	[48]	RS	PCR; [49]
Mogador	SP	PSB Producción Vegetal S.L.		R	M*	[28,49]	RS	PCR; [28]
Moixent (=GM961)	SP	IVIA, Valencia	Goldrich × Mitger [18]	R	D	IVIA	RS	PCR; [49]
Murciana	SP	CEBAS-CSIC, Murcia	Orange Red × Currot [73]	R	D / M	[75] / [15]	RS	WGS; PCR; [49]
Nikitskii	UKR			R	M	[15]	RS	WGS
NJA42	USA	New Jersey	NJA12 × NJA13 [76]	R	?	[77]	RS	PCR
Orange Red (=Barth; NJA-32)	USA	New Jersey	Lasgerdi Mashhad × NJA2 (= Morden 604 OP) [78]	R	D / M	[68] / [79]	RS	PCR; [21]
Pandora	GR		SEO × Tirynthos [18]	R	M / D	[74] / [47]	RS	PCR

Table 1. *Cont.*

Name	Country [a]	Origin	Pedigree	PPV Resistance Phenotype [b]	PPV Strain Used	First Phenotype Ref	PMC2 Genotype [c]	PMC2 Genotype Ref
Pelese di Giovanniello	IT			Tolerant	D	[64]	SS	[49]
Perla	SP	Murcia		R	D	[64]	SS	PCR
Petra (BO8617102)	IT	University of Bologna and Milan, Italy, by D Bassi	Goldrich × Pelese di Giovanniello [73]	R	M*	[28]	RS	[28]
Precoce d'Imola	IT			tolerant	D	[64]	SS	WGS
Priboto (=Zebra)	FR		bud mutation of Goldrich [80]	R	M	[15]	RS	WGS; [49]
Pricia	FR	Marie-France BOIS, France (IPS)		R	M*	[28,49]	RS	[28]
Pseudo Royal	USA			R	M	[15]	RS	WGS
Robada (= K106-2)	USA	Parlier, California	Orange Red × K113-40 (ancestry includes Blenheim, Blush and Perfection) [81]	R	M	[82]	RS	WGS
Rojo Pasión	SP	CEBAS-CSIC	Orange Red × Currot [83]	R	D	[83]	RS	PCR; [49]
Rosa	SP	CEBAS-CSIC, Murcia	Orange Red × Palsteyn [73]	R / Tolerant	D	[41] / [23]	RS	[49]
Rubista	FR	Marie-France BOIS, France (IPS)		R	M*	[28,49]	RS	[28]
Sabbatani (= Selezione Sabbatani?)	IT			R	D	[64]	SS	[49]
Selene	SP	CEBAS-CSIC	Goldrich × A2564 (=Screara × SEO) [18]	R	D	[84]	RS	PCR; [49]
SEOP934	SP	IVIA	SEO × Palau [18]	R	D	IVIA	RS	PCR
Spring Blush (= EA3126TH)	FR	Escande EARL		R	M*	[57]	RR	[49]
Stark Early Orange (= SEO, Earle Orange)	USA	Grandview, Washington, by WL Roberts	Unknown [66]	R	M / D	[85] / [70]	RS	PCR; [20,21,24, 28,61]
Stella	USA		Unknown [18]	R	M / D	[85] / [70]	RR	PCR; [21]
Sunglo (= Sun Glo)	USA	Columbia & Okanogan Nursery Co.	Unknown [66]	R	M / D	[45] / [47]	RS / SS	PCR / WGS

Table 1. *Cont.*

Name	Country [a]	Origin	Pedigree	PPV Resistance Phenotype [b]	PPV Strain Used	First Phenotype Ref	PMC2 Genotype [c]	PMC2 Genotype Ref
Sunnycot (= 97-3-203)	USA	SDR FRUIT LLC – USA		R		[62]	RS	[49]
Traian	RO			R	D	[86]	RS	PCR; [87]
Tsunami (= EA 5016)	FR	Escande EARL		R	M*	[28]	RS	[28]
Wonder Cot (= RM 7)	USA	SDR FRUIT LLC – USA		R	M*	[28]	RS	[28]
Zard	CA			R	T	[61]	RS	[61]

M *: strain likely used for phenotyping by the Phytosanitary Service, Emilia-Romagna (Italy). [a] Countries: C: Canada, CA: Central Asia, CR: Czech Republic, FR: France, GR: Greece, IR: Iran, IT: Italy, RO: Romania, SP: Spain, TU: Tunisia, TR: Turkey, UKR: Ukraine, US: United States of America; [b] Phenotype: R: Resistant, S: Susceptible; [c] Genotype: RR: homozygous for PMC2 resistant allele, SS: homozygous for PMC2 susceptible allele, RS: heterozygous.

Table 2. Apricot PPV susceptible accessions genotyped for *ParPMC2*-del.

Cultivar	Country [a]	Origin	Pedigree	PPV Resistance Phenotype [b]	PPV Strain Used	First Phenotype Ref	PMC2 Genotype [c]	PMC2 Genotype Ref
A3521	IR			S	M	[15]	SS	WGS
A3522	IR			S	M	[15]	SS	WGS
Amabile Vecchoni	IT	Seedling by Prof. F. Scaramuzzi	Unknown [67]	S	M	[45]	SS	[49]
Aprikoz	TR			S	M	[88]	SS	PCR
Arrogante	SP	Murcia		S	D	[89]	SS	[21]
Avikaline	FR			S	M	[15]	SS	WGS
Bebecou (Bebeco)	GR		Unknown [18]	S	M/D	[90]	SS	PCR; [21,28]
Bella Di Imola	IT		Spontaneous seedling [23]	S	D	[64]	SS	[28]
Bergeron	FR	Saint-Cyr-au-Mont-d'Or, Lyon	Spontaneous seedling [23]	S	M	[90]	SS	PCR; [21]
Big Red (EA4006)	FR	Escande EARL, France		S	M	[57]	RS	[28]
BO04624042	IT		Portici × Goldrich [28]	S	M*	[28]	SS	[28]
BO04624043	IT		Portici × Goldrich [28]	S	M*	[28]	SS	[28]
BO06609003	IT		Silvercot × Bora [28]	S	M*	[49]	RS	[49]
BO08160431l	IT		San Castrese × Reale di Imola [73]	S	D	[91]	SS	[24]
BO96621021	IT		Goldrich × Lito [28]	S	M*	[28]	RS	[28]
Boucheran Boutard	FR			S	M	[15]	SS	WGS
Búlida	SP	Murcia	Unknown [73]	S	D	[92]	SS	PCR; [21]
				S	M	[93]		

Table 2. *Cont.*

Cultivar	Country [a]	Origin	Pedigree	PPV Resistance Phenotype [b]	PPV Strain Used	First Phenotype Ref	PMC2 Genotype [c]	PMC2 Genotype Ref
Cafona	IT	Vesuvian area		S	M / D	[94] / [64]	SS	WGS
CAID AGDZ n2	MO			S	M	[15]	SS	WGS
Canino	SP	Valencia	Unknown [18]	S	D / M	[95] / [90]	SS	PCR; [20,21]
Castlebrite (=K111-6)	USA	USDA, Fresno, California	OP seedling of B60-12 (= Perfection × Castleton) [66]	S	M	[45]	SS	PCR
Cegléd Bibor	HU	Cegléd Horticultural Research Institute	Chance seedling [96]	S	M	[46]	SS	[28]
Colorado (Colorao 43-15)	SP	PSB Producción Vegetal SL	Unknown	S	M* / D	[49] / [89]	SS	PCR; [28]
Corbató	SP	Valencia		S	D / M	[95] / [46]	SS	PCR
Currot	SP	Valencia	Unknown [18]	S	D / M	[95] / [46]	SS	PCR
Estrella	SP	CEBAS-CSIC	Orange Red × Z211-18 (= Goldrich × Pepito del Rubio) [23]	S	D	[23]	SS	PCR; [49]
Faralia	FR	Marie-France BOIS, IPS		S	M*	[28]	SS	[28]
Farclo	FR	Marie-France BOIS, IPS		S	M	[57]	SS	[28]
Favorit	RO			S	M	[94,97]	SS	[49]
Geç Abligoz	TR			S	T	[61]	SS	[61]
Ginesta	SP	Valencia	Unknown [18]	S	D / M	[95] / [46]	SS	PCR
Dama Vermella (HG9869)	SP	IVIA	Harcot × Ginesta [18]	S	D	IVIA	SS	PCR; [49]
Hachaliloğlu	TR			S	T	[61]	SS	[61]
Hargrand (= HW410)	C	Richard EC Layne, Agr. Canada, Res. Station	V51092 ((Reliable × OP) × OP) × NJA1 (Phelps × Perfection) [66]	S	M	[45]	SS	[21]
Hasanbey	TR			S	M	[45][1]	SS	PCR
Hungarian Best = (Best of Hungary?)	HU/RO			S	T / M	[61] / [94]	SS	[61]
Katy	USA	Zaiger's Genetics[4]		S	D	[18]	SS	PCR; [21]
Krasnoshchekii	UKR	Advanced/improved cultivar		S	D	[20]	SS	[20,21]

Table 2. *Cont.*

Cultivar	Country[a]	Origin	Pedigree	PPV Resistance Phenotype[b]	PPV Strain Used	First Phenotype Ref	PMC2 Genotype[c]	PMC2 Genotype Ref
Kyoto (= Kioto)	FR	Escande	Unknown [73]	S	M*	[28]	SS	[28]
Lambertin-1	USA	USDA, Fresno, California	A95-45 × B69-85 (=Perfection × Royal) [98]	S	M	[45]	SS	[21]
Larclyd (= F168 cv; Jenny Cot)	NZ	Central Otago	Sundrop × Moorpark [99]	S	M	[15]	SS	WGS
Le-3218	CR	Faculty of Horticulture in Lednice	Vestar × SEO [51]	S	M	[51]	SS	[24]
Luizet (= Suchet; Hatif du clos; Abricot du Clos)	FR		Spontaneous seedling [71]	S	M	[93]	SS	WGS
Luna	IT			S	M*	[28]	RS	[28]
Madarska Narijlepsia	SL			S	M	[15]	SS	WGS
Magic cot (= RM 22)	USA	SDR FRUIT LLC - USA	Unknown [23]	S	D	[23]	SS	[49]
Manicot	FR			S	D	[97,100]	SS	WGS
					M	[15]		
Maravilla	SP	CEBAS-CSIC, Murcia	Orange Red × Z211-18 (= Goldrich × Pepito) [23]	S	D	[23]	SS	PCR; [49]
Mari de Cenad	RO	Unknown		S		[86]	RS	PCR
Markulești	TR			S	T	[61]	SS	PCR
Marlén	CR	Horticulture Faculty, Lednice	clone of Hungarian Best [59]	S	Rec	[14]	SS	PCR; [24]
Marouch 14	MO		Local landrace	S	M	[15]	SS	WGS
Marouch 4	MO			S	M	[15]	SS	WGS
Mei Hwang	CH		Traditional cultivar/landrace	S	M	[15]	SS	WGS
Mektep	TR			S	T	[61]	SS	[61]
Mektep 8	TR			S	T	[61]	SS	[61]
Mitger	SP	Castellón [30]	Unknown [18]	S	D	[95]	SS	PCR
					M	[46]		
Monaco Bello	IT			S	M	[97]	SS	WGS
Moniqui	SP	Murcia	Unknown	S	M	[90]	SS	PCR; [21,24]
Mono	USA	Le Grand, California, by FW Anderson	Perfection OP [66]	S	M	[93]	SS	[49]
Moongold (= Moongola?)	USA	University of Minnesota		S	-	[77]	SS	PCR
Moorpark (=Moor Park)	USA			S	M	[46]	SS	WGS
Morden 604	C	Morden, Manitoba, by Canada Dept. Agr. Res. Sta.	Scout × McClure [66]	S	M	[15]	SS	WGS
Ninfa (BO8162075)	IT	University of Bologna and Milan, by D. Bassi	Ouardy × Tyrinthos [55]	S	M*	[28]	SS	PCR; [28,61]
					T	[61]		

Table 2. *Cont.*

Cultivar	Country[a]	Origin	Pedigree	PPV Resistance Phenotype[b]	PPV Strain Used	First Phenotype Ref	PMC2 Genotype[c]	PMC2 Genotype Ref
Olimp	RO			S	M	[45]	SS	WGS; [49]
Orange Rubis (=Couloumine)	FR	Mallard		S	M	[57]	SS	[28]
Ordubat B.	TR			S	T	[61]	SS	[61]
Ouardi	TU	INRAT, Ariana	Canino × Hamidi [101]	S	M	[46]	SS	[49]
Palsteyn (Palstein)	SA		Blenhein × Canino [73]	S	M	[102]	SS	WGS
Palabras	SP			S	D / M	[95] / [46]	SS	PCR
Palau	SP		Unknown [18]	S	D	[95]	SS	PCR
Paviot	FR			S	M	[93]	SS	WGS
Peche De Nancy	FR			S	M	[15]	SS	WGS
Perfection	USA	Waterville, Washington	Unknown [66]	S	M	[46]	SS	[21]
Piera				S	M	[65]	RS	PCR
Poizat	FR			S	M	[15]	SS	WGS
Polonais	FR		Spontaneous seedling [23]	S	M	[93]	SS	[24]
Poppy	USA	Zaiger Genetics, Inc., Modesto, CA	78EB575 × 123GD161 [58]	S	D	[23]	SS	[49]
Portici (= Pertini)	IT	Vesuvian area	Unknown; Local selection [23]	S	M / D	[46] / [64]	SS	PCR; [28]
Precoce Ampuis	FR			S	M	[15]	SS	WGS
Reale d'Imola	IT		Luizet OP [23]	S	M / D	[46] / [64]	SS	[21,24,49]
Rojo de Carlet	SP	Valencia		S	D / M	[95] / [46]	SS	PCR
Rouge Du Roussillon	FR			S	M	[45]	SS	WGS
Rouge De Fournes	FR			S	M	[15]	SS	WGS
Saturn	RO			S	M	[45]	SS	WGS
Screara	FR			S	D / M	[70] / [45]	SS	WGS
Şekerpare B.	TR			S	T	[61]	SS	[61]
Shalakh (=Yerevani, Erevani)	AR		Local selection [23]	S	M	[93]	SS	WGS; [20,21]
Silistra × Ananas (Marculesti 43/1)	RO			S	M	[15]	SS	WGS

Table 2. *Cont.*

Cultivar	Country [a]	Origin	Pedigree	PPV Resistance Phenotype [b]	PPV Strain Used	First Phenotype Ref	PMC2 Genotype [c]	PMC2 Genotype Ref
Sucre De Holub	HU	Bohème, by M. Holub		S	M	[15]	SS	WGS
Sublime	SP	CEBAS-CSIC	Orange Red × Z211-18 (= Goldrich × Pepito del Rubio) [103]	S	D	[103]	SS	PCR; [49]
Super Rouge	FR			S	M	[15]	SS	WGS
Sweet Red	FR			S	M	[57]	SS	[49]
Szegedi mamut (=Szegadti Mamut?)	HU	Foki István and Kovács Imre	Hybrid of Cegledi orias, "Giant" group [96]	S	M	[94]	SS	[49]
Tabriz	TR			S		[86]	SS	PCR
Tadeo (= Taddeo)	SP	Valencia		S	D M	[95] [45]	SS	PCR
Tardif De Bordaneil	FR		Unknown [23]	S	M D	[46] [64]	SS	WGS
Tardif De Tain	FR			S	M	[15]	SS	WGS
Tonda di costigliole	IT	Piedmont		S		[104]	SS	[49]
Trevatt	AU			S	M	[45]	SS	PCR
Tyrinthos	GR		Unknown [18]	S	D M	[70] [97]	SS	PCR; WGS; [49]
Uleanos	SP	Ulea, Murcia		S	D	[89]	SS	[49]
Velázquez	SP	Murcia		S	D	[89]	SS	PCR; [21]
Venus (= Venus 1414?)	RO		(Umberto × Ananas) × (Luizet × Umberto) [96]	S	M	[46]	SS	[49]
Vestar	CR		Hungarian Best × mixture of pollen from Chinese cultivars [55]	S	M	[105]	RS	WGS; [24]
Vivagold	C	Vineland Station, Ontario	Veecot × V49024 (= Geneva × Gibb) [66]	S	M	[15]	SS	WGS
Xirivello (=Chirivello)	SP	Valencia	Unknown	S	M	[46]	SS	PCR
Yilbat (=M243)	TR			S	T	[61]	RS	[61]

M *: strain likely used for phenotyping by the Phytosanitary Service, Emilia-Romagna (Italy). [a] Countries: AR: Armenia, AU: Australia, C: Canada, CH: China, CR: Czech Republic, FR: France, GR: Greece, HU: Hungary, IR: Iran, IT: Italy, MO: Morocco, NZ: New Zealand, RO: Romania, SA: South Africa, SL: Slovakia, SP: Spain, TR: Turkey, UKR: Ukraine and US: United States of America; [b] Phenotype: R: Resistant, S: Susceptible; [c] Genotype: RR: homozygous for PMC2 resistant allele, SS: homozygous for PMC2 susceptible allele and RS: heterozygous.

Table 3. Apricot accessions with uncertain PPV resistance phenotype genotyped for *ParPMC2*-del.

Cultivar	Country[a]	Origin	Pedigree	PPV Resistance Phenotype[b]	PPV Strain Used	First Phenotype Ref	PMC2 Genotype[c]	PMC2 Genotype Ref
Badami	IR			S	M	[102]	SS	WGS
				T	D			
Farbaly	FR	Marie-France BOIS, IPS		S	M*	[28]	RS	[28]
				R	M	[56]		
Goldrich	USA	USDA and Washington State University, Prosser, Washington	Sun Glo × Perfection [73]	R	D	[68]	RS	PCR; [20,21,24,28]
					M	[45]		
				uncertain	M	[106]		
					D	[68]		
				S	D	[64]		
Harcot	C	Agr. Canada, Res. Station, Harrow, Ontario, by REC Layne	(T2 (Geneva × Naramata) × Morden 604 (Scout × McClure)) × NJA1 (Phelps × Perfection) [66]	T?	-	[28]	RS	PCR; [21,24,28]
				R	M	[90]		
					D	[70]		
				S	T	[61]		
Incomparable de Malissard (= Valssard)	FR	Malissard, Valence		R	M	[15]	SS	[61]
				S		[57]		
Pisana	IT		ICAPI 26/5 OP [55]	S	M*	[28]	RS	WGS
				R	M	[65]		
				R	D	[64]		
				S	D	[28]		
Pieve (BO89608015)	IT	University of Bologna and Milan, by D. Bassi	Harcot × Reale di Imola [73]	S	M*	[28]	SS	PCR; [28]
				R	M	[65]		
San Castrese	IT	Naples	Unknown [73]	T	D	[64]	SS	WGS; [49]
				S	M	[46]		
Sulmona	RO		(Luizet × Re Umberto) × (Ananas × Ananas) [71]	S	M	[45]	SS	[49]
				R	-	[77]		
Veecot	C	Ontario Dept Agr Res Inst, Vineland Station, Ontario, by OA Bradt	Reliable OP [18]	R	M	[45]	RS	PCR; [21]
				S		[105]		
				T	D	[47]		
Viceroy (=Viceroy_603_G?)	RO			R	-	[77]	SS	PCR
				S		[86]		

M *: strain likely used for phenotyping by the Phytosanitary Service, Emilia-Romagna (Italy). [a] Countries: C: Canada, FR: France, IR: Iran, IT: Italy and RO: Romania; [b] Phenotype: R: Resistant and S: Susceptible; and [c] Genotype: RR: homozygous for PMC2 resistant allele, SS: homozygous for PMC2 susceptible allele and RS: heterozygous.

Accurate evaluation of PPV resistance is a complex process, and results obtained by different researchers sometimes are contradictory, as exemplified by Farbaly and Pieve (Table 3), which may lead to GPIs. This problem is also observed in well-known accessions. For instance, Goldrich, usually classified as resistant against both PPV-D and M strains, has also been classified as uncertain or even as susceptible at least once (Table 3). Moreover, the effect of the PPV strain used [9,24] has also been observed, as at least 5 accessions showed different behaviour against PPV-M, D or T infection (Table 3). In addition, the environmental effect on symptoms and the different PPV detection techniques employed could also been involved in GPIs [9].

On the other hand, PPV resistance has been related with the downregulation of both *ParPMC2* and, especially, *ParPMC1*, putatively due to an RNA silencing mechanism triggered by the pseudogenization of *ParPMC2res* [21]. Notwithstanding, the presence of epigenetic changes has also been suggested as a possible cause [22]. In any case, resistant cultivars show residual expression levels that could somehow be influenced by environmental conditions. This might explain sporadic symptoms that eventually lead to GPI classification. Moreover, the role of additional PPV resistance loci or genes may also contribute to GPIs. In this sense, Gallois et al. [105] pointed out that a large part of a resistant phenotype conferred by a given QTL depends on the genetic background due to frequent epistatic effects between resistance genes. In fact, other minor loci, linked or not to *PPVres*, have been suggested to underlie PPV resistance in apricot [13–16]. Altogether, the identification and/or confirmation of GPIs in this work pave the way for future studies to unravel the PPV resistance mechanism.

The handful of North American cultivars originally described as PPV resistant [9] have been extensively used as donors in all breeding programs currently in progress. As a result, the *PPVres* locus has been introduced in different genetic backgrounds. In order to complete our survey, genotypic information was compiled from other 96 accessions without available PPV phenotype data (Table S1, [107–113]). In summary, 152 accessions (46.8%) have at least one copy of the *ParPMC2*-del (Figure 3) and 15 out of them are homozygous for *ParPMC2-del*, including the North American PPV-resistant cultivar Stella [114]. Those materials derived from crosses with North American PPV-resistant cultivars represent an opportunity to accelerate the development of new varieties better adapted to the Mediterranean basin conditions [9]. In this context, it should be highlighted that MAS allows to improve cost efficiency and/or genetic gain in apricot breeding programs aimed to select PPV-resistant seedlings. This improvement is highly significant even if some PPV susceptible individuals among those with *ParPMC2*-del are dragged, since they will be later identified by PPV phenotyping. Similarly, Tartarini et al. [115] underlined the advantage of the identification of homozygous *Rvi6* scab-resistant plants using MAS, despite segregating progenies showing at least 5% of GPIs.

4. Conclusions

Here, we present a high-throughput method to quickly perform DNA testing for PPV resistance that may greatly improve the efficiency of apricot breeding programs. The long-lasting PPV phenotyping process will only be performed with those advanced selections showing promising agronomic behaviour in advanced stages to guarantee the selection of PPV-resistant individuals. Additionally, a wide survey over 300 accessions has been made to identify PPV-resistant sources that could also be useful in apricot breeding programs.

Author Contributions: Conceptualization: C.R., M.L.B. and E.Z.; experimental procedures: Á.P.-O., I.L. and E.Z.; bioinformatics: E.Z.; funding acquisition: M.L.B.; writing—original draft, E.Z.; writing—review and editing, Á.P.-O., C.R., I.L., M.L.B. and E.Z. All authors have read and agreed to the published version of the manuscript.

Acknowledgments: The authors would like to express their gratitude to Bassi (University of Milan, Italy) for providing pedigree information from their apricot breeding program.

References

1. Rehder, A. *Manual of Cultivated Trees and Shrubs Hardy in North America*, 2nd ed.; The Macmillan Company: New York, NY, USA, 1940.
2. Zhebentyayeva, T.N.; Ledbetter, C.; Burgos, L.; Llácer, G. Apricots. In *Fruit Breeding*, 1st ed.; Badenes, M.L., Byrne, D.H., Eds.; Springer: New York, NY, USA, 2012; Volume 3, pp. 415–458. [CrossRef]
3. Moreno, M.A. Breeding and selection of *Prunus* rootstocks at the Aula Dei experimental station, Zaragoza, Spain. *Acta Hortic.* **2004**, *658*, 519–528. [CrossRef]
4. García, J.A.; Cambra, M. Plum pox virus and sharka disease. *Plant. Viruses* **2007**, *1*, 69–79.
5. García, J.A.; Glasa, M.; Cambra, M.; Candresse, T. Plum pox virus and sharka: A model potyvirus and a major disease. *Mol. Plant. Pathol.* **2014**, *15*, 226–241. [CrossRef]
6. Chirkov, S.; Ivanov, P.; Sheveleva, A.; Zakubanskiy, A.; Osipov, G. New highly divergent Plum pox virus isolates infecting sour cherry in Russia. *Virology* **2017**, *502*, 56–62. [CrossRef]
7. James, D.; Varga, A.; Sanderson, D. Genetic diversity of Plum pox virus: Strains, disease and related challenges for control. *Can. J. Plant. Pathol.* **2013**, *35*, 431–441. [CrossRef]
8. Sihelská, N.; Glasa, M.; Šubr, Z.W. Host preference of the major strains of Plum pox virus—Opinions based on regional and world-wide sequence data. *J. Integr. Agric.* **2017**, *16*, 510–515. [CrossRef]
9. Martínez-Gómez, P.; Dicenta, F.; Audergon, J.M. Behaviour of apricot (*Prunus armeniaca* L.) cultivars in the presence of Sharka (Plum pox potyvirus): A review. *Agronomie* **2000**, *20*, 407–422. [CrossRef]
10. Dondini, L.; Lain, O.; Vendramin, V.; Rizzo, M.; Vivoli, D.; Adami, M.; Guidarelli, M.; Gaiotti, F.; Palmisano, F.; Bazzoni, A.; et al. Identification of QTL for resistance to Plum pox virus strains M and D in Lito and Harcot apricot cultivars. *Mol. Breed.* **2011**, *27*, 289–299. [CrossRef]
11. Hurtado, M.A.; Romero, C.; Vilanova, S.; Abbott, A.G.; Llácer, G.; Badenes, M.L. Genetic linkage maps of two apricot cultivars (*Prunus armeniaca* L.) and mapping of PPV (sharka) resistance. *Theor. Appl. Genet.* **2002**, *105*, 182–191. [CrossRef]
12. Lalli, D.A.; Abbott, A.G.; Zhebentyayeva, T.N.; Badenes, M.L.; Damsteegt, V.; Polák, J.; Krška, B.; Salava, J. A genetic linkage map for an apricot (*Prunus armeniaca* L.) BC1 population mapping Plum pox virus resistance. *Tree Genet. Genomes* **2008**, *4*, 481–493. [CrossRef]
13. Lambert, P.; Dicenta, F.; Rubio, M.; Audergon, J.M. QTL analysis of resistance to sharka disease in the apricot (*Prunus armeniaca* L.) 'Polonais' x 'Stark Early Orange' F1 progeny. *Tree Genet. Genomes* **2007**, *3*, 299–309. [CrossRef]
14. Marandel, G.; Pascal, T.; Candresse, T.; Decroocq, V. Quantitative resistance to Plum pox virus in *Prunus davidiana* P1908 linked to components of the eukaryotic translation initiation complex. *Plant. Pathol.* **2009**, *58*, 425–435. [CrossRef]
15. Mariette, S.; Wong Jun Tai, F.; Roch, G.; Barre, A.; Chague, A.; Decroocq, S.; Groppi, A.; Laizet, Y.; Lambert, P.; Tricon, D.; et al. Genome-wide association links candidate genes to resistance to Plum pox virus in apricot (*Prunus armeniaca*). *New Phytol.* **2016**, *209*, 773–784. [CrossRef]
16. Pilarova, P.; Marandel, G.; Decroocq, V.; Salava, J.; Krška, B.; Abbott, A.G. Quantitative trait analysis of resistance to Plum pox virus in the apricot F1 progeny 'Harlayne' x 'Vestar'. *Tree Genet. Genomes* **2010**, *6*, 467–475. [CrossRef]
17. Soriano, J.M.; Vera-Ruiz, E.; Vilanova, S.; Martínez-Calvo, J.; Llácer, G.; Badenes, M.L.; Romero, C. Identification and mapping of a locus conferring Plum pox virus resistance in two apricot-improved linkage maps. *Tree Genet. Genomes* **2008**, *4*, 391–402. [CrossRef]
18. Soriano, J.M.; Domingo, M.L.; Zuriaga, E.; Romero, C.; Zhebentyayeva, T.; Abbott, A.G.; Badenes, M.L. Identification of simple sequence repeat markers tightly linked to Plum pox virus resistance in apricot. *Mol. Breed.* **2012**, *30*, 1017–1026. [CrossRef]
19. Zhebentyayeva, T.N.; Reighard, G.L.; Lalli, D.; Gorina, V.M.; Krška, B.; Abbott, A.G. Origin of resistance to *Plum pox virus* in apricot: What new AFLP and targeted SSR data analyses tell. *Tree Genet. Genomes* **2008**, *4*, 403–417. [CrossRef]
20. Zuriaga, E.; Soriano, J.M.; Zhebentyayeva, T.; Romero, C.; Dardick, C.; Cañizares, J.; Badenes, M.L. Genomic analysis reveals MATH gene(s) as candidate(s) for Plum pox virus (PPV) resistance in apricot (*Prunus armeniaca* L.). *Mol. Plant Pathol.* **2013**, *14*, 663–677. [CrossRef]

21. Zuriaga, E.; Romero, C.; Blanca, J.M.; Badenes, M.L. Resistance to Plum pox virus (PPV) in apricot (*Prunus armeniaca* L.) is associated with down-regulation of two MATHd genes. *BMC Plant Biol.* **2018**, *18*, 25. [CrossRef]

22. Rodamilans, B.; Valli, A.; García, J.A. Molecular plant-Plum pox virus interactions. *Mol. Plant. Microbe Interact.* **2020**, *33*, 6–17. [CrossRef]

23. Rubio, M.; Ruiz, D.; Egea, J.; Martínez-Gómez, P.; Dicenta, F. Opportunities of marker assisted selection for Plum pox virus resistance in apricot breeding programs. *Tree Genet. Genomes* **2014**, *10*, 513–525. [CrossRef]

24. Decroocq, S.; Chague, A.; Lambert, P.; Roch, G.; Audergon, J.M.; Geuna, F.; Chiozzotto, R.; Bassi, D.; Dondini, L.; Tartarini, S.; et al. Selecting with markers linked to the *PPVres* major QTL is not sufficient to predict resistance to Plum pox virus (PPV) in apricot. *Tree Genet. Genomes* **2014**, *10*, 1161–1170. [CrossRef]

25. Moustafa, T.A.; Badenes, M.L.; Martínez-Calvo, J.; Llácer, G. Determination of resistance to sharka (*plum pox*) virus in apricot. *Sci. Hort.* **2001**, *91*, 59–70. [CrossRef]

26. Lommel, S.A.; McCain, A.H.; Morris, T.J. Evaluation of indirect-linked immunosorbent assay for the detection of plant viruses. *Phytopathology* **1982**, *72*, 1018–1022. [CrossRef]

27. Wetzel, T.; Candresse, T.; Ravelonandro, M.; Dunez, J. A polymerase chain reaction assay adapted to plum pox potyvirus detection. *J. Virol. Methods* **1991**, *33*, 355–365. [CrossRef]

28. Passaro, M.; Geuna, F.; Bassi, D.; Cirilli, M. Development of a high-resolution melting approach for reliable and cost-effective genotyping of *PPVres* locus in apricot (*P. armeniaca*). *Mol. Breed.* **2017**, *37*, 74. [CrossRef]

29. Ru, S.; Main, D.; Evans, K.; Peace, C. Current applications, challenges, and perspectives of marker-assisted seedling selection in Rosaceae tree fruit breeding. *Tree Genet. Genomes* **2015**, *11*, 8. [CrossRef]

30. Doyle, J.J.; Doyle, J.L. A rapid isolation procedure for small quantities of fresh leaf tissue. *Phytochem. Bull.* **1987**, *19*, 11–15.

31. Muñoz-Sanz, J.V.; Zuriaga, E.; Badenes, M.L.; Romero, C. A disulfide bond A-like oxidoreductase is a strong candidate gene for self-incompatibility in apricot (*Prunus armeniaca*) pollen. *J. Exp. Bot.* **2017**, *68*, 5069–5078. [CrossRef]

32. Haas, B.J.; Papanicolaou, A.; Yassour, M.; Grabherr, M.; Blood, P.D.; Bowden, J.; Couger, M.B.; Eccles, D.; Li, B.; Lieber, M.; et al. De novo transcript sequence reconstruction from RNA-seq using the trinity platform for reference generation and analysis. *Nat. Protoc.* **2013**, *8*, 1494–1512. [CrossRef]

33. Verde, I.; Jenkins, J.; Dondini, L.; Micali, S.; Pagliarani, G.; Vendramin, E.; Paris, R.; Aramini, V.; Gazza, L.; Rossini, L.; et al. The Peach v2.0 release: High-resolution linkage mapping and deep resequencing improve chromosome-scale assembly and contiguity. *BMC Genom.* **2017**, *18*, 225. [CrossRef] [PubMed]

34. Langmead, B.; Salzberg, S.L. Fast gapped-read alignment with bowtie 2. *Nat. Meth.* **2012**, *9*, 357–359. [CrossRef] [PubMed]

35. Thorvaldsdóttir, H.; Robinson, J.T.; Mesirov, J.P. Integrative Genomics Viewer (IGV): High-performance genomics data visualization and exploration. *Brief. Bioinform.* **2013**, *14*, 178–192. [CrossRef] [PubMed]

36. Peace, C. DNA-informed breeding of rosaceous crops: Promises, progress and prospects. *Hortic. Res.* **2017**, *4*, 17006. [CrossRef]

37. Edge-Garza, D.A.; Luby, J.J.; Peace, C. Decision support for cost-efficient and logistically feasible marker-assisted seedling selection in fruit breeding. *Mol. Breed.* **2015**, *35*, 223. [CrossRef]

38. Audergon, J.M.; Blanc, A.; Gilles, F.; Broquaire, J.M.; Clauzel, G.; Gouble, B.; Grotte, M.; Reich, M.; Bureau, S.; Pitiot, C. New recent selections issued from INRA's apricot breeding programme. *Acta Hortic.* **2010**, *862*, 179–182. [CrossRef]

39. Bassi, D.; Audergon, J.M. Apricot breeding: Update and perspectives. *Acta Hortic.* **2006**, *701*, 279–294. [CrossRef]

40. Bassi, D.; Bellini, E.; Guerriero, R.; Monastra, F.; Pennone, F. Apricot breeding in Italy. *Acta Hortic.* **1995**, *384*, 47–54. [CrossRef]

41. Egea, J.; Dicenta, F.; Burgos, L.; Martínez-Gómez, P.; Rubio, M.; Campoy, J.A.; Ortega, E.; Patiño, J.L.; Nortes, L.; Molina, A.; et al. New apricot cultivars from CEBAS-CSIC (Murcia, Spain) breeding programme. *Acta Hortic.* **2010**, *862*, 113–118. [CrossRef]

42. Martínez-Calvo, J.; Font, A.; Llácer, G.; Badenes, M.L. Apricot and Peach breeding programs from the IVIA. *Acta Hortic.* **2009**, *814*, 185–188. [CrossRef]

43. Gürcan, K.; Ceylan, A. Strain identification and sequence variability of Plum pox virus in Turkey. *Turkish J. Agric.* **2016**, *40*, 746–760. [CrossRef]

44. Decroocq, S.; Cornille, A.; Tricon, D.; Babayeva, S.; Chague, A.; Eyquard, J.P.; Karychev, R.; Dolgikh, S.; Kostritsyna, T.; Li, S.; et al. New insights into the history of domesticated and wild apricots and its contribution to Plum pox virus resistance. *Mol. Ecol.* **2016**, *25*, 4712–4729. [CrossRef] [PubMed]

45. Dosba, F.; Orliac, S.; Dutranoy, F.; Maison, P.; Massonie, G.; Audergon, J.M. Evaluation of resistance to Plum pox virus in apricot trees. *Acta Hortic.* **1992**, *309*, 211–219. [CrossRef]

46. Karayiannis, I.; Mainou, A. Resistance to Plum pox virus in apricots. *EPPO Bull.* **1994**, *24*, 761–766. [CrossRef]

47. Martínez-Gómez, P.; Rubio, M.; Dicenta, F. Evaluation of resistance to Plum pox virus of North American and European apricot cultivars. *HortScience* **2003**, *38*, 568–569. [CrossRef]

48. Egea, J.; Rubio, M.; Campoy, J.A.; Dicenta, F.; Ortega, E.; Nortes, M.D.; Martínez-Gómez, P.; Molina, A.; Molina, A., Jr.; Ruiz, D. 'Mirlo Blanco', 'Mirlo Anaranjado', and 'Mirlo Rojo': Three new very early-season apricots for the fresh market. *HortScience* **2010**, *45*, 1893–1894. [CrossRef]

49. Passaro, M. Cost-Effective Use of Molecular Markers in the Practical Resolution of Common Horticultural Challenges. Ph.D. Thesis, Agriculture, Environment and Bioenergy in Universitá Degli Studi Di Milano, Milan, Italy, 2016.

50. Krška, B.; Salava, J.; Polák, J. Breeding for resistance: Breeding for Plum pox virus resistant apricots (*Prunus armeniaca* L.) in the Czech Republic. *EPPO Bull.* **2006**, *36*, 330–331. [CrossRef]

51. Krška, B.; Salava, J.; Polák, J.; Komínek, P. Genetics of resistance to Plum pox virus in apricot. *Plant Protect. Sci.* **2002**, *38*, 180–182. [CrossRef]

52. Krška, B.; Vachun, Z.; Nečas, T.; Ondrásek, I. New Sharka resistant apricots at the Horticultural Faculty in Lednice. *Acta Hortic.* **2015**, *1063*, 105–110. [CrossRef]

53. Rankovic, M.; Duli-Markovic, I.; Paunovic, S. Sharka virus in apricot and its diagnosis. *Acta Hortic.* **1999**, *488*, 783–786. [CrossRef]

54. CEP INNOVATION Website. Available online: https://cepinnovation-novadi.com/variete/anegat/ (accessed on 29 July 2020).

55. Milatović, D.; Nikolić, D.; Krška, B. Testing of self-(in)compatibility in apricot cultivars from European breeding programmes. *HortScience* **2013**, *40*, 65–71. [CrossRef]

56. Brans, Y. Evaluation de la sensibilité de cultivars d'abricotier à la Sharka en zone confinée. Présentation de l'essai Ctifl 2012–2015. In Proceedings of the Rencontres Phytosanitaires Ctifl/DGAL—SDQPV Fruits à Noyau Ctifl Balandran, Bellegarde, France, 16 October 2016.

57. Babini, A.R.; Vicchi, V.; Missere, D. L'esame delle cultivar tolleranti alla sharka. *Ermes Agricoltura* 2010. Available online: http://www.crpv.it/doc/549738/DLFE-9612.pdf (accessed on 29 August 2020).

58. Finn, C.E.; Clark, J.R. Register of New Fruit and Nut Cultivars List 44. *HortScience* **2008**, *43*, 1321–1343. [CrossRef]

59. Krška, B.; Vachůn, Z. Apricot Breeding at the Faculty of Horticulture in Lednice. *Agronomy* **2016**, *6*, 27. [CrossRef]

60. Krška, B. Genetic resources of apricot for adaptability improvement and breeding. *Acta Hortic.* **2010**, *862*, 203–208. [CrossRef]

61. Gürcan, K.; Çetinsağ, N.; Pınar, H.; Macit, T. Molecular and biological assessment reveals sources of resistance to Plum pox virus—Turkey strain in Turkish apricot (*Prunus armeniaca*) germplasm. *Sci. Hortic.* **2019**, *252*, 348–353. [CrossRef]

62. Drogoudi, P. ΕΥΠΑΘΕΙΑ ΠΟΙΚΙΛΙΩΝ ΒΕΡΙΚΟΚΙΑΣ ΣΤΟΝ ΙΟ ΤΗΣ ΕΥΛΟΓΙΑΣ ΤΗΣ ΔΑΜΑΣΚΗΝΙΑΣ (PPV) (Sensitivity of apricot varieties against PPV). Π. Δρογούδη 28ο Συνέδριο Ελληνικής Εταιρείας πιστήμης Οπωροκηπευτικών '50 χρόνια από την ίδρυση της ΕΕΕΟ'. In Proceedings of the 28th Conference of the Hellenic Fruit and Vegetable Credit Society '50 Years Since the Founding of EEEO', Thessaloniki, Greece, 16–20 October 2017.

63. Goffreda, J.C.; Voordeckers, A.; Butenis-Vorsa, L.; Cowgill, W.P., Jr.; Maletta, M.H.; Frecon, J.L. NJA53 apricot. *HortScience* **1995**, *30*, 389–390. [CrossRef]

64. Faggioli, F.; Barba, M. Screening of stone fruit germplasm for resistance to Plum Pox Potyvirus. *Adv. Hortic. Sci.* **1996**, *10*, 91–94.

65. Poggi Pollini, C.; Bianchi, L.; Babini, A.; Vicchi, V.; Liverani, A.; Brandi, F.; Giunchedi, L.; Autonell, C.; Ratti, C. Evaluation of Plum pox virus infection on different stone fruit tree varieties. *J. Plant. Pathol.* **2008**, *90*, S27–S31.

66. Brooks, R.M.; Olmo, H.P. *The Brooks and Olmo Register of Fruit and Nut Varieties*, 3rd ed.; ASHS Press: Alexandria, VA, USA, 1997.

67. Leccese, A.; Bartolini, S.; Viti, R. Genotype harvest season, and cold storage influence on fruit quality and antioxidant properties of apricot. *Int. J. Food Prop.* **2012**, *15*, 864–879. [CrossRef]

68. Fuchs, E.; Grünzig, M.; Kegler, H. Investigation on the Plum pox virus resistance in different apricot genotypes. *Acta Virol.* **1998**, *42*, 222–225.

69. Polák, J.; Oukropec, I.; Krška, B.; Pívalová, J.; Miller, W. Difference in reactions of apricot and peach cultivars to Plum pox virus: Serological and symptomatological evaluation. *HortScience* **2003**, *30*, 129–134. [CrossRef]

70. Audergon, J.M.; Dosba, F.; Karayiannis, I.; Dicenta, F. Amélioration de l'abricotier pour la résistance à la sharka. *EPPO Bull.* **1994**, *24*, 741–748. [CrossRef]

71. Mesarović, J.; Trifković, J.; Tosti, T.; Akšić, M.F.; Milatović, D.; Ličina, V.; Milojković-Opsenica, D. Relationship between ripening time and sugar content of apricot (*Prunus armeniaca* L.) kernels. *Acta Physiol. Plant.* **2018**, *40*, 157. [CrossRef]

72. Krška, B.; Oukropec, I.; Polak, J.; Kominek, P. The evaluation of apricot (*Prunus armeniaca* L.) cultivars and hybrids resistant to Sharka. *Acta Hortic.* **2000**, *538*, 143–146. [CrossRef]

73. Salazar, J.A.; Rubio, M.; Ruiz, D.; Tartarini, S.; Martínez-Gómez, P.; Dondini, L. SNP development for genetic diversity analysis in apricot. *Tree Genet. Genomes* **2015**, *11*, 15. [CrossRef]

74. Syrgiannidis, G.; Mainou, A. Two new apricot varieties resistant to Sharka (Plum pox virus) disease created by crossing. In *Agriculure, Proceedings of the Programme de recherche Agrimed. Deuxiemes Rencontres sur L'abricotier, Avignon, France, 27–31 May 1991*; CEC Commission of the European Communities: Luxembourg, 1993.

75. Egea, J.; Ruiz, D.; Dicenta, F.; Burgos, L. Murciana apricot. *HortScience* **2005**, *40*, 254–255. [CrossRef]

76. Ledbetter, C.A.; Peterson, S.J. 'Apache' and 'Kettleman': Two early season apricots for the fresh market. *HortScience* **2005**, *40*, 2202–2203. [CrossRef]

77. Trandafirescu, M.; Dumitru, L.M.; Trandafirescu, I. Evaluating the Resistance to the Plum pox virus of Some Apricot Tree Cultivars and Hybrids in South-Eastern Romania. *Proc. Latvian Acad. Sci.* **2013**, *67*, 203–206. [CrossRef]

78. Egea, J.; Ruiz, D.; Martínez-Gómez, P. Influence of rootstock on the productive behaviour of 'Orange Red' apricot under Mediterranean conditions. *Fruits* **2004**, *59*, 367–373. [CrossRef]

79. Karayannis, I.; Di Terlizzi, B.; Audergon, J.M. Susceptibility of apricot cultivars to Plum pox virus. *Acta Hortic.* **1999**, *488*, 753–760. [CrossRef]

80. Halász, J. Molecular Background of the S-locus Controlled Self-Incompatibility in Apricot. Ph.D. Thesis, Department of Genetics and Plant Breeding, Corvinus University, Budapest, Hungary, 2007.

81. Ledbetter, C.A.; Ramming, D.W. Apricot cv. Robada. United. States Patent USPP9890P, 13 May 1997.

82. Karayiannis, I.; Ledbetter, C.A. Susceptibility of certain apricot and plumcot cultivars to Plum pox virus infection. *Acta Hortic.* **2009**, *825*, 153–156. [CrossRef]

83. Egea, J.; Dicenta, F.; Burgos, L. 'Rojo Pasión' Apricot. *HortScience* **2004**, *39*, 1490–1491. [CrossRef]

84. Egea, J.; Dicenta, F.; Martínez-Gómez, P.; Burgos, L. Selene apricot. *HortScience* **2004**, *39*, 192–1493. [CrossRef]

85. Syrgiannidis, G. Selection of two apricot varieties resistant to Sharka virus. *Acta Phytopathol. Acad. Sci. Hung.* **1980**, *15*, 85–87. [CrossRef]

86. Adascalului, M.; Hoza, D.; Ion, L. Behaviour study for pollination a Romanian apricot varieties using different source of resistance to Sharka. *J. Hortic. For. Biotechnol.* **2014**, *18*, 13–17.

87. Ion, L.; Asănică, A.; Moale, C. Studies of resistance to Sharka in several Romanian apricot progenies. In Proceedings of the International Conference on Chemical, Agricultural and Biological Sciences (CABS-2015), Istanbul, Turkey, 4–5 September 2015.

88. Elibüyük, S.; Erdiller, G. The susceptibility of some apricot and plum varieties to Plum pox (sharka) virus. *Acta Hortic.* **1995**, *384*, 549–552. [CrossRef]

89. Rodríguez, J.; Andrés, V.; Gil, L.; Martínez, J.; Hita, I. *Sensibilidad a Sharka en Variedades de Albaricoquero de Murcia*; Frutales Hueso Fund. La Caixa: Barcelona, Spain, 1995; pp. 56–64.

90. Dosba, F.; Lansac, M.; Maison, P.; Massonie, G.; Audergon, J.M. Tolerance to Plum pox virus in apricot. *Acta Hort.* **1988**, *235*, 275–281. [CrossRef]

91. Dondini, L.; Lain, O.; Geuna, F.; Banfi, R.; Gaiotti, F.; Tartarini, S.; Bassi, D.; Testolin, R. Development of a new SSR-based linkage map in apricot and analysis of synteny with existing *Prunus* maps. *Tree Genet. Genomes* **2007**, *3*, 239–249. [CrossRef]

92. Martínez-Gómez, P.; Dicenta, F. Evaluation of resistance of apricot cultivars to a Spanish isolate of plum pox potyvirus (PPV). *Plant. Breed.* **2000**, *119*, 179–181. [CrossRef]

93. Karayiannis, I. Susceptibility of apricots cultivars to Plum pox virus in Greece. *Acta Hortic.* **1989**, *235*, 271–274. [CrossRef]

94. Balan, V.; Stoian, E. Susceptibility of certain apricot-tree to the Plum pox virus pathogenic action. *Acta Hortic.* **1995**, *384*, 565–569. [CrossRef]

95. Avinent, L.; Hermoso de Mendoza, A.; Llácer, G.; García, S. Transmisión del virus de la sharka y sensibilidad varietal en albaricoquero. In Proceedings of the II Congreso Ibérico Ciencias Hortícolas, Zaragoza, Spain, 27–30 April 1993.

96. Maghuly, F.; Borroto-Fernandez, E.; Ruthner, S.; Pedryc, A.; Laimer, M. Microsatellite variability in apricots (*Prunus armeniaca* L.) reflects their geographic origin and breeding history. *Tree Genet. Genomes* **2005**, *1*, 151–165. [CrossRef]

97. Audergon, J.M.; Morvan, G.; Dicenta, F.; Chastelliere, G.; Karayiannis, I. A method to determine the susceptibility of apricot to Plum pox virus. *Acta Hortic.* **1995**, *384*, 575–579. [CrossRef]

98. Egea, J.; Burgos, L. Detecting cross-incompatibility of three North American apricot cultivars and establishing the first incompatibility group in apricot. *J. Am. Soc. Hort. Sci.* **1996**, *121*, 1002–1005. [CrossRef]

99. McLaren, J. Apricot tree, 'F168 cv'. United. States Patent USPP16071P2, 25 October 2005.

100. Dicenta, F.; Audergon, J.M. Localization of Plum pox virus (PPV) in tissues of susceptible and resistant apricot cultivars. *Phytopathol. Med.* **1995**, *34*, 83–87.

101. Lachkar, A.; Mlika, M. New apricot varieties selected from the Tunisian breeding programme. *Acta Hort.* **2006**, *717*, 189–192. [CrossRef]

102. Dosba, F.; Denise, F.; Audergon, J.M.; Maison, P.; Massonie, G. Plum pox virus resistance of apricot. *Acta Hortic.* **1991**, *293*, 569–580. [CrossRef]

103. Egea, J.; Campoy, J.A.; Dicenta, F.; Burgos, L.; Patiño, J.L.; Ruiz, D. 'Estrella' and 'Sublime' apricot cultivars. *HortScience* **2009**, *44*, 469–470. [CrossRef]

104. Eynard, A.; Roggero, P.; Lenzi, R.; Conti, M.; Milne, R.G. Test for pollen and seed transmission on Plum pox virus (Sharka) in two apricot cultivars. *Adv. Hortic. Sci.* **1991**, *5*, 104–106.

105. Gallois, J.L.; Moury, B.; German-Retana, S. Role of the Genetic Background in Resistance to Plant Viruses. *Int. J. Mol. Sci.* **2018**, *19*, 2856. [CrossRef]

106. Polák, J.; Kominek, P.; Jokes, M.; Oukropec, I.; Krška, B. The evaluation of resistance of apricots to Plum pox virus by ELISA and ISEM. *Acta Hortic.* **1995**, *386*, 285–289. [CrossRef]

107. Layne, R.E.C.; Hunter, D.M. 'AC Harostar' apricot. *HortScience* **2003**, *38*, 140–141. [CrossRef]

108. Hegedűs, A.; Lénárt, J.; Halász, J. Sexual incompatibility in *Rosaceae* fruit tree species: Molecular interactions and evolutionary dynamics. *Biol. Plant* **2012**, *56*, 201–209. [CrossRef]

109. Austin, P.T. Pollination of Sundrop Apricot. Ph.D. Thesis, Massey University, Auckland, New Zealand, 1995.

110. Egea, J.; Ruiz, D.; Burgos, L. "Dorada" apricot. *HortScience* **2005**, *40*, 1919–1920. [CrossRef]

111. Muñoz-Sanz, J.V.; Zuriaga, E.; López, I.; Badenes, M.L.; Romero, C. Self-(in)compatibility in apricot germplasm is controlled by two major loci, *S* and *M*. *BMC Plant. Biol.* **2017**, *17*, 82. [CrossRef]

112. Corrin, A.A. "Ruby" Apricot Tree. United. States Patent USPP8177, 16 March 1993.

113. Zaiger, C.F. Apricot Tree (Spring Giant). United. States Patent USPP5138, 15 November 1983.

114. Dicenta, F.; Audergon, J.M. Inheritance of resistance to plum pox potyvirus (PPV) in 'Stella' apricot seedlings. *Plant. Breed.* **1998**, *117*, 579–581. [CrossRef]

115. Tartarini, S.; Sansavini, S.; Vinatzer, B.; Gennari, F.; Domizi, C. Efficiency of marker assisted selection (MAS) for the Vf scab resistance gene. *Acta Hortic.* **2000**, *538*, 549–552. [CrossRef]

11

Identification and Verification of Quantitative Trait Loci Affecting Milling Yield of Rice

Hui Zhang [1,2,3], Yu-Jun Zhu [2], An-Dong Zhu [2], Ye-Yang Fan [2], Ting-Xu Huang [3], Jian-Fu Zhang [3,*], Hua-An Xie [1,3,*] and Jie-Yun Zhuang [2,*]

[1]   College of Crop Science, Fujian Agriculture and Forestry University, Fuzhou 350002, China; hzhangfafu@163.com
[2]   State Key Laboratory of Rice Biology and Chinese National Center for Rice Improvement, China National Rice Research Institute, Hangzhou 310006, China; yjzhu2013@163.com (Y.-J.Z.); zhuandong1234@163.com (A.-D.Z.); fanyeyangcnrri@163.com (Y.-Y.F.)
[3]   Rice Research Institute and Fuzhou Branch of the National Center for Rice Improvement, Fujian Academy of Agricultural Sciences, Fuzhou 350018, China; txhuang@sina.com
*   Correspondence: jianfzhang@163.com (J.-F.Z.); huaanxie@163.com (H.-A.X.); zhuangjieyun@caas.cn (J.-Y.Z.)

Abstract: Rice is generally consumed in the form of milled rice. The yield of total milled rice and head mill rice is affected by both the paddy rice yield and milling efficiency. In this study, three recombinant inbred line (RIL) populations and one $F_{4:5}$ population derived from a residual heterozygous (RH) plant were used to determine quantitative trait loci (QTLs) affecting milling yield of rice. Seven traits were analyzed, including recovery of brown rice (BR), milled rice (MR) and head rice (HR); grain yield (GY); and the yield of brown rice (BRY), milled rice (MRY) and head rice (HRY). A total of 77 QTLs distributed on 35 regions was detected in the three RIL populations. Four regions, where qBR5, qBR7, qBR10, and qBR12 were located, were validated in the RH-derived $F_{4:5}$ population. In the three RIL populations, all the 11 QTLs for GY detected were accompanied with QTLs for two or all the three milling yield traits. Not only the allele direction for milling yield traits was unchanged, but also the effects were consistent with GY. In the RH-derived $F_{4:5}$ population, regions controlling GY also affected all three milling yield traits. Results indicated that variations of BRY and MRY were mainly ascribed to GY, but HRY was determined by both GY and HR. Results also showed that the regions covering GW5–Chalk5 and Wx loci had major effects on milling quality and milling yield of rice. These two regions, which have been known to affect multiple traits determining grain quality and yield of rice, provide good candidates for milled yield improvement.

Keywords: brown rice recovery; milled rice recovery; head rice recovery; milling yield traits; QTL mapping; rice (*Oryza sativa* L.)

1. Introduction

As a major cereal crop, rice (*Oryza sativa* L.) provides staple food for at least half of the global population. Rice food is mainly consumed in the form of cooked milled rice. Farmer's incomes are based on both the paddy rice yield and the milling efficiency. In the postharvest processing, paddy grains are firstly de-hulled into brown rice and then milled into milled rice. Milled rice is separated into head rice (also called whole rice) and broken rice. Milled rice whose length is longer than or equal to 3/4 of its unbroken length falls into the category of head rice, and the rest is called broken rice. Head rice has a higher price than broken rice [1]. Three parameters, recovery of brown rice (BR), milled rice (MR) and head rice (HR), are used to evaluate rice milling quality and efficiency of the milling processing [2–4]. Generally, BR is defined as the percentage of brown rice to grains, MR the percentage of milled rice to grains, and HR the percentage of head rice to grains. Alternatively, MR is measured as the percentage of milled rice to brown rice, and HR the percentage of head rice to milled rice [5]. In some reports, these traits are called percent milling yield, brown rice yield, milled rice yield, total milled yield, or head rice yield [3,6–8].

In the past two decades, analysis of quantitative trait loci (QTLs) was employed to study the genetic basis of rice milling quality. A larger number of QTLs were identified using various populations developed from crosses of the same subspecies [3,9], between the *indica* and *japonica* subspecies [6,10–12], or between different species [2]. In these studies, clustering of QTLs for different milling traits was commonly observed. Some of them detected only one cluster. For example, Tan et al. [9] located QTLs for MR and HR in the C1087–RZ403 region on chromosome 3; Aluko et al. [2] detected *br8* and *hr8* in RM126–RM137 on chromosome 8; and Lou et al. [12] found *qMRR-3* and *qHRR-3* in RM3204–RM6283 on chromosome 3. Other studies detected more clusters. Li et al. [6] located *qBR-4* and *qMR-4* in the C975–C734 interval on chromosome 4, *qBR-9* and *qMR-9* in R1751–R2272 on chromosome 9, *qBR-10* and *qMR-10* in C488–R716 on chromosome 10, and *qBR11* and *qMR11* in R728–G202 on chromosome 11. Zheng et al. [11] detected *QBr6* and *QMr6* in the *Wx* region on chromosome 6, and *QBr7* and *QMr7* in RM505–RM118 on chromosome 7.

When QTL mapping for milling quality traits and components of grain yield was performed using the same population, a proportion of genomic regions were found to be associated with both types of traits. Using 240 backcross introgression lines derived from the Ce258/IR75862 cross, three QTL regions were found to simultaneously affect components of milling quality and grain yield [5]. In the RM71–RM300 interval harboring *qBR2*, the Ce258 allele increased BR and panicle weight. In RM348–RM349 harboring *qBR4*, the Ce258 allele increased BR and grain weight but decreased spikelet number. In RM250–RM482 harboring *qHR2*, the Ce258 allele increased HR, grain number and spikelet number but decreased grain weight. Using 205 recombinant inbred lines (RILs) developed from the L-204/01Y110 cross, two QTL regions were found to simultaneously affect milling traits and grain weight [8]. For QTLs located in RM5638–RM1361 on chromosome 1, the L-204 allele increased BR and HR but decreased grain weight. For QTLs linked to RM3283 on chromosome 10, the L-204 allele increased HR but decreased grain weight. These results provide evidences for genetic association between milling quality and grain yield in rice, but it remains unknown whether this association has a consequence on the yield of milled and head rice.

In the present study, QTL analysis was employed to determine the dependence of milled and head rice yield on milling quality and grain yield. Firstly, three RIL populations were used to detect QTLs for brown, milled and head rice recovery; grain yield; and brown, milled and head rice yield. Then, QTL validation was performed using one secondary population derived from a residual heterozygote (RH) identified from one of the RIL populations.

2. Materials and Methods

2.1. Plant Materials

Four populations of *indica* rice (*Oryza sativa* subsp. *indica*) were used, including three RIL populations and one RH-derived F$_{4:5}$ population.

The three RIL populations were previously used to detect QTLs for components of appearance quality and physiochemical traits for eating and cooking quality [13–15]. The TI population consisting of 204 lines was constructed from crosses between Teqing (TQ) and IRBB lines. The IRBB lines are near isogenic lines carrying different bacterial blight resistance genes [16] in the background of IR24, including IRBB50, IRBB51, IRBB52, IRBB54, IRBB55, and IRBB59. The ZM population consisting of 230 lines was constructed from a cross between Zhenshan 97 and Milyang 46 (MY46). The XM population consisting of 209 lines was constructed from a cross between Xieqingzao and MY46. All the parental lines of these populations have been widely used in the breeding and production of three-line hybrid rice in China.

The RH-derived $F_{4:5}$ population was previously used to validate minor QTL for gel consistency [15]. It was derived from one RH plant that was an F_7 progeny of the cross TQ/IRBB52. Of the 135 polymorphic markers included in the TI map, 33 were heterozygous and 102 were homozygous in the RH plant. This plant was selfed to produce an F_2-type population consisting of 250 individuals. Single seed descent was applied to advance the population to F_4. Seeds of the F_4 plants were harvested and a population consisting of 250 $F_{4:5}$ families was constructed.

2.2. Field Experiment and Trait Measurement

All the populations were planted in the middle rice growing season (from May to October) at the China National Rice Research Institute (CNRRI), Hangzhou (30°04′ N, 119°54′ E), China. The three RIL populations were tested for 2 years, including 2008 and 2009 for TI, 2009 and 2010 for ZM, and 2003 and 2009 for XM. The RH-derived $F_{4:5}$ population was tested for 1 year in 2017. The experiments followed a randomized complete block design. Twelve plants per line were transplanted in a single row with 16.7 cm between plants and 26.7 cm between rows. Field management followed common practice in rice production.

At maturity, five plants from the 10 middle plants of each line were randomly sampled and harvested. The grains were dried and weighted to calculate grain yield per plant (GY, g). Dried grains were stored at room temperature for three months. Then, two replicates of filled grains were processed independently. Filled grain of 100 g was de-husked using a Satake Rice Machine (Suzhou, China). The brown rice was milled using a JNMJ3 rice miller (Taizhou, China). Head rice of which the length was longer than or equal to 3/4 of the full length was separated from broken rice. Three traits for milling quality, i.e., BR, MR and HR, were calculated as the percentage of the weight of brown, milled and head rice to grain weight, respectively. Three traits for milling yield, i.e., brown, milled and head rice yield per plant, were calculated as follows:

$$\text{Brown rice yield per plant (BRY, g)} = \text{BR} \times \text{GY} \tag{1}$$

$$\text{Milled rice yield per plant (MRY, g)} = \text{MR} \times \text{GY} \tag{2}$$

$$\text{Head rice yield per plant (HRY, g)} = \text{HR} \times \text{GY} \tag{3}$$

2.3. Marker Data and Genetic Maps

Marker data and genetic maps of the four populations have been available [15]. The TI, ZM and XM maps were 1345.3, 1814.7 and 2080.4 cM in length, consisting of 135, 256 and 240 DNA markers, respectively. Genomic coverage and distances between neighboring markers are satisfactory for primary QTL mapping in the ZM and XM populations. A number of large homozygous segments remain in the TI map due to low polymorphism between the female and male parents of the TI population. For the RH-derived $F_{4:5}$ population, the map consisted of 35 markers, including 28 simple sequence repeats, six InDels and one single nucleotide polymorphism.

2.4. Data Analysis

For the three RIL populations that were tested for 2 years, phenotypic data averaged over 2 years were used for computing the descriptive statistics, plotting the frequency distribution and calculating the Pearson correlation coefficient, and the data of each year were used for QTL mapping. QTL analysis was performed using the default setting of the MET (multi-environmental trials) approach in IciMapping V4.1 [17], taking the 2 years for each population as two environments. *LOD* thresholds for genome-wide type I error of $p < 0.05$ were calculated with 1000 permutation test and used to claim a putative QTL. QTL effects and the proportions of phenotypic variance explained (R^2) were estimated. When a QTL was shown to have a significant genotype-by-environment (GE) interaction, the effect and R^2 due to GE interaction were also measured. QTLs were designated as proposed by McCouch and CGSNL [18].

For the RH-derived $F_{4:5}$ population that was tested for 1 year, QTLs were determined with the BIP (bi-parental populations) approach in IciMapping V4.1 [17]. *LOD* > 2.0 was used as the threshold to claim a putative QTL. QTLs were designated as proposed by McCouch and CGSNL [18].

3. Results

3.1. Phenotypic Performance of the Three RIL Populations

Descriptive statistics of the seven traits in the three RIL populations are presented in Table S1. Two of the seven traits, BR and MR, showed similar coefficients of variation (CV) among the three populations, ranging from 0.0116 to 0.0119 and 0.0140 to 0.0180, respectively. The CV of the five other traits were higher in the ZM and XM populations than in the TI population, ranging from 0.1266 to 0.2642 in ZM, 0.1288 to 0.2942 in XM, and 0.0967 to 0.1871 in TI (Table S1). Continuous distributions were observed for all the traits in the three populations (Figure S1), suggesting polygenic inheritance of these traits.

Correlations between the seven traits were either non-significant or positive highly significant ($p < 0.01$) (Table 1). Regarding the three traits for milling quality, the correlation was strong between BR and MR but weak between these two traits and HR. The correlation coefficients (*r*) between BR and MR ranged from 0.764 to 0.815 in the three populations, but their correlations with HR were either non-significant or had low *r* values (0.230–0.363). These results suggest that the control of properties for maintaining whole milled rice may differ greatly from that for achieving high brown and milled rice recovery.

Table 1. Simple correlation coefficients between seven traits in three RIL populations of rice.

Population	Trait	MR	HR	GY	BRY	MRY	HRY
TI	BR	0.764 **	−0.157	0.229 **	0.292 **	0.287 **	0.107
	MR		0.160	0.205 **	0.253 **	0.287 **	0.275 **
	HR			−0.064	−0.074	−0.046	0.505 **
	GY				0.998 **	0.996 **	0.816 **
	BRY					0.998 **	0.809 **
	MRY						0.826 **
ZM	BR	0.815 **	0.148	−0.006	0.055	0.054	0.063
	MR		0.230 **	0.063	0.112	0.136	0.157
	HR			0.354 **	0.362 **	0.368 **	0.710 **
	GY				0.998 **	0.997 **	0.902 **
	BRY					0.999 **	0.905 **
	MRY						0.907 **
XM	BR	0.784 **	0.233 **	0.246 **	0.292 **	0.296 **	0.300 **
	MR		0.363 **	0.365 **	0.398 **	0.425 **	0.449 **
	HR			0.224 **	0.232 **	0.241 **	0.605 **
	GY				0.999 **	0.998 **	0.900 **
	BRY					0.999 **	0.903 **
	MRY						0.907 **

TI = Teqing/IRBB lines; ZM = Zhenshan 97/Milyang 46; XM = Xieqingzao/Milyang 46; BR = Brown rice recovery; MR = Milled rice recovery; HR = Head rice recovery; GY = Grain yield per plant; BRY = Brown rice yield per plant; MRY = Milled rice yield per plant; HRY = Head rice yield per plant; **, $p < 0.01$.

Regarding the four yield traits, GY, BRY, MRY and HRY, not only the correlations were all significant but also the coefficients were all high. Near-perfect correlation was observed between GY, BRY and MRY, with the *r* values ranging from 0.996 to 0.999. Lower *r* values were found between these three traits and HRY, ranging from 0.816 to 0.826, 0.902 to 0.907 and 0.900 to 0.907 in the TI, ZM and XM populations, respectively. These results suggest that GY is the main source of variations for BRY,

MRY and HRY, and the postharvest processing has a more significant influence on HRY than on BRY and MRY.

Regarding the three pairs of corresponding traits for milling quality and yield, the correlation was stronger between HR and HRY than between the other two pairs of traits. The r values ranged from 0.505 to 0.710 between HR and HRY in the three populations and decreased to 0.292–0.055 between BR and BRY, and 0.136–0.425 between MR and MRY. These results also suggest that the postharvest processing has a more significant influence on HRY than on BRY and MRY.

3.2. QTL Detected in the Three RIL Populations

In the TI, ZM and XM populations, a total of 27, 33 and 17 QTLs were detected for the seven traits analyzed, of which one, four and none showed significant GE interactions, respectively (Figure 1; Tables 2–4). Based on the physical position of DNA markers, it was found that all the five QTLs having significant GE effects were located in the region covering the Wx locus on the short arm of chromosome 6.

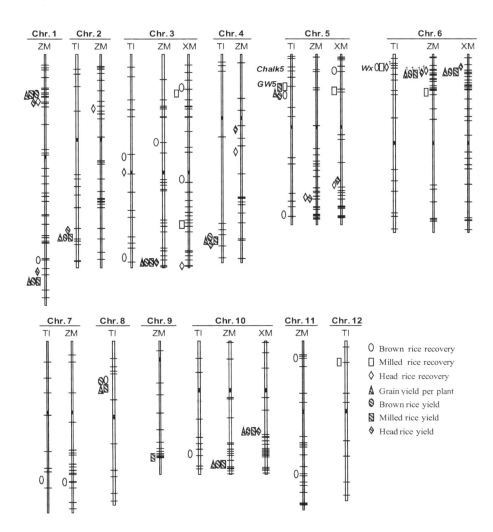

Figure 1. Genomic distribution of QTLs for seven traits detected in three RIL populations. TI = Teqing/IRBB lines; ZM = Zhenshan 97/Milyang 46; XM = Xieqingzao/Milyang 46. Marker positions in each chromosome are indicated by solid lines and the distances are in proportion to the physical length. Solid rectangles refer to the approximate positions of centromeres. QTLs are drawn on the left side of the corresponding interval. Significant genotype-by-environment interaction is indicated by the number "1".

Table 2. QTLs for seven traits detected in the TI population.

Chr	Interval	QTL	*LOD*	*LOD* (A)	*LOD* (ge)	*A*	*ge*	R^2 (A)	R^2 (ge)
2	RM6–RM240	*qGY2*	8.84	8.25		−1.89		10.75	
		qBRY2	8.76	8.09		−1.54		10.37	
		qMRY2	8.15	7.61		−1.30		9.78	
		qHRY2	7.44	6.57		−1.21		8.38	
3	RM15139–RM15303	*qBR3.1*	26.64	25.34		0.41		14.16	
		qHR3	14.07	13.17		−2.51		17.81	
	RM16048–RM16184	*qBR3.2*	8.02	7.46		0.21		3.75	
4	RM6992–RM349	*qGY4*	4.90	4.88		1.44		6.23	
		qBRY4	4.59	4.58		1.14		5.64	
		qMRY4	4.53	4.52		0.99		5.69	
		qHRY4	3.90	3.85		0.89		4.55	
5	RM437–RM18189	*qBR5.1*	35.75	34.08		−0.51		20.20	
		qMR5	11.22	10.51		−0.39		11.99	
		qGY5	4.18	3.52		−1.30		4.52	
		qBRY5	5.66	4.81		−1.25		6.02	
		qMRY5	6.17	5.06		−1.13		6.48	
	RM274–RM334	*qBR5.2*	3.11	2.93		−0.13		1.38	
6	RM190–RM587	*qBR6*	15.18	14.32		−0.30		7.45	
		qMR6	6.30	5.96		−0.28		6.44	
		qHR6	10.08	4.35	5.73	−1.40	−1.20	5.59	4.28
7	RM70–RM18	*qBR7*	5.82	5.54		0.19		2.75	
8	RM547–RM22755	*qBR8*	8.00	6.82		−0.20		3.31	
		qGY8	2.85	2.82		−1.10		3.55	
		qBRY8	3.42	3.33		−0.98		4.09	
		qMRY8	2.85	2.82		−0.79		3.53	
10	RM6100–RM3773	*qBR10*	14.93	12.44		−0.28		6.60	
12	RM20–RM27610	*qMR12*	3.48	3.36		0.21		3.50	

QTLs are designated as proposed by McCouch and CGSNL [18]. *A*: additive effect of replacing a maternal with a paternal allele; *ge*: effect due to genotype-by-environment interaction; R^2: percentage of phenotypic variance explained by the additive or GE effect.

Table 3. QTLs for seven traits detected in the ZM population.

Chr	Interval	QTL	*LOD*	*LOD* (A)	*LOD* (ge)	*A*	*ge*	R^2 (A)	R^2 (ge)
1	RG532–RM5359	*qHR1*	9.34	9.14		2.21		8.95	
		qGY1.1	5.25	4.35		0.89		4.79	
		qBRY1.1	5.53	4.56		0.74		4.93	
		qMRY1.1	5.68	4.73		0.69		5.01	
		qHRY1.1	6.16	4.60		0.70		4.59	
	RZ730–RG381	*qBR1*	7.34	6.67		−0.24		5.91	
		qGY1.2	6.63	6.61		1.12		7.51	
		qBRY1.2	5.96	5.93		0.86		6.59	
		qMRY1.2	5.58	5.58		0.76		6.02	
		qHRY1.2	4.69	4.30		0.67		4.34	
2	A5–RM71	*qHR2*	3.30	3.30		−1.29		3.00	
3	RM251–RG393	*qBR3*	6.00	5.77		−0.22		5.02	
	RZ613–RG418A	*qGY3*	4.17	3.14		−0.76		3.42	
		qBRY3	4.17	3.04		−0.61		3.24	
		qMRY3	4.00	2.98		−0.56		3.18	
		qHRY3	5.06	4.08		−0.65		4.03	
4	RZ69–RM3317	*qHRY4*	5.72	5.68		0.76		5.52	
	RM401–RM3643	*qHR4*	3.76	3.51		1.31		3.20	
5	CDO348–RG480	*qHR5*	4.95	4.94		−1.61		4.69	
		qHRY5	5.80	5.79		−0.79		5.81	

Table 3. *Cont.*

Chr	Interval	QTL	LOD	LOD (A)	LOD (ge)	A	ge	R^2 (A)	R^2 (ge)
6	RZ516–RM197	qHR6	5.72	4.78		−1.56		4.51	
		qGY6	7.97	3.30	4.67	−0.79	0.80	3.72	3.72
		qBRY6	7.74	3.22	4.52	−0.63	0.63	3.54	3.54
		qMRY6	7.38	3.13	4.24	−0.58	0.54	3.39	3.03
		qHRY6	7.41	4.15	3.27	−0.66	−0.64	4.11	3.89
	RM276–RZ667	qMR6	3.16	3.01		0.20		3.25	
7	RG650–RZ395	qBR7	4.08	3.50		0.18		3.25	
9	RG667–RM201	qMRY9	4.13	3.59		0.60		3.77	
10	RZ811–RZ583	qGY10	4.46	3.68		−0.83		4.13	
		qBRY10	5.21	4.34		−0.73		4.78	
		qMRY10	5.17	4.45		−0.68		4.82	
11	RZ816–RM332	qBR11.1	3.12	3.10		0.16		2.63	
	RM187–RM254	qBR11.2	3.20	3.18		−0.16		2.73	

QTLs are designated as proposed by McCouch and CGSNL [18]. *A*: additive effect of replacing a maternal with a paternal allele; *ge*: effect due to genotype-by-environment interaction; R^2: percentage of phenotypic variance explained by the additive or GE effect.

Table 4. QTLs for seven traits detected in the XM population.

Chr	Interval	QTL	LOD	LOD (A)	A	R^2 (A)
3	RM6849–RM14629	qBR3.1	3.97	3.58	0.21	4.57
		qMR3.1	3.76	3.01	0.28	2.96
	RZ696–RG445A	qBR3.2	3.14	2.99	−0.19	3.58
	RZ519–RZ328	qMR3.2	3.35	3.26	−0.28	3.10
	RM85–RG418A	qHR3	3.39	3.12	−1.61	4.28
5	RM13–RM267	qBR5	3.94	3.02	0.19	3.53
	RG182–RG413	qMR5	6.86	6.86	0.43	7.05
	RM163–RG470	qHR5	3.78	3.28	−1.62	4.58
		qHRY5	3.02	3.00	−0.80	3.09
6	RM190–RM204	qGY6	7.14	6.64	−1.74	8.46
		qBRY6	7.14	6.65	−1.43	8.25
		qMRY6	7.31	6.73	−1.32	8.53
		qHRY6	4.97	4.30	−0.99	4.85
10	RM1859–RM184	qGY10	4.56	4.06	−1.36	5.15
		qBRY10	4.51	4.00	−1.11	4.96
		qMRY10	4.70	4.17	−1.04	5.28
		qHRY10	6.11	5.39	−1.11	6.05

QTLs are designated as proposed by McCouch and CGSNL [18]. *A*: additive effect of replacing a maternal with a paternal allele; R^2: percentage of phenotypic variance explained by the additive effect.

3.3. QTLs Detected in the TI Population

The 27 QTLs identified in the TI population were distributed across nine of the 12 rice chromosomes (Figure 1, Table 2). Numbers of QTLs detected for BR, MR, HR, GY, BRY, MRY, and HRY were 8, 3, 2, 4, 4, 4, and 2, having overall R^2 of 59.55%, 21.93%, 23.40%, 25.05%, 26.13%, 25.49%, and 12.93%, respectively. Twenty-two of these QTLs formed six clusters distributed on chromosomes 2, 3, 4, 5, 6, and 8.

The largest cluster consisted of five QTLs, followed by three clusters of four QTLs. It was found that the 14 QTLs detected for grain yield and the three traits for milling yield were all included in these four clusters. In the RM437–RM18189 region on chromosome 5, the TQ allele increased BR, MR, GY, BRY, and MRY by 0.51%, 0.39%, 1.30 g, 1.25 g, and 1.13 g, respectively (Table 2). In the RM6–RM240 region on chromosome 2, the TQ allele increased GY, BRY, MRY, and HRY by 1.89 g, 1.54 g, 1.30 g, and 1.21 g, respectively. In the RM6992–RM349 region on chromosome 4, the TQ allele decreased

GY, BRY, MRY, and HRY by 1.44 g, 1.14 g, 0.99 g, and 0.89 g, respectively. In the RM547–RM22755 region on chromosome 8, the TQ allele increased BR, GY, BRY, and MRY by 0.20%, 1.10 g, 0.98 g, and 0.79 g, respectively.

The fifth cluster consisted of three QTLs, which were located in the RM190–RM587 region covering the *Wx* locus [19] on chromosome 6. The TQ allele increased BR, MR and HR by 0.30%, 0.28% and 1.40%, respectively. The sixth cluster consisted of two QTLs, which were located in the RM15139–RM15303 region covering the *GS3* locus [20] on chromosome 3. The TQ allele decreased BR by 0.41 g but increased HR by 2.51 g.

In the other five regions, one QTL was detected in each region. Included were *qBR3.2* located in the interval RM16048–RM16184 on chromosome 3, *qBR5.2* in RM274–RM334 on chromosome 5, *qBR7* in RM70–RM18 on chromosome 7, *qBR10* in RM6100–RM3773 on chromosome 10, and *qMR12* in RM20–RM27610 on chromosome 12.

3.4. QTLs Detected in the ZM Population

The 33 QTLs identified in the ZM population were distributed across 10 of the 12 rice chromosomes (Figure 1, Table 3). Numbers of QTLs detected for BR, MR, HR, GY, BRY, MRY, and HRY were 5, 1, 5, 5, 5, 6, and 6, having overall R^2 of 19.55%, 3.25%, 24.34%, 23.57%, 23.08%, 26.20%, and 28.39%, respectively. Twenty-four of these QTLs formed six clusters distributed on chromosomes 1, 3, 5, 6, and 10.

Two clusters on chromosome 1 and one on chromosome 6 were the three largest clusters consisting of five QTLs. Each of them affected one milling quality trait and all the four yield traits. In the RG532–RM5359 region on the short-arm chromosome 1, the MY46 allele increased HR, GY, BRY, MRY, and HRY by 2.21%, 0.89 g, 0.74 g, 0.69 g, and 0.70 g, respectively (Table 3). In the RZ730–RG381 region on the long arm of chromosome 1, the MY46 allele decreased BR by 0.24% but increased GY, BRY, MRY, and HRY by 1.12 g, 0.86 g, 0.76 g, and 0.67 g, respectively. In the RZ516–RM197 region covering the *Wx* locus on chromosome 6, the MY46 allele decreased HR, GY, BRY, MRY, and HRY by 1.56%, 0.79 g, 0.63 g, 0.58 g, and 0.66 g, respectively.

The other three clusters consisted of four, three and two QTLs, respectively. The RZ613–RG418A region on chromosome 3 affected all four yield traits, with the MY46 allele decreasing GY, BRY, MRY, and HRY by 0.76 g, 0.61 g, 0.56 g, and 0.65 g, respectively. The RZ811–RZ583 region on chromosome 10 affected three yield traits, with the MY46 allele decreasing GY, BRY and MRY by 0.83 g, 0.73 g and 0.68 g, respectively. The CDO348–RG480 region on chromosome 5 affected the recovery and yield of head rice, with the MY46 allele decreasing HR and HRY by 1.61% and 0.79 g, respectively.

Two other QTLs, *qHRY4* and *qHR4*, were mapped in close positions on chromosome 4 (Figure 1). The MY46 allele increased HRY and HR by 0.76 g and 1.31%, respectively (Table 3). In the other seven regions, one QTL was detected in each region. Included were *qHR2* located in the interval A5–RM71 on chromosome 2, *qBR3* in RM251–RG393 on chromosome 3, *qMR6* in RM276–RZ667 on chromosome 6, *qBR7* in RG650–RZ395 on chromosome 7, *qMRY9* in RG667–RM201 on chromosome 9, *qBR11.1* in RZ816–RM332, and *qBR11.2* in RM187–RM254 on chromosome 11.

3.5. QTLs Detected in the XM Population

The 17 QTLs identified in the XM population were distributed on four of the 12 rice chromosomes (Figure 1, Table 4). Numbers of QTLs detected for BR, MR, HR, GY, BRY, MRY, and HRY were 3, 3, 2, 2, 2, 2, and 3, having overall R^2 of 11.68%, 13.11%, 8.86%, 13.62%, 13.21%, 13.80%, and 13.98%, respectively. Twelve of these QTLs formed four clusters distributed on chromosomes 3, 5, 6, and 10.

The clusters on chromosomes 6 and 10 were the two largest clusters consisting of four QTLs, both of which affected all four yield traits. In the RM190–RM204 region covering the *Wx* locus on chromosome 6, the MY46 allele decreased GY, BRY, MRY, and HRY by 1.74 g, 1.43 g, 1.32 g, and 0.99 g, respectively (Table 4). In the RM1859–RM184 region on chromosome 10, the MY46 allele decreased GY, BRY, MRY, and HRY by 1.36 g, 1.11 g, 1.04 g, and 1.11 g, respectively.

The other two clusters each consisted of two QTLs. In the RM6849–RM14629 region on the short arm of chromosome 3, the MY46 allele increased BR and MR by 0.21% and 0.28%, respectively. In the RM163–RG470 on the long arm of chromosome 5, the MY46 allele decreased HR and HRY by 1.62% and 0.80 g, respectively. Two other QTLs, *qBR5* and *qMR5*, were mapped in close positions on the short arm of chromosome 5 (Figure 1). The MY46 allele increased BR and MR by 0.19% and 0.43%, respectively (Table 4). The remaining three QTLs were loosely linked on the long arm of chromosome 3, including *qBR3.2* for brown rice recovery, *qMR3.2* for milled rice recovery, and *qHR3* for head rice recovery.

3.6. Validation of Five QTL Regions in an RH-Derived F$_{4:5}$ Population

In the Ti52-3 population that was derived from an RH-plant of TQ/IRBB52, correlations between the seven traits (Table S2) are much the same as in the three RIL populations. Regarding the three traits for milling quality, the correlation between BR and MR (r = 0.773) was much stronger than between these two traits and HR (r = 0.243 and 0.266). Regarding the four yield traits, near-perfect correlation was observed between GY, BRY and MRY (r values ranging as 0.993–0.998), and their correlations with HRY were slightly weaker (r values ranging as 0.923–0.929). Regarding the three pairs of traits for milling quality and yield, the correlation between HR and HRY (r = 0.440) was much stronger than between the two others (r = 0.065 and 0.159).

Among the 16 segregating regions distributed on 12 chromosomes in the Ti52-3 population, QTLs were detected in 10 regions across nine chromosomes (Figure 2; Table 5). A total of 26 QTLs were found, including 6, 6, 1, 3, 3, 3, and 4 for BR, MR, HR, GY, BRY, MRY, and HRY, which had overall R^2 of 33.39%, 45.99%, 6.01%, 22.75%, 22.93%, 23.73%, and 23.27%, respectively.

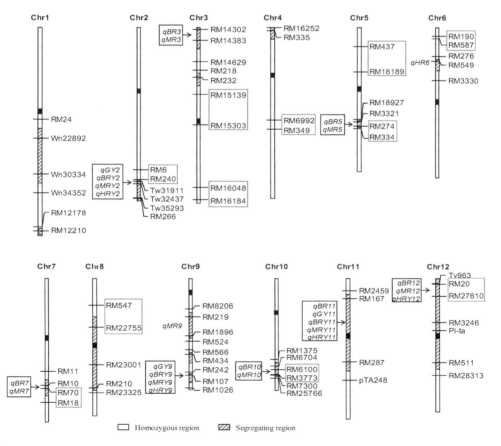

Figure 2. Genomic distribution of QTLs for seven traits detected in the RH-F$_{4:5}$ population. BR = Brown rice recovery; MR = Milled rice recovery; HR = Head rice recovery; GY = Grain yield per plant; BRY = Brown rice yield per plant; MRY = Milled rice yield per plant; HRY = Head rice yield per plant. Markers within the blue rectangle are flanking markers of QTLs detected in the TI population.

Table 5. QTLs for seven traits detected in the RH-$F_{4:5}$ population.

Chr	Interval	QTL	LOD	A	D	R^2 (%)
2	Tw31911–Tw32437	qGY2	8.93	1.87	0.41	11.60
		qBRY2	8.88	1.52	0.29	11.59
		qMRY2	9.25	1.40	0.20	12.03
		qHRY2	7.35	1.29	0.16	10.67
3	RM14302–RM14383	qBR3	2.34	−0.13	0.05	2.82
		qMR3	6.41	−0.31	0.08	7.30
5	RM3321–RM274	qBR5	4.88	−0.17	−0.18	6.25
		qMR5	5.92	−0.26	−0.33	6.54
6	RM549	qHR6	8.05	1.34	0.31	6.01
7	RM10–RM70	qBR7	7.23	0.23	−0.09	9.28
		qMR7	12.89	0.44	−0.15	15.51
9	RM219–RM1896	qMR9	3.06	0.20	0.12	3.34
9	RM107	qGY9	2.83	−0.93	−0.84	3.35
		qBRY9	2.81	−0.76	−0.65	3.35
		qMRY9	3.09	−0.73	−0.52	3.68
		qHRY9	2.04	−0.62	−0.28	2.67
10	RM6704–RM7300	qBR10	2.20	−0.10	−0.21	2.52
		qMR10	2.92	−0.18	−0.24	3.29
11	RM167–RM287	qBR11	2.20	0.10	0.20	2.52
		qGY11	3.30	0.76	3.05	7.80
		qBRY11	3.48	0.63	2.52	8.00
		qMRY11	3.46	0.55	2.30	8.03
		qHRY11	2.43	0.36	2.34	7.19
12	Tv963–RM3246	qBR12	2.84	0.10	−0.57	9.99
		qMR12	3.74	0.12	−0.88	10.01
		qHRY12	2.04	0.62	−0.32	2.73

QTLs are designated as proposed by McCouch and CGSNL [18]. *A*: additive effect of replacing a maternal with a paternal allele; *D*: dominance effect; R^2: proportion of the phenotypic variance explained by the QTL.

Of the QTLs detected in the TI population, four were covered by the segregating regions of the Ti52-3 population (Figure 2), including *qBR5.2* located in the interval RM274–RM334 on chromosome 5, *qBR7* in RM70–RM18 on chromosome 7, *qBR10* in RM6100–RM3773 on chromosome 10, and *qMR12* in RM20–RM27610 on chromosome 12. They were all well validated. The TQ alleles consistently increased BR in the *qBR5.2* and *qBR10* regions, decreased BR in the *qBR7* region, and decreased MR in the *qMR12* region (Tables 2 and 5). Additionally, significant effects were newly detected on MR in the *qBR5.2*, *qBR7* and *qBR10* regions, and on BR and HRY in the *qMR12* region. The QTL region *qBR8/qGY8/qBRY8/qMRY8* found in the TI population was overlapped with the segregating region RM22755–RM23001 in the Ti52-3 population (Figure 2). These QTLs were not detected in Ti52-3. Since one side of this putative QTL region was homozygous in the new population, it is possible that the QTLs were not segregated in Ti52-3.

The other six QTL regions found in the Ti52-3 population were not detected in the TI population. One of them, Tw31911–Tw32437 on chromosome 2, showed significant effects on four traits. In the neighboring region RM6–RM240, QTLs for the same four traits were detected in the TI population. However, the QTL directions were opposite between the two regions. It is noted that RM6–RM240 segregated in the TI population was homozygous in the Ti52-3 population (Figure 2). The gene underlying this QTL cluster may be located between RM6–RM240 and Tw31911–Tw32437, and crossover may have occurred between the gene and Tw31911.

Five other QTL regions detected in the Ti52-3 population included three QTL clusters and two regions affecting a single trait. The RM14302–RM14383 region on chromosome 3 affected two traits, in which the TQ allele increased BR and MR by 0.13% and 0.31%, respectively. The RM107 region on chromosome 9 affected four traits, in which the TQ allele increased GY, BRY, MRY, and HRY by 0.93 g, 0.76 g, 0.73 g, and 0.62 g, respectively. The RM167–RM287 region on chromosome 11 affected five traits, in which the TQ allele decreased BR, GY, BRY, MRY, and HRY by 0.10%, 0.76 g, 0.63 g, 0.55 g, and 0.36 g, respectively. The remaining two QTLs were qHR6 and qMR9, of which the TQ allele decreased HR and MR by 1.34% and 0.20%, respectively.

4. Discussion

Milled and head rice yield, two of the most important commercial traits in rice production, are determined by grain yield and milling quality. Understanding the genetic relationship among these traits is critical for the improvement of milled and head rice yield in breeding. In this study, QTL analysis for seven traits—brown, milled and head rice recovery, grain yield, and brown, milled and head rice yield—was performed using three RIL populations and one RH-derived $F_{4:5}$ population. New knowledge on the genetic basis underlying the control of brown, milled and head rice yield is provided.

In the four populations investigated in this study, correlations between the four yield traits were all highly significant. Near-perfect correlations were observed between GY, BRY and MRY, and their correlations with HRY were slightly weaker. These results were supported by QTLs detected for the four traits. Four, five, two, and three QTLs were detected for grain yield in the TI, ZM, XM, and Ti52-3 populations. It is worth noting that each of these QTL regions had significant effects on all or two of the three traits for milling yield. For multiple QTLs accompanied in the same region, not only the allelic direction remained unchanged, but also the effects were consistent. Of the four regions controlling GY in TI, the qGY2 and qGY4 regions controlled all four traits, but the qGY5 and qGY8 regions were non-significant for HRY (Table 2). No other QTLs for these traits were detected in TI. Of the five regions controlling GY in ZM, the qGY1.1, qGY1.2, qGY3, and qGY6 regions controlled all four traits, but the qGY10 region was non-significant for HRY (Table 3). One more QTL for MRY, qMRY9, was detected alone. One more QTL for HRY, qHRY4, was detected and accompanied with a QTL for HR, qHR4. In XM, the two regions controlling GY both affected all four traits (Table 4). One more QTL for HRY, qHRY5, was detected and accompanied with a QTL for HR, qHR5. Similarly, all three regions controlling GY in Ti52-3 affected all four traits (Table 5). These results have two implications. Firstly, variation on paddy grain yield might be the only main source of variation for brown and milled rice yield. Secondly, variations on the paddy grain yield and head rice recovery both make important contributions to the variation of head rice yield.

Significant correlations between different quality traits in rice have been commonly observed [7,8, 12,21], which could be partly ascribed to the influence of a QTL region on multiple traits [7,12]. By comparing the locations of genes or QTLs reported for various grain quality traits in rice, it is found that some regions harboring QTLs for milling quality are associated with other traits that determine appearance quality or eating and cooking characteristics. Two typical examples are the GW5–Chalk5 region on chromosome 5 [22,23] and the Wx region on chromosome 6 [19]. In the GW5–Chalk5 region, QTLs having major effects for MR were detected in the TI and XM populations. In a previous study reported by Zheng et al. [11], one QTL for MR, QMr5, was also detected in this region, having an additive effect of 1.10% and R^2 of 11.5%. In addition, this region was reported to affect various traits for grain chalkiness, endosperm transparency and grain size in the TI and XM populations [13,14].

The Wx gene not only plays a key role in controlling eating and cooking quality of rice, but also influences other traits including protein content, head rice recovery, grain chalkiness, and grain weight [2,3,8,9,24]. The Wx locus was segregated in the TI, ZM and XM populations, having major effects on amylose content and gel consistency [15]. The Wx region also showed significant effects on grain chalkiness, grain width and endosperm transparency in TI; on grain chalkiness and grain length in ZM; and on grain chalkiness, grain length and endosperm transparency in XM [13,14]. In the present study, this region was found to have significant effects on BR, MR and HR in TI; on HR, GY, BRY, MRY, and HRY in ZM; and on GY, BRY, MRY, and HRY in XM.

In conclusion, the *GW5–Chalk5* and *Wx* regions are good targets for studying the genetic control of multiple traits determining grain yield, appearance quality, eating and cooking quality, milling quality, and milling yield.

5. Conclusions

A total of 77 QTLs for seven traits affecting milling yield in rice were detected using three RIL populations. All the regions harboring QTLs for grain yield were found to affect two or all three milling yield traits. QTLs for head rice yield were usually accompanied with grain yield and head rice recovery. Variations of brown and milled rice yield were mainly ascribed to grain yield, but head rice yield was determined by both grain yield and head rice recovery. Two regions covering *GW5–Chalk5* and *Wx* loci, respectively, had a major contribution to milling quality and milling yield of rice.

Author Contributions: Conceptualization, J.-Y.Z., H.-A.X. and J.-F.Z.; investigation, H.Z., Y.-J.Z., A.-D.Z., Y.-Y.F. and T.-X.H.; writing—original draft preparation, H.Z.; writing—review and editing, J.-Y.Z. All authors have read and agreed to the published version of the manuscript.

Acknowledgments: The authors would like to thank D.-P. Li for his assistance in field work. We acknowledge Y.-F. Sun and H.-Z. Lin for their technical assistance in laboratory works.

References

1.   Nalley, L.; Tack, J.; Barkley, A.; Jagadish, K.; Brye, K. Quantifying the agronomic and economic performance of hybrid and conventional rice varieties. *Agron. J.* **2016**, *108*, 1514–1523. [CrossRef]

2.   Aluko, G.; Martinez, C.; Tohme, J.; Castano, C.; Bergman, C.; Oard, J.H. QTL mapping of grain quality traits from the interspecific cross *Oryza sativa* x *O. glaberrima*. *Theor. Appl. Genet.* **2004**, *109*, 630–639. [CrossRef] [PubMed]

3.   Kepiro, J.L.; McClung, A.M.; Chen, M.H.; Yeater, K.M.; Fjellstrom, R.G. Mapping QTLs for milling yield and grain characteristics in a tropical *japonica* long grain cross. *J. Cereal Sci.* **2008**, *48*, 477–485. [CrossRef]

4.   Bao, J. Genes and QTLs for rice grain quality improvement. In *Rice-Germplasm, Genetics and Improvement*; Yan, W., Bao, J., Eds.; Intech: Rijeka, Croatia, 2014; pp. 239–278.

5.   Hu, X.; Shi, Y.-M.; Jia, Q.; Xu, Q.; Wang, Y.; Chen, K.; Sun, Y.; Zhu, L.-H.; Xu, J.-L.; Li, Z.-K. Analyses of QTLs for rice panicle and milling quality traits and their interaction with environment. *Acta Agron. Sin.* **2011**, *37*, 1175–1185.

6.   Li, Z.F.; Wan, J.M.; Xia, J.F.; Zhai, H.Q.; Ikehashi, H. Identification of quantitative trait loci underlying milling quality of rice (*Oryza sativa*) grains. *Plant Breed.* **2004**, *123*, 229–234. [CrossRef]

7.   Nelson, J.C.; McClung, A.M.; Fjellstrom, R.G.; Moldenhauer, K.A.K.; Boza, E.; Jodari, F.; Oard, J.H.; Linscombe, S.; Scheffler, B.E.; Yeater, K.M. Mapping QTL main and interaction influences on milling quality in elite US rice germplasm. *Theor. Appl. Genet.* **2011**, *122*, 291–309. [CrossRef]

8.   Nelson, J.C.; Jodari, F.; Roughton, A.I.; McKenzie, K.M.; McClung, A.M.; Fjellstrom, R.G.; Scheffler, B.E. QTL mapping for milling quality in elite western U.S. rice germplasm. *Crop Sci.* **2012**, *52*, 242–252. [CrossRef]

9.   Tan, Y.F.; Sun, M.; Xing, Y.Z.; Hua, J.P.; Sun, X.L.; Zhang, Q.F.; Corke, H. Mapping quantitative trait loci for milling quality, protein content and color characteristics of rice using a recombinant inbred line population derived from an elite rice hybrid. *Theor. Appl. Genet.* **2001**, *103*, 1037–1045. [CrossRef]

10.   Dong, Y.; Tsuzuki, E.; Lin, D.; Kamiunten, H.; Terao, H.; Matsuo, M.; Cheng, S. Molecular genetic mapping of quantitative trait loci for milling quality in rice (*Oryza sativa* L.). *J. Cereal Sci.* **2004**, *40*, 109–114. [CrossRef]

11.   Zheng, T.Q.; Xu, J.L.; Li, Z.K.; Zhai, H.Q.; Wan, J.M. Genomic regions associated with milling quality and grain shape identified in a set of random introgression lines of rice (*Oryza sativa* L.). *Plant Breed.* **2007**, *126*, 158–163. [CrossRef]

12.   Lou, J.; Chen, L.; Yue, G.; Lou, Q.; Mei, H.; Xiong, L.; Luo, L. QTL mapping of grain quality traits in rice. *J. Cereal Sci.* **2009**, *50*, 145–151. [CrossRef]

13. Mei, D.-Y.; Zhu, Y.-J.; Yu, Y.-H.; Fan, Y.-Y.; Huang, D.-R.; Zhuang, J.-Y. Quantitative trait loci for grain chalkiness and endosperm transparency detected in three recombinant inbred line populations of *indica* rice. *J. Integr. Agric.* **2013**, *12*, 1–11. [CrossRef]
14. Wang, Z.; Chen, J.-Y.; Zhu, Y.-J.; Fan, Y.-Y.; Zhuang, J.-Y. Validation of *qGS10*, a quantitative trait locus for grain size on the long arm of chromosome 10 in rice (*Oryza sativa* L.). *J. Integr. Agric.* **2017**, *16*, 20–30. [CrossRef]
15. Zhang, H.; Zhu, Y.-J.; Fan, Y.-Y.; Huang, T.-X.; Zhang, J.-F.; Xie, H.-A.; Zhuang, J.-Y. Identification and verification of quantitative trait loci for eating and cooking quality of rice (*Oryza sativa*). *Plant Breed.* **2019**, *138*, 568–576. [CrossRef]
16. Huang, N.; Angeles, E.R.; Domingo, J.; Magpantay, G.; Singh, S.; Zhang, G.; Kumaravadivel, N.; Bennett, J.; Khush, G.S. Pyramiding of bacterial blight resistance genes in rice: Marker-assisted selection using RFLP and PCR. *Theor. Appl. Genet.* **1997**, *95*, 313–320. [CrossRef]
17. Meng, L.; Li, H.; Zhang, L.; Wang, J. QTL IciMapping: Integrated software for genetic linkage map construction and quantitative trait locus mapping in biparental populations. *Crop J.* **2015**, *3*, 269–283. [CrossRef]
18. McCouch, S.R.; CGSNL (Committee on Gene Symbolization, Nomenclature and Linkage, Rice Genetic Cooperative). Gene nomenclature system for rice. *Rice* **2008**, *1*, 72–84. [CrossRef]
19. Wang, Z.Y.; Wu, Z.L.; Xing, Y.Y.; Zheng, F.G.; Guo, X.L.; Zhang, W.G.; Hong, M.M. Nucleotide sequence of rice *waxy* gene. *Nucleic Acids Res.* **1990**, *18*, 5898. [CrossRef]
20. Fan, C.; Xing, Y.; Mao, H.; Lu, T.; Han, B.; Xu, C.; Li, X.; Zhang, Q. *GS3*, a major QTL for grain length and weight and minor QTL for grain width and thickness in rice, encodes a putative transmembrane protein. *Theor. Appl. Genet.* **2006**, *112*, 1164–1171. [CrossRef]
21. Wang, D.-Y.; Zhang, X.-F.; Zhu, Z.-W.; Chen, N.; Min, J.; Yao, Q.; Yan, J.-L.; Liao, X.-Y. Correlation analysis of rice grain quality characteristics. *Acta Agron. Sin.* **2005**, *31*, 1086–1091. (In Chinese)
22. Weng, J.; Gu, S.; Wan, X.; Gao, H.; Guo, T.; Su, N.; Lei, C.; Zhang, X.; Cheng, Z.; Guo, X.; et al. Isolation and initial characterization of *GW5*, a major QTL associated with rice grain width and weight. *Cell Res.* **2008**, *18*, 1199–1209. [CrossRef] [PubMed]
23. Li, Y.; Fan, C.; Xing, Y.; Yun, P.; Luo, L.; Yan, B.; Peng, B.; Xie, W.; Wang, G.; Li, X.; et al. *Chalk5* encodes a vacuolar H+-translocating pyrophosphatase influencing grain chalkiness in rice. *Nat. Genet.* **2014**, *46*, 398–404. [CrossRef] [PubMed]
24. Zhou, P.H.; Tan, Y.F.; He, Y.Q.; Xu, C.G.; Zhang, Q. Simultaneous improvement for four quality traits of Zhenshan 97, an elite parent of hybrid rice, by molecular marker-assisted selection. *Theor. Appl. Genet.* **2003**, *106*, 326–331. [CrossRef] [PubMed]

12

SSR Marker-Assisted Management of Parental Germplasm in Sugarcane (*Saccharum* spp. hybrids) Breeding Programs

Jiantao Wu [1,*,†], **Qinnan Wang** [1,†], **Jing Xie** [1], **Yong-Bao Pan** [2], **Feng Zhou** [1], **Yuqiang Guo** [1], **Hailong Chang** [1], **Huanying Xu** [1], **Wei Zhang** [1], **Chuiming Zhang** [1] and **Yongsheng Qiu** [1,*]

[1]   Guangdong Provincial Bioengineering Institute (Guangzhou Sugarcane Industry Research Institute), Guangzhou 510316, China

[2]   Sugarcane Research Unit, USDA-ARS, Houma, LA 70360, USA

*   Correspondence: wujiantao2010@163.com (J.W.); siriwu@126.com (Y.Q.)

†   These authors contributed equally to this work.

Abstract: Sugarcane (*Saccharum* spp. hybrids) is an important sugar and bioenergy crop with a high aneuploidy, complex genomes and extreme heterozygosity. A good understanding of genetic diversity and population structure among sugarcane parental lines is a prerequisite for sugarcane improvement through breeding. In order to understand genetic characteristics of parental lines used in sugarcane breeding programs in China, 150 of the most popular accessions were analyzed with 21 fluorescence-labeled simple sequence repeats (SSR) markers and high-performance capillary electrophoresis (HPCE). A total of 226 SSR alleles of high-resolution capacity were identified. Among the series obtained from different origins, the YC-series, which contained eight unique alleles, had the highest genetic diversity. Based on the population structure analysis, the principal coordinate analysis (PCoA) and phylogenetic analysis, the 150 accessions were clustered into two distinct sub-populations (Pop1 and Pop2). Pop1 contained the majority of clones introduced to China (including 28/29 CP-series accessions) while accessions native to China clustered in Pop2. The analysis of molecular variance (AMOVA), fixation index (*Fst*) value and gene flow (*Nm*) value all indicated the very low genetic differentiation between the two groups. This study illustrated that fluorescence-labeled SSR markers combined with high-performance capillary electrophoresis (HPCE) could be a very useful tool for genotyping of the polyploidy sugarcane. The results provided valuable information for sugarcane breeders to better manage the parental germplasm, choose the best parents to cross, and produce the best progeny to evaluate and select for new cultivar(s).

Keywords: sugarcane; parental line; population structure; plant breeding; genetic diversity; simple sequence repeats (SSR)

1. Introduction

Sugarcane cultivars are allopolyploids with highly heterozygous and complex genomes, which render a slow progress in breeding. To date, most commercial sugarcane varieties can be traced back to a limited number of popular cultivars belonging to either the POJ- or Co-series, which represent a very narrow genetic base [1]. Therefore, it is important for sugarcane breeders to fully understand the genetic relationship among parental lines and to choose elite parents of different genetic background for crossing in order to broaden the genetic diversity of sugarcane population [2].

Hainan sugarcane breeding station (HSBS) is the primary sugarcane crossing facility in Mainland China. It produces nearly all the seeds for sugarcane breeders in China every year [3]. HSBS has more than 2000 germplasm materials. Currently, thousands of new elite sugarcane genotypes are created by breeders each year. The utilization of these ever-increasing germplasm materials is a daunting challenge. Parental selection is a crucial step for good quality cross-breeding. Therefore, breeding materials should be adequately evaluated by different analytical methods to ensure their genetic suitability.

In the past, sugarcane breeders studied the genetic differences of parents mainly from the aspects of the genetic relationship, geographical origin and morphology. The genetic differences of sugarcane parents cannot really be reflected by pedigree because of mixed pollen, selfing and seed admixture [4]. Although morphological traits can be evaluated, these traits are easily influenced by the environment and may not reflect the real genetic diversity of sugarcane germplasm resources [5]. DNA molecular markers with high stability, multiple quantity and high polymorphism are more suitable for evaluating sugarcane germplasm collection [1]. With the rapid development of biotechnology, sugarcane researchers have utilized different types of DNA molecular markers, including amplified fragment length polymorphisms (AFLP) [1,5], restriction fragment length polymorphisms (RFLP) [6,7], random amplification of polymorphic DNAs (RAPD) [8,9], single nucleotide polymorphism (SNP) [10], simple sequence repeats (SSRs) [11], inter simple sequence repeat (ISSRs) [12,13], expressed sequence tag-simple sequence repeat (EST-SSRs) [14–16], 5S rRNA intergenic spacers [17], start codon targeted (SCoT) [18], target region amplification polymorphism (TRAP) [5,19,20], and cleaved amplified polymorphism sequences (CAPS) [21] for evaluating sugarcane germplasm.

Among PCR-based markers, SSR (microsatellite) markers are considered one of the most efficient markers for plant breeding due to large quantity, low dosage, co-dominant, reliability and multi-allelic detecting [22]. SSR markers have been used widely to study sugarcane genetic diversity and population structure [22–24], variety identity [25], genetic map [26,27], and genetic association [28–30]. Furthermore, fluorescence-labeled SSR markers combined with high-performance capillary electrophoresis (HPCE) have manifested better performance in genotyping of polyploid sugarcane, due to higher accuracy and better detection power [22–24,31–37].

Now, this paper reports a study that was designed to manage the parental germplasm of the sugarcane breeding programs in China through the microsatellite (SSR) DNA fingerprinting using fluorescence-labeled SSR primers and the high-performance capillary electrophoresis (HPCE) system. The results will help sugarcane breeders better manage the parental germplam, choose cross parents, design cross combinations, and produce high quality seedlings for the selection and development of elite varieties.

2. Materials and Methods

2.1. Plant Materials

One hundred and fifty parental clones were chosen for this study, based on the number of lines used most often in crossing from 2014 to 2018 in all Chinese sugarcane breeding programs (Table 1 and S1). These included 32 of clones from foreign origin, 109 clones from the China Mainland, and nine ROC-series clones from China Taiwan. Among the 32 foreign clones, one was from India (Co-series), 29 were from the U.S. (CP-series) and two were from Thailand (K-series). Among the 109 clones from China Mainland, four were from the Dehong Sugarcane Research Institute, Yunnan Province (DZ-series); 11 were from the Fujian Agriculture and Forestry University, Fujian Province (FN-series); two were from the Jiangxi Sugarcane Research Institute, Jiangxi Province (GN-series); 21 were from the Guangxi Academy of Agricultural Sciences, Guangxi Province (GT-series); six were from the Liucheng Academy of Agricultural Sciences, Guangxi Province (LC-series); six were from the Neijiang Academy of Agricultural Sciences, Sichuan Province (NJ-series); 18 were from the Hainan Sugarcane Breeding Station of Guangzhou Sugarcane Industry Research Institute, Hainan Province (YC-series); 29 were from the Guangzhou Sugarcane Industry Research Institute, Guangdong Province (YT-series); 10 were from the Yunnan Academy of Agricultural Sciences, Yunnan Province (YZ-series) and two were from other breeding units in China Mainland (one from Sichuan Research Institute of Sugar Crops, Sichuan Province and one from the Guangdong Academy of Agricultural Sciences, Guangdong Province).

Table 1. The 150 sugarcane accessions used in the experiment.

No.	Accession	Series	No.	Accession	Series	No.	Accession	Series
1	Co1001	Co	51	GZ75-65	GN	101	YC06-92	YC
2	CP57-614	CP	52	HoCP00-1142	CP	102	YC07-65	YC
3	CP67-412	CP	53	HoCP00-2218	CP	103	YC07-71	YC
4	CP72-1210	CP	54	HoCP01-517	CP	104	YC09-13	YC
5	CP72-2086	CP	55	HoCP01-564	CP	105	YC71-374	YC
6	CP80-1827	CP	56	HoCP02-610	CP	106	YC94-46	YC
7	CP81-1254	CP	57	HoCP02-623	CP	107	YC97-24	YC
8	CP84-1198	CP	58	HoCP03-704	CP	108	YC97-40	YC
9	CP89-2143	CP	59	HoCP03-708	CP	109	YC98-2	YC
10	CP93-1382	CP	60	HoCP03-716	CP	110	YC98-27	YC
11	CP93-1634	CP	61	HoCP05-902	CP	111	YN73-204	YN
12	CP94-1100	CP	62	HoCP07-612	CP	112	YT00-236	YT
13	CT89-103	CT	63	HoCP07-613	CP	113	YT00-318	YT
14	DZ03-83	DZ	64	HoCP07-617	CP	114	YT00-319	YT
15	DZ05-61	DZ	65	HoCP91-555	CP	115	YT01-120	YT
16	DZ06-51	DZ	66	HoCP92-648	CP	116	YT01-125	YT
17	DZ93-88	DZ	67	HoCP93-746	CP	117	YT01-71	YT
18	FN02-6404	FN	68	HoCP95-988	CP	118	YT03-373	YT
19	FN02-6427	FN	69	K5	K	119	YT03-393	YT
20	FN05-2848	FN	70	K86-110	K	120	YT85-177	YT
21	FN0711	FN	71	LC03-1137	LC	121	YT86-368	YT
22	FN0712	FN	72	LC03-182	LC	122	YT89-240	YT
23	FN0713	FN	73	LC04-256	LC	123	YT91-976	YT
24	FN0717	FN	74	LC05-128	LC	124	YT92-1287	YT
25	FN91-23	FN	75	LC05-136	LC	125	YT93-124	YT
26	FN92-4621	FN	76	LC05-291	LC	126	YT93-159	YT
27	FN95-1702	FN	77	LCP85-384	CP	127	YT94-128	YT
28	FN99-20169	FN	78	NJ00-118	NJ	128	YT96-86	YT
29	GN95-108	GN	79	NJ00-15	NJ	129	YT97-20	YT
30	GT00-122	GT	80	NJ03-218	NJ	130	YT97-76	YT
31	GT02-1156	GT	81	NJ07-13	NJ	131	YT99-66	YT
32	GT02-208	GT	82	NJ86-117	NJ	132	YZ02-2540	YZ
33	GT02-281	GT	83	NJ92-244	NJ	133	YZ02-588	YZ
34	GT02-467	GT	84	ROC1	ROC	134	YZ03-194	YZ
35	GT02-761	GT	85	ROC10	ROC	135	YZ07-100	YZ
36	GT02-901	GT	86	ROC16	ROC	136	YZ07-49	YZ
37	GT03-11	GT	87	ROC20	ROC	137	YZ89-7	YZ
38	GT03-1403	GT	88	ROC22	ROC	138	YZ94-343	YZ
39	GT03-2112	GT	89	ROC23	ROC	139	YZ94-375	YZ
40	GT03-3005	GT	90	ROC25	ROC	140	YZ99-601	YZ
41	GT03-3089	GT	91	ROC26	ROC	141	YZ99-91	YZ
42	GT03-8	GT	92	ROC28	ROC	142	ZZ33	YT
43	GT03-91	GT	93	YC04-55	YC	143	ZZ41	YT
44	GT05-3084	GT	94	YC05-64	YC	144	ZZ43	YT
45	GT05-3595	GT	95	YC06-111	YC	145	ZZ45	YT
46	GT73-167	GT	96	YC06-140	YC	146	ZZ49	YT
47	GT89-5	GT	97	YC06-166	YC	147	ZZ50	YT
48	GT92-66	GT	98	YC06-61	YC	148	ZZ80-101	YT
49	GT94-119	GT	99	YC06-63	YC	149	ZZ90-76	YT
50	GT96-154	GT	100	YC06-91	YC	150	ZZ92-126	YT

2.2. SSR Genotyping

Young leaf tissues were collected from three individual clones, rinsed with 75% ethanol, and kept at −80 °C prior to DNA extraction. The genomic DNA was extracted from leaf tissues using the cetyl trimethyl ammonium bromide (CTAB) method [38] with minor modifications. The quality and concentration of DNA were measured using the UV-Vis Spectrophotometer Q5000 of Quawell (Quawell Technology, Inc. San Jose, CA, USA) and diluted to 20 ng/μL. A set of 21 SSR primer pairs (Table 1) with stable and clear amplification was selected from previous reports [3,11,33,39–42]. All forward primers were labeled with a fluorescence dye, 6-carboxy-fluorescein (FAM) or Hexachlorofluorescein (HEX). PCR reactions were performed with the following cycling condition: 95 °C for 2 min, followed by 40 cycles of 94 °C for 30 s, then primer-specific annealing temperature (Tm) for 90 s, 65 °C for 30 s, followed by one cycle at 65 °C for 10 min. The annealing temperatures for the 21 primer pairs were optimized separately, ranging from 49 °C to 62 °C (Table 2). The amplified PCR products were checked by a 3% agarose gel electrophoresis. High-performance capillary electrophoreses (HPCE) was conducted on the ABI 3730XL DNA analyzer (Applied Biosystems, Inc. Foster City, CA, USA) to generate GeneScan files. The GeneScan files were analyzed using the GeneMarker V2.2 software (SoftGenetics, LLC. State College, PA, USA) to show SSR DNA fragments (alleles) and the sizes of these fragments were calibrated automatically against the GeneScan500 size standards. Due to the polyploidy nature of sugarcane, the SSR alleles had to be manually called first and the score sheet was manually rechecked according to Pan [43]. The presence of an allele was scored as "1" and its absence scored as "0". SSR alleles were named using a combination of primer name and allele size.

Table 2. The 21 simple sequence repeat (SSR) markers used in this study.

Primer Name	Type [a]	Repeat Motif	Primer Sequence (5'-3')	Annealing Temperatures (°C)
mSSCIR36	G-SSR	$(GA)_{18}GT$ $(GA)_4$	CAACAATAACTTAACTGGTA CTGTCCTTTTTATTCTCTTT	52
mSSCIR46	G-SSR	$(GT)_{10}$	ATGCTCCGCTTCTCACTC AAGGGGAAAATGAAAACC	52
mSSCIR74	G-SSR	$(CGC)_9$	GCGCAAGCCACACTGAGA ACGCAACGCAAAACAACG	56
SCM4	E-SSR	$(CGGAT)_4$	CATTGTTCTGTGCCTGCT CCGTTTCCCTTCCTTCCC	52
SCM7	E-SSR	$(GCAC)4$	ACGGTGCTCTTCACTGCT GGGCATACTTCCTCCTCTAC	60
SCM18	E-SSR	$(ATAC)_3$	CATCAGTATCATTTCATCTTGG CAGTCACAGTCGGGTAGA	60
SMC1825LA	G-SSR	$(TG)_{11}$	CACGTCCTTCCGCCTTGA TCATCGTTCGTCGCACTG	56
SMC286CS	G-SSR	$(TG)_{43}$	TCAAATGGGACCTTATTGGAG TCCCTCGATCTCCGTTGTT	52
SMC477CG	G-SSR	$(CA)_{31}$	CCAACAACGAATTGTGCATGT CCTGGTTGGCTACCTGTCTTCA	60
SMC486CG	G-SSR	$(CA)_{14}$	GAAATTGCCTCCCAGGATTA CCAACTTGAGAATTGAGATTCG	60
SMC569CS	G-SSR	$(TG)_{37}$	GCGATGGTTCCTATGCAACTT TTCGTGGCTGAGATTCACACTA	60
SMC597CS	G-SSR	$(AG)_{31}$	GCACACCACTCGAATAACGGAT AGTATATCGTCCCTGGCATTCA	52
SMC334BS	G-SSR	$(TG)_{36}$	CAATTCTGACCGTGCAAAGAT CGATGAGCTTGATTGCGAATG	60
SMC36BUQ	G-SSR	$(TTG)_7$	GGGTTTCATCTCTAGCCTACC TCAGTAGCAGAGTCAGACGCTT	56

Table 2. *Cont.*

Primer Name	Type [a]	Repeat Motif	Primer Sequence (5′-3′)	Annealing Temperatures (°C)
SMC7CUQ	G-SSR	$(CA)_{10}(C)_4$	GCCAAAGCAAGGGTCACTAGA AGCTCTATCAGTTGAAACCGA	60
SEGM285	G-SSR	$(GCAC)_4$	AAGAAGAAGACTGAGAAGAACACT TAGCAACAACTTAATTTAGCAATC	56
UGSM345	E-SSR	$(TG)_6$	CTGTACTGGTATTACATGTGACCT TCTACTAATCACAAGAGAAGATGC	60
UGSM10	E-SSR	$(GGC)_{11}$	GCTACTATGGACAACAGGG ATGAAGAGACGAGACGAAGA	56
UGSuM50	E-SSR	$(TC)_{14}$	CTACTGCCGAGGAAAGATCG GGAAAAGTTTGTGGCAAGGA	56
MCSA068G08	E-SSR	$(CAG)_6$	CTAATGCCATGCCCCAGAGG GCTGGTGATGTCGCCCATCT	56
MCSA176C01	E-SSR	$(GGT)_5$	GAGTCAGTTGGTGCCGAGATTG GAACAGGTTAAAGCCCATGTC	56

[a] G-SSR: SSR primer pair designed from genomic sequence; E-SSR: SSR primer pair designed from UniGene or cDNA sequences.

2.3. Genetic Diversity Analysis

Qualitative allelic data matrix was constructed and formatted using the DataFormatter software [44]. The PowerMarker v3.25 software [45] was used to calculate allele frequency, number of alleles per locus, polymorphism information content (PIC), the gene diversity index (h), Shannon's information index (I), and percentage of polymorphic loci (PPL) of each marker. The resolving power of the primer (Rp) [46] was calculated using allele frequencies. The probability of identity (PI) [23] was computed using the CERVUS v3.0 software [47]. Unique (Series-specific) alleles were estimated using GeneALEx v6.502 [48,49].

2.4. Population Structure Analysis

The model-based program Structure v2.3.4 [50] was used to analyze the population structure involving the 226 alleles amplified by the 21 SSR primer pairs. The number of populations (K) was set from one to 10, and at each K value, ten runs were conducted separately with 50,000 iterations of burn-in length and 50,000 Markov Chain Monte Carlo (MCMC). Then, the best K value was estimated using Evanno's ΔK method [51] with an online tool, Structure Harvester [52]. An individual Q matrix was generated by CLUMPP v1.1.2 [53]. Parental clones with membership probabilities greater than 0.5 were identified as the same group [54]. A Principal Coordinate Analysis (PCoA) map was generated based on the genetic distances between pairs of clones by GeneALEx v6.502 [48,49]. An unrooted phylogenetic tree was constructed based on the neighbor-joining (NJ) method and the genetic distance matrix using PowerMarker v3.25 [45] and adjusted with MEGA v6.06 [55].

2.5. Differentiation Analysis and Genetic Diversity Indices

Analysis of Molecular Variance (AMOVA) was conducted to find the genetic differentiation within and among subpopulations using GeneALEx v6.502 [48,49]. From AMOVA, the fixation index (Fst) and gene flow (Nm) within the population was also acquired. In addition, genetic diversity indices, including number of different alleles (Na), number of effective alleles (Ne), Shannon's information index (I), observed heterozygosity (Ho), expected heterozygosity (He), unbiased expected heterozygosity (uHe), and percentage of polymorphic loci (PPL) of different sub-groups were also calculated using GeneALEx v6.502 [48,49].

3. Results

3.1. Polymorphism Revealed by SSR Genotyping

The 21 SSR primer pairs amplified a total of 226 alleles with an average of 10.8 alleles per primer pair (Table 2). Of the 226 alleles, 220 alleles were polymorphic and the other six alleles could be amplified in each clone. The number of alleles amplified by one primer pair ranged from five by MCSA176C01 to 25 by SCM4. The mean PIC value of each SSR primer pair ranged from 0.15 to 0.29 with an average of 0.23. The probability of identity (*PI*) of the 21 markers was all very low, which ranged from 0.000001 (mSSCIR36) to 0.071332 (SMC569CS) with an average of 0.015532. For the 21 primers pairs, the resolving power of the primer (*Rp*) was relatively high, ranging from 3.68 (SMC569CS) to 21.01 (mSSCIR36) with an average of 9.14. The mean number of alleles and the mean PIC value of genomic SSRs were 10.6 and 0.23, and were 9.8 and 0.23 for EST SSRs, respectively (Table 3).

Table 3. Genetic diversity parameters of 150 of the most popular parental clones from sugarcane hybrid breeding programs.

Primer Name	Allele (No.)	Product Size (bp)	Range of PIC [a] Values	Mean of PIC Values	PI [b]	RP [c]
mSSCIR36	21	127–168	0.01–0.38	0.15	0.000001	7.09
mSSCIR46	12	146–177	0.01–0.37	0.15	0.002858	13.04
mSSCIR74	6	215–228	0.00–0.37	0.17	0.042135	4.69
SCM4	25	92–209	0.01–0.37	0.17	0.000087	4.16
SCM7	7	155–188	0.03–0.37	0.18	0.048672	3.68
SCM18	9	226–251	0.00–0.38	0.19	0.010157	8.67
SMC1825LA	10	91–119	0.01–0.37	0.20	0.001240	6.53
SMC286CS	13	128–152	0.01–0.37	0.21	0.000411	7.31
SMC477CG	15	115–134	0.00–0.36	0.21	0.000125	4.11
SMC486CG	7	222–243	0.06–0.36	0.22	0.051066	4.88
SMC569CS	6	166–220	0.04–0.38	0.24	0.071332	14.05
SMC597CS	14	143–166	0.03–0.37	0.24	0.000034	10.99
SMC334BS	12	145–163	0.01–0.38	0.24	0.000140	6.27
SMC36BUQ	12	103–251	0.00–0.37	0.25	0.010448	7.49
SMC7CUQ	7	156–170	0.00–0.37	0.26	0.024118	9.76
SEGM285	13	306–389	0.03–0.38	0.26	0.000143	21.01
UGSM345	8	320–334	0.01–0.38	0.27	0.005772	13.68
UGSM10	10	97–125	0.00–0.38	0.28	0.005289	9.31
UGSuM50	6	123–139	0.05–0.38	0.28	0.023095	6.24
MCSA068G08	8	179–202	0.06–0.38	0.29	0.003035	15.57
MCSA176C01	5	427–440	0.11–0.38	0.29	0.026013	13.31

[a] PIC: Polymorphism information content; [b] PI: Probability of identity; [c] RP: Resolving power.

3.2. Genetic Diversity

The gene diversity (*h*) of the polymorphic allele ranged from 0.013 to 0.500 with an average of 0.282. The Shannon's information index (*I*) of the polymorphic allele ranged from 0.010 to 0.534 with an average of 0.261. Among the different series of sugarcane parental lines, the highest values of both gene diversity (*h*) and Shannon's information index (*I*) were found in the YC-series (0.261, 0.397), followed by the YT-series (0.254, 0.386,) and the GT-series (0.251, 0.376) (Table 3), indicating that the YC-series is genetically more diverse than the other series. The average percentages of polymorphic allele for the YT-, YC-, and CP-series were 0.814, 0.805 and 0.743, respectively. Alleles were identified that were unique to the 12 distinct germplasm groups (Table 4).

Table 4. Gene diversity, Shannon's information index, percentage of polymorphic loci and series-specific alleles of different series.

Series	Sample Size	h [a]	I [b]	PPL [c]	Series-Specific Alleles
CP	29	0.239	0.361	0.743	SCM7-188, SCM18-238, SMC486CG-225, SMC486CG-233
DZ	4	0.235	0.341	0.562	
FN	11	0.245	0.365	0.677	mSSCIR46-153
GN	2	0.148	0.205	0.296	
GT	21	0.251	0.376	0.721	
LC	6	0.197	0.293	0.522	
NJ	6	0.205	0.302	0.527	SMC36BUQ-125
ROC	9	0.201	0.301	0.558	SMC36BUQ-184, SEGM285-359
K	2	0.164	0.227	0.327	
YC	18	0.261	0.397	0.805	mSSCIR46-146, mSSCIR46-149, SCM7-175, SMC569CS-174, SMC569CS-202, SMC36BUQ-106, SMC36BUQ-132, UGSM10-113
YT	29	0.254	0.386	0.814	SMC36BUQ-105, SMC36BUQ-139
YZ	10	0.241	0.358	0.650	
Mean		0.176	0.261	0.480	

[a] h, Gene diversity; [b] I, Shannon's information index; [c] PPL, percentage of polymorphic loci.

3.3. Population Structure and Phylogeny

The K-value was used to estimate the number of clusters of the clones based on the genotypic data. A continuous gradual increase was observed in the log-likelihood of K-value (LnP(K)) with the increase of K-value (Figure 1B and Table S2). The number of clusters (K) was plotted against Delta K (ΔK), which revealed a sharp peak at $K = 2$ (Figure 1A and Table S2). The optimal K-value was $K = 2$, which revealed that the highest probability for the presence of two sub-populations (Pop1 and Pop2) among the 150 sugarcane clones (Figure 1C); Pop1 consisted of 50 clones and Pop2 contained 100 clones (Table S3). Pop1 clones were mainly introduction accessions and most of the Pop2 clones were from Mainland China.

In accordance with the population structure results, PCoA also showed two clusters with the first three axes together explained 20.04% of cumulative variation. In the PCoA plot, the first and second principal coordinates accounted for 8.41% and 6.71% of the total variations, respectively (Figure 2). Furthermore, the unrooted neighbor-joining phylogenetic tree (Figure 3) also showed two clusters. One cluster contained most of the clones of Pop1; the other cluster contained most of the clones of Pop2. However, the admixture of clones between the two sub-populations does exist. Few accessions (YC98-27, GT03-2112 and FN0717) native to China were clustered into Pop1 while several others (HoCP01-517, ROC10, ROC16, K5, ROC25, ROC22, ROC1) introduced to China Mainland were grouped into Pop2.

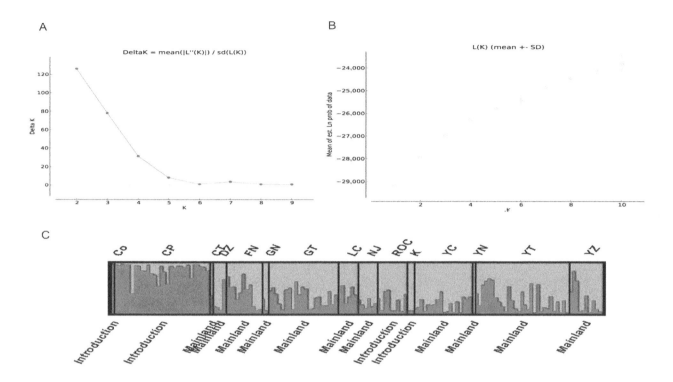

Figure 1. (**A**) Delta K (ΔK) for different numbers of subpopulations (K); (**B**) average log-likelihood K-value (LnP(K)) against the number of K; (**C**) the population structure of 150 most popular parental clones in the hybrid breeding programs in China based on the distribution of 226 SSR alleles among these clones. Pop1 clones are coded in red and Pop2 clones in green.

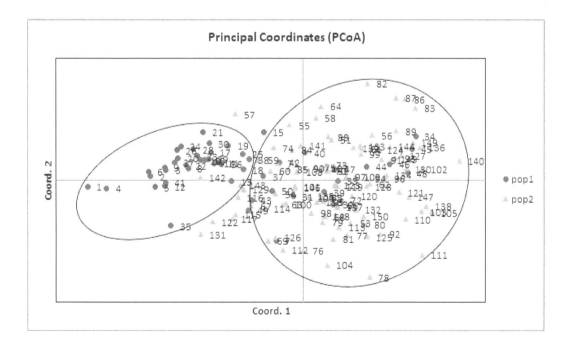

Figure 2. Principal coordinates analysis (PCoA) scatter plots. Red circles represent the Pop1 clones and green triangles the Pop2 clones.

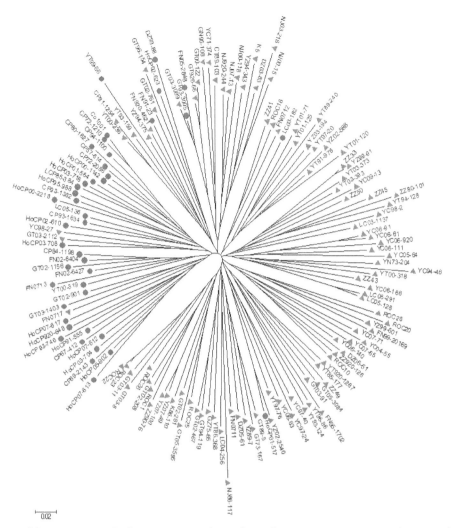

Figure 3. A neighbor-joining phylogenetic tree based on the pair-wise genetic distance between 150 most popular parental clones from hybrid breeding programs in China. Red circles represent the Pop1 clones and green triangles the Pop2 clones.

3.4. Genetic Differentiation and Allelic Pattern Across Populations

The two sub-populations Pop1 and Pop2 identified by the Structure analysis were subjected to the GeneALEx analysis to calculate the values of Analysis of Molecular Variance (AMOVA), *Nei*'s genetic distance and genetic diversity indices (Table 5). The variation value within the sub-populations (95% of total variation) was significantly higher than that between the sub-populations (5% of total variation). In addition, a high gene flow ($Nm = 4.981$) and a low fixation index value ($Fst = 0.048$) were obtained on the basis of *Nei*'s genetic distance analysis.

Table 5. Analysis of molecular variance (AMOVA) of SSR-based genetic variation between and within two sub-populations of Pop1 and Pop2.

Source of Variation	Degrees of Freedom	Sum of Squares	Mean Sum of Squares	Estimated Variance	Percentage of Variation
Between sub-Pops	1	546.240	546.240	6.308	5%
Within sub-Pop	148	18,601.600	125.686	125.686	95%
Total	149	19,147.840		131.995	100%
Fixation Index	$Fst = 0.048$				
Gene Flow	$Nm = 4.981$				

The mean value of the number of different alleles (Na) and effective alleles (Ne) of the two sub-populations were 1.885 ± 0.015 and 1.462 ± 0.017, respectively. The mean values for I, He and uHe among the 150 parental clones were 0.413 ± 0.011, 0.272 ± 0.008 and 0.274 ± 0.009, respectively. Pop2 ($I = 0.423 \pm 0.016$, $He = 0.278 \pm 0.012$, and $uHe = 0.278 \pm 0.012$) showed higher levels of genetic diversity than Pop1 ($I = 0.403 \pm 0.017$, $He = 0.267 \pm 0.012$, and $uHe = 0.269 \pm 0.012$). The percentage of polymorphic loci per population (PPL) ranged from 83.63% (Pop1) to 93.36% (Pop2) with an average of 88.50% (Figure 4).

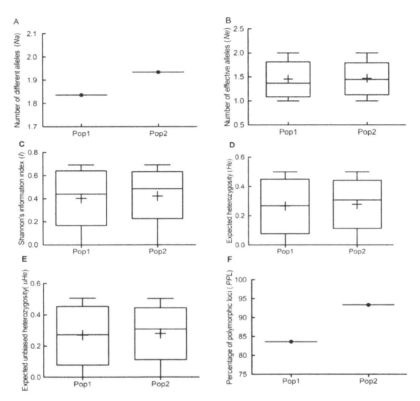

Figure 4. Allelic pattern of SSR across the two sub-populations Pop1 and Pop2. (**A**) Number of SSR alleles (Na); (**B**) number of effective SSR alleles (Ne); (**C**) Shannon's information index (I); (**D**) expected heterozygosity (He); (**E**) expected unbiased heterozygosity (uHe); and (**F**) percentage of polymorphic loci (PPL).

4. Discussion

Cross hybridization has become the main breeding method for the sugarcane variety improvement. In the traditional sugarcane cross-breeding process, selecting parental clones for crossing is the most important step. Only parental clones sharing a highly level of genetic diversity and complementarity can generate high quality seedling populations [56,57]. Since the 1950s, some sugarcane cultivars from America and China Taiwan have played a very important role in China's sugarcane cross-breeding programs [3]. Meanwhile, some new elite sugarcane parents are being created and utilized by the breeders every year. To make informed crossing choices, the genetic relationship among the parental clones involved in the latest sugarcane cross-breeding programs should be clarified.

In this study, we used 21 pairs of SSR primers to investigate the genetic diversity and population structure of 150 of the most commonly used parental clones. These primer pairs amplified 226 alleles, of which 97.3% were polymorphic. The mean PIC and the gene diversity (h) of the polymorphic alleles were 0.23 and 0.28, respectively, which were lower than the values reported on the "World Collections of Sugarcane and Related Grasses" (WGSRG) (PIC = 0.2568, h = 0.310) [23]. This may be largely due to the number of accessions involved in the world collection study. The WCSRG study involved 1002 highly diverse accessions, belonging to nine species, whereas only 150 clones were used in this study.

Since 2000, a large number of genomic SSR and EST-SSR markers has been developed and applied effectively in estimating genetic diversity in the sugarcane [16,35,39,41,58]. After a lot of screening and identification (unpublished), we selected the best 21 primer pairs from these reports, including eight EST-SSR and 13 genomic SSR. We found that the number and mean PIC value of the EST-SSR alleles were lower than those of the genomic SSR alleles (Table 2). This can be due to the fact that the EST-SSR alleles are located in more conserved regions of the genome [16].

The probability of identity (PI) is an individual identification estimator that shows the probability of two different accessions sharing the same genotypes at one specific locus in a population [23]. In this study, the PI values of all SSR primer pairs were very low, ranging from 0.000001 (mSSCIR36) to 0.071332 (SMC569CS) (Table 2). The combined PI value for all markers was 9.04×10^{-57}, indicating that these 21 SSR primer pairs are able to distinguish the 150 parental clones. The resolving power of the primer pair (Rp) is an index, which explains the primer pair's ability to identify different genotypes. Rp is related to the distribution of alleles within the sampled genotypes [46] and has been found to correlate strongly with the genotype in evaluating 34 potato cultivars using four primers [46]. The mean Rp value (9.135) of the 21 SSR primer pairs is much higher than other studies, such as 2.37 by [59] and 2.2 by [12], indicating these primer pairs are more informative and could identify more cultivars.

Based on geographic origin, the 150 clones were sorted into 15 series. Among these series, the genetic diversity (h) indices ranged from 0 to 0.261 and the Shannon's information index (I) ranged from 0 to 0.397. At the series level, the YC-series had the highest genetic diversity ($h = 0.261$, $I = 0.397$), which was similar to the previous results reported by You et al. [35,60]. The YC-series clones are from the Hainan Sugarcane Breeding Station of Guangzhou Sugarcane Industry Research Institute in Sanya city, Hainan province, where the primary sugarcane crossing facility of China is located. The YC-series clones were selected from crosses involving indigenous clones, foreign clones, and clones of closely related *Saccharum* species and genera [35]. Furthermore, the YC-series also had the greatest number of eight series-specific alleles. Only four, two, one, and one unique alleles were found in the CP-series, YT-series, ROC-series, FN-series and NJ-series clones, respectively. Series-specific alleles are the alleles found only in a single population among a broader collection of populations [61,62]. These alleles have been proven to be informative for population genetic studies [63,64] and we may use these alleles for variety identification and marker assisted selection.

The 150 parental clones were classified into two groups (Pop1 and Pop2) based on the PCoA, phylogenetic analysis and population structure analysis. Pop1 contained the majority of foreign accessions with the membership probabilities of >0.5, while most accessions from Mainland China were assigned to Pop2. Certain specific target traits intentionally selected by different germplasm collectors or breeders might also contribute to the population structure [54]. However, admixture of clones between the two sub-populations do exist (Figures 1–3). For example, one out of the 29 CP-series clones, nine ROC-series clones and two K-series clones clustered into Pop2, but the majority of introduction clones clustered into Pop1. Likewise, one out of four DZ-series, five out of 11 FN-series, four out of 21 GT-series, two out of six LC-series, seven out of 29 YT-series, and two out 10 YZ-series clones clustered into Pop1, while the majority of the clones from Mainland China clustered into Pop2. This might be due to genetic exchange among different series, or the similar threshold (Pop1: 0.5098, Pop2: 0.4902) (Table S3) resulting in several clones to be clustered completely into a certain group (Pop1 or Pop1), while others being clustered into both groups.

The utilization data was based the most widely used 150 parental clones of sugarcane breeding programs in China during the recent five years. These included 32 of clones from foreign origin, 109 clones from the China Mainland, and nine ROC-series clones from China Taiwan. Among the 32 foreign clones, only one was from India (Co1001), two were from Thailand (K5 and K86-110) while the majority of them (29/32) were from the US (CP-series). Co1001 has been used as parental line extensively in the sugarcane breeding programs in the world. Some sugarcane cultivars, including the CP-series and China Mainland clones, were the progenies of Co-series varieties. Compared to clones from China Mainland, the CP-series clones may have closer genetic distance with the Co-series. So CP-series clones and Co-series clone can be clustered into Pop1. K5 and K86-110, which were from Thailand, were two of the most widely used parental clones in China. Some clones from China Mainland were the progenies of K5 and K86-110. Clones from China mainland may have the closer genetic distance with the two clones to be clustered into Pop2. The ROC-series varieties have been used as major cultivars in China Mainland accounting for greater than 80% of sugarcane planting areas [24]. In addition, the ROC-series accessions were also the most widely used parents in China Mainland during the recent five years (Table S1). In our study, the ROC-series accessions were clustered into Pop2 because of their closer genetic distance with China Mainland's clones. It is suggested that less attention be continually paid on the utilization of ROC-series accessions in China Mainland's sugarcane breeding programs.

Fixation index (Fst) measures the genetic distance between populations. An Fst value of zero indicates no differentiation between the sub-populations, while one indicates complete differentiation [65]. An Fst value less than 0.05 is considered no differentiation, while an Fst value greater than 0.15 is considered significant in differentiating populations [66]. In this study, the Fst value between the two sub-populations was 0.048 (Table 5), which was low and would indicate a very low genetic differentiation. This is consistent with the results obtained from the AMOVA, where the genetic variation within sub-populations (95%) was significantly higher than between sub-populations (5%). Gene flow (Nm) is the transfer of genetic variation from one population to another. If the value is less than one, then the gene exchange would be limited between sub-populations [67].In this study, the Nm value was high, 4.981 suggesting that a high level of genetic exchange may have occurred and this can result in a low genetic differentiation between the two sub-populations. Since the genetic diversity indices of Pop2, such as the number of different alleles (Na), effective alleles (Ne), I, He and uHe, were all higher than those of Pop1, Pop2 is more diverse than Pop1.

Selecting genetically distant accessions from Pop1 and Pop2 for crossing parents in sugarcane breeding programs will potentially lead to elite varieties with broadened genetic bases. Almost all the CP-series clones from the US were clustered into Pop1. These clones have been used extensively as parental lines in the sugarcane breeding programs in China; some have become or are elite progenitors of Chinese cultivars [67]. In addition, this study shows that several YC-series clones are also good crossing parents with a high level of genetic diversity.

5. Conclusions

Using a high-performance capillary electrophoresis (HPCE) detection system, the most widely used 150 sugarcane parental clones from 15 different series were fingerprinted with 21 SSR primer pairs. A total of 226 SSR alleles were identified and the distribution of these SSR alleles were subjected to genetic variation, phylogeny, population structure, and principal coordinate analyses. The results showed that the parental lines were clustered into two distinct groups, Pop1 and Pop2. Pop1 contained the majority of foreign clones, while Pop2 consisted of the majority of accessions from Mainland China. Genetic differentiation between the two groups was low. The YC-series clones of Pop2 displayed a high level of genetic diversity and the CP-series clones were elite parents of several Chinese cultivars. The introduction and utilization of more clones of the YC- and CP-series into China's sugarcane breeding programs will broaden the genetic base of breeding germplasm and produce high quality seedlings for selection and development of elite varieties.

Author Contributions: Methodology, J.W.; Validation, J.W. and Q.W.; Formal Analysis, J.W. and Y.-B.P.; Investigation, J.X., H.X. and J.W.; Resources, F.Z., C.Z., and W.Z.; Data Curation, J.X., Y.G. and H.C.; Writing—Original Draft Preparation, J.W., Y.-B.P. and J.X.; Writing—Review and Editing, Q.W. and Y.-B.P. Funding Acquisition, Q.W. and Y.Q.

Acknowledgments: We thank Perng-Kuang Chang, James Todd and Yunlin Jia for their review comments and language editing.

References

1.   Lima, M.L.A.; Garcia, A.A.F.; Oliveira, K.M.; Matsuoka, S.; Arizono, H.; de Souza, C.L.; de Souza, A.P. Analysis of genetic similarity detected by AFLP and coefficient of parentage among genotypes of sugar cane (*Saccharum* spp.). *Theor. Appl. Genet.* **2002**, *104*, 30–38. [CrossRef] [PubMed]

2.   Ming, R.; Moore, P.H.; Wu, K.; D'Hont, A.; Glaszmann, J.C.; Tew, T.L.; Mirkov, T.E.; Da Silva, J.; Jifon, J.; Rai, M.; et al. *Sugarcane Improvement through Breeding and Biotechnology*; John Wiley & Sons, Ltd.: New York, NY, USA, 2010; pp. 15–118.

3.   Qi, Y.W.; Pan, Y.B.; Fang, Y.L.; Zhang, C.M.; Fan, L.N.; He, H.Y.; Liu, R.; Wang, Q.N.; Liu, S.M.; Liu, F.Y.; et al. Genetic structure and diversity of parental cultivars involved in China mainland sugarcane breeding programs as inferred from DNA microsatellites. *J. Integr. Agric.* **2012**, *11*, 1794–1803. [CrossRef]

4.   Nair, N.V.; Selvi, A.; Sreenivasan, T.V.; Pushpalatha, K.N. Molecular diversity in Indian sugarcane cultivars as revealed by Randomly Amplified DNA polymorphisms. *Euphytica* **2002**, *127*, 219–225. [CrossRef]

5.   Creste, S.; Sansoli, D.M.; Tardiani, A.C.S.; Silva, D.N.; Gonçalves, F.K.; Fávero, T.M.; Medeiros, C.N.F.; Festucci, C.S.; Carlini-Garcia, L.A.; Landell, M.G.A.; et al. Comparison of AFLP, TRAP and SSRs in the estimation of genetic relationships in sugarcane. *Sugar Tech.* **2010**, *12*, 150–154. [CrossRef]

6.   Jannoo, N.; Grivet, L.; Seguin, M.; Paulet, F.; Domaingue, R.; Rao, P.S.; Dookun, A.; D'Hont, A.; Glaszmann, J.C. Molecular investigation of the genetic base of sugarcane cultivars. *Theor. Appl. Genet.* **1999**, *99*, 171–184. [CrossRef]

7.   Silva, J.A.G.D.; Sorrells, M.E.; Burnquist, W.L.; Tanksley, S.D. RFLP linkage map and genome analysis of *Saccharum spontaneum*. *Genome* **1993**, *36*, 782–791. [CrossRef]

8.   Pan, Y.B.; Burner, D.M.; Legendre, B.L.; Grisham, M.P.; White, W.H. An assessment of the genetic diversity within a collection of *Saccharum spontaneum* L. with RAPD-PCR. *Genet. Resour. Crop Evol.* **2004**, *51*, 895–903. [CrossRef]

9.   Singh, P.; Singh, S.P.; Tiwari, A.K.; Sharma, B.L. Genetic diversity of sugarcane hybrid cultivars by RAPD markers. *3 Biotech* **2017**, *7*, 222. [CrossRef]

10. Garcia, A.A.F.; Mollinari, M.; Marconi, T.G.; Serang, O.R.; Silva, R.R.; Vieira, M.L.C.; Vicentini, R.; Costa, E.A.; Mancini, M.C.; Garcia, M.O.S.; et al. SNP genotyping allows an in-depth characterisation of the genome of sugarcane and other complex autopolyploids. *Sci. Rep. UK* **2013**, *3*, 3399. [CrossRef]

11. Parida, S.K.; Kalia, S.K.; Kaul, S.; Dalal, V.; Hemaprabha, G.; Selvi, A.; Pandit, A.; Singh, A.; Gaikwad, K.; Sharma, T.R.; et al. Informative genomic microsatellite markers for efficient genotyping applications in sugarcane. *Theor. Appl. Genet.* **2009**, *118*, 327–338. [CrossRef]

12. Devarumath, R.M.; Kalwade, S.B.; Kawar, P.G.; Sushir, K.V. Assessment of Genetic Diversity in Sugarcane Germplasm Using ISSR and SSR Markers. *Sugar Tech.* **2012**, *14*, 334–344. [CrossRef]

13. Oliveira, L.A.R.; Machado, C.A.; Cardoso, M.N.; Oliveira, A.C.A.; Amaral, A.L.; Rabbani, A.R.C.; Silva, A.V.C.; Ledo, A.S. Genetic diversity of Saccharum complex using ISSR markers. *Genet. Mol. Res.* **2017**, *16*, 1–9. [CrossRef]

14. Pinto, L.R.; Oliveira, K.M.; Marconi, T.; Garcia, A.A.F.; Ulian, E.C.; de Souza, A.P. Characterization of novel sugarcane expressed sequence tag microsatellites and their comparison with genomic SSRs. *Plant Breed.* **2006**, *125*, 378–384. [CrossRef]

15. James, B.T.; Chen, C.; Rudolph, A.; Swaminathan, K.; Murray, J.E.; Na, J.; Spence, A.K.; Smith, B.; Hudson, M.E.; Moose, S.P.; et al. Development of microsatellite markers in autopolyploid sugarcane and comparative analysis of conserved microsatellites in sorghum and sugarcane. *Mol. Breed.* **2012**, *30*, 661–669. [CrossRef]

16. Parthiban, S.; Govindaraj, P.; Senthilkumar, S. Comparison of relative efficiency of genomic SSR and EST-SSR markers in estimating genetic diversity in sugarcane. *3 Biotech* **2018**, *8*, 144. [CrossRef]

17. Pan, Y.B.; Burner, D.M.; Legendre, B.L. An Assessment of the Phylogenetic Relationship Among Sugarcane and Related Taxa Based on the Nucleotide Sequence of 5S rRNA Intergenic Spacers. *Genetica* **2000**, *108*, 285–295. [CrossRef]

18. Que, Y.X.; Pan, Y.B.; Lu, Y.H.; Yang, C.; Yang, Y.T.; Huang, N.; Xu, L.P. Genetic analysis of diversity within a Chinese local sugarcane germplasm based on start codon targeted polymorphism. *Biomed Res. Int.* **2014**, *2014*, 468375. [CrossRef]

19. Que, Y.X.; Chen, T.; Xu, L.; Chen, R.K. Genetic diversity among key sugarcane clones revealed by TRAP markers. *J. Agric. Biotechnol.* **2009**, *17*, 496–503.

20. Alwala, S.; Suman, A.; Arro, J.A.; Veremis, J.C.; Kimbeng, C.A. Target region amplification polymorphism (TRAP) for assessing genetic diversity in sugarcane germplasm collections. *Crop Sci.* **2006**, *46*, 448. [CrossRef]

21. Khan, M.; Pan, Y.B.; Iqbal, J. Development of an RAPD-based SCAR marker for smut disease resistance in commercial sugarcane cultivars of Pakistan. *Crop Prot.* **2017**, *94*, 166–172. [CrossRef]

22. Ali, A.; Pan, Y.; Wang, Q.; Wang, J.; Chen, J.; Gao, S. Genetic diversity and population structure analysis of Saccharum and Erianthus genera using microsatellite (SSR) markers. *Sci. Rep. UK* **2019**, *9*, 395. [CrossRef]

23. Nayak, S.N.; Song, J.; Villa, A.; Pathak, B.; Ayala-Silva, T.; Yang, X.; Todd, J.; Glynn, N.C.; Kuhn, D.N.; Glaz, B.; et al. Promoting Utilization of *Saccharum* spp. Genetic Resources through Genetic Diversity Analysis and Core Collection Construction. *PLoS ONE* **2014**, *9*, e110856. [CrossRef]

24. Liu, H.L.; Yang, X.P.; You, Q.; Song, J.; Wang, L.P.; Zhang, J.S.; Deng, Z.H.; Ming, R.; Wang, J.P. Pedigree, marker recruitment, and genetic diversity of modern sugarcane cultivars in China and the United States. *Euphytica* **2018**, *214*, 48. [CrossRef]

25. Pan, Y. Development and Integration of an SSR-Based Molecular Identity Database into Sugarcane Breeding Program. *Agronomy* **2016**, *6*, 28. [CrossRef]

26. Marconi, T.G.; Costa, E.A.; Miranda, H.R.; Mancini, M.C.; Cardososilva, C.B. Functional markers for gene mapping and genetic diversity studies in sugarcane. *BMC Res. Notes* **2011**, *4*, 264. [CrossRef]

27. Andru, S.; Pan, Y.; Thongthawee, S.; Burner, D.M.; Kimbeng, C.A. Genetic analysis of the sugarcane (*Saccharum* spp.) cultivar 'LCP 85-384'. I. Linkage mapping using AFLP, SSR, and TRAP markers. *Theor. Appl. Genet.* **2011**, *123*, 77–93. [CrossRef]

28. Banerjee, N.; Siraree, A.; Yadav, S.; Kumar, S.; Singh, J.; Kumar, S.; Pandey, D.K.; Singh, R.K. Marker-trait association study for sucrose and yield contributing traits in sugarcane (*Saccharum* spp. hybrid). *Euphytica* **2015**, *205*, 185–201. [CrossRef]

29. Racedo, J.; Gutiérrez, L.; Perera, M.F.; Ostengo, S.; Pardo, E.M.; Cuenya, M.I.; Welin, B.; Castagnaro, A.P. Genome-wide association mapping of quantitative traits in a breeding population of sugarcane. *BMC Plant Biol.* **2016**, *16*, 142. [CrossRef]

30. Ukoskit, K.; Posudsavang, G.; Pongsiripat, N.; Chatwachirawong, P.; Klomsa-ard, P.; Poomipant, P.; Tragoonrung, S. Detection and validation of EST-SSR markers associated with sugar-related traits in sugarcane using linkage and association mapping. *Genomics* **2019**, *111*, 1–9. [CrossRef]

31. Pan, Y.B. Highly polymorphic microsatellite DNA markers for sugarcane germplasm evaluation and variety identity testing. *Sugar Tech.* **2006**, *8*, 246–256. [CrossRef]

32. Chen, P.H.; Pan, Y.B.; Chen, R.K.; Xu, L.P.; Chen, Y.Q. SSR marker-based analysis of genetic relatedness among sugarcane cultivars (*Saccharum* spp. hybrids) from breeding programs in China and other countries. *Sugar Tech.* **2009**, *11*, 347–354. [CrossRef]

33. Liu, P.W.; Que, Y.X.; Pan, Y.B. Highly Polymorphic Microsatellite DNA Markers for Sugarcane Germplasm Evaluation and Variety Identity Testing. *Sugar Tech.* **2011**, *13*, 129–136. [CrossRef]

34. Pan, Y.B.; Liu, P.W.; Que, Y.X. Independently Segregating Simple Sequence Repeats (SSR) Alleles in Polyploid Sugarcane. *Sugar Tech.* **2015**, *17*, 235–242. [CrossRef]

35. You, Q.; Pan, Y.; Xu, L.; Gao, S.; Wang, Q.; Su, Y.; Yang, Y.; Wu, Q.; Zhou, D.; Que, Y. Genetic Diversity Analysis of Sugarcane Germplasm Based on Fluorescence-Labeled Simple Sequence Repeat Markers and a Capillary Electrophoresis-based Genotyping Platform. *Sugar Tech.* **2016**, *18*, 380–390. [CrossRef]

36. Ali, A.; Wang, J.; Pan, Y.; Deng, Z.; Chen, Z.; Chen, R.; Gao, S. Molecular identification and genetic diversity analysis of Chinese sugarcane (*Saccharum* spp. Hybrids) varieties using SSR markers. *Trop. Plant Biol.* **2017**, *10*, 194–203. [CrossRef]

37. Fu, Y.; Pan, Y.; Lei, C.; Grisham, M.P.; Yang, C.; Meng, Q. Genotype-Specific Microsatellite (SSR) Markers for the Sugarcane Germplasm from the Karst Region of Guizhou, China. *Am. J. Plant Sci.* **2016**, *7*, 2209–2220. [CrossRef]

38. Rogers, S.O.; Bendich, A.J. *Extraction of DNA from plant tissues. Plant Molecular Biology Manual*; Gelvin, S.B., Schilperoort, R.A., Verma, D.P.S., Eds.; Springer: Dordrecht, The Netherlands, 1989.

39. Pan, Y.B.; Scheffler, B.E.; Richard, E., Jr. High throughput genotyping of commercial sugarcane clones with microsatellite (SSR) DNA markers. *Sugar Tech.* **2007**, *9*, 176–181.

40. Singh, R.K.; Mishra, S.K.; Singh, S.P.; Mishra, N.; Sharma, M.L. Evaluation of microsatellite markers for genetic diversity analysis among sugarcane species and commercial hybrids. *Aust. J. Crop Sci.* **2010**, *4*, 116–125.

41. Parida, S.K.; Pandit, A.; Gaikwad, K.; Sharma, T.R.; Srivastava, P.; Singh, N.K.; Mohapatra, T. Functionally relevant microsatellites in sugarcane unigenes. *BMC Plant Biol.* **2010**, *10*, 251. [CrossRef]

42. Singh, R.K.; Singh, R.B.; Singh, S.P.; Sharma, M.L. Identification of sugarcane microsatellites associated to sugar content in sugarcane and transferability to other cereal genomes. *Euphytica* **2011**, *182*, 335–354. [CrossRef]

43. Pan, Y.B.; Tew, T.L.; Schnell, R.J.; Viator, R.P.; Richard, E.P.; Grisham, M.P.; White, W.H. Microsatellite DNA marker-assisted selection of *Saccharum spontaneum* cytoplasm-derived germplasm. *Sugar Tech.* **2006**, *8*, 23–29. [CrossRef]

44. Fan, W.Q.; Ge, H.M.; Sun, X.S.; Yang, A.G.; Zhang, Z.F.; Ren, M. DataFormater, A software for SSR data formatting to develop population genetics analysis. *Mol. Plant Breed.* **2016**, *14*, 1029–1034.

45. Liu, K.; Muse, S.V. PowerMaker: An integrated analysis environment for genetic maker analysis. *Bioinformatics* **2005**, *21*, 2128–2129. [CrossRef]

46. Prevost, A.; Wilkinson, M.J. A new system of comparing PCR primers applied to ISSR fingerprinting of potato cultivars. *Theor. Appl. Genet.* **1999**, *98*, 107–112. [CrossRef]

47. Kalinowski, S.T.; Taper, M.L.; Marshall, T.C. Revising how the computer program CERVUS accommodates genotyping error increases success in paternity assignment. *Mol. Ecol.* **2007**, *16*, 1099–1106. [CrossRef]

48. Peakall, R.; Smouse, P.E. GENALEX 6: Genetic analysis in Excel. Population genetic software for teaching and research. *Mol. Ecol. Notes* **2006**, *6*, 288–295. [CrossRef]

49. Peakall, R.; Smouse, P.E. GenAlEx 6.5: genetic analysis in Excel. Population genetic software for teaching and research-an update. *Bioinformatics* **2012**, *28*, 2537–2539. [CrossRef]

50. Falush, D.; Stephens, M.; Pritchard, J.K. Inference of Population Structure Using Multilocus Genotype Data: Linked Loci and Correlated Allele Frequencies. *Genetics* **2003**, *164*, 1567–1587.

51. Evanno, G.; Regnaut, S.; Goudet, J. Detecting the number of clusters of individuals using the software STRUCTURE: A simulation study. *Mol. Ecol.* **2010**, *14*, 2611–2620. [CrossRef]

52. Earl, D.A.; VonHoldt, B.M. Structure harvester: A website and program for visualizing structure output and implementing the Evanno method. *Conserv. Genet. Resour.* **2012**, *4*, 359–361. [CrossRef]

53. Jakobsson, M.; Rosenberg, N.A. Clumpp: A cluster matching and permutation program for dealing with label switching and multimodality in analysis of population structure. *Bioinformatics* **2007**, *23*, 1801–1806. [CrossRef]

54. Luo, Z.; Brock, J.; Dyer, J.M.; Kutchan, T.; Schachtman, D.; Augustin, M.; Ge, Y.; Fahlgren, N.; Abdel-Haleem, H. Genetic Diversity and Population Structure of a Camelina sativa Spring Panel. *Front. Plant Sci.* **2019**, *10*, 184. [CrossRef]

55. Tamura, K.; Stecher, G.; Peterson, D.; Filipski, A.; Kumar, S. MEGA6: Molecular Evolutionary Genetics Analysis Version 6.0. *Mol. Biol. Evol.* **2016**, *30*, 2725–2729. [CrossRef]

56. Jackson, P.A. Breeding for improved sugar content in sugarcane. *Field Crop. Res.* **2005**, *92*, 277–290. [CrossRef]

57. Stevenson, G.C. *Genetic and Breeding of Sugarcane*; Longmans: London, UK, 1965.

58. Cordeiro, G.M.; Casu, R.; McIntyre, C.L.; Manners, J.M.; Henry, R.J. Microsatellite markers from sugarcane (*Saccharum* spp.) ESTs cross transferable to erianthus and sorghum. *Plant Sci.* **2001**, *160*, 1115–1123. [CrossRef]

59. Hameed, U.; Pan, Y.; Muhammad, K.; Afghan, S.; Iqbal, J. Use of simple sequence repeat markers for DNA fingerprinting and diversity analysis of sugarcane (*Saccharum* spp.) cultivars resistant and susceptible to red rot. *Genet. Mol. Res.* **2012**, *11*, 1195. [CrossRef]

60. You, Q.; Xu, L.P.; Zheng, Y.F.; Que, Y.X. Genetic diversity analysis of sugarcane parents in Chinese breeding programmes using gSSR markers. *Sci. World J.* **2013**, *2013*, 613062. [CrossRef]

61. Szpiech, Z.A.; Rosenberg, N.A. On the size distribution of private microsatellite alleles. *Theor. Popul. Biol.* **2011**, *80*, 100–113. [CrossRef]

62. Slatkin, M. Rare alleles as indicators of gene flow. *Evolution* **1985**, *39*, 53–65. [CrossRef]

63. Kalinowski, S.T. Counting Alleles with Rarefaction: Private Alleles and Hierarchical Sampling Designs. *Conserv. Genet.* **2004**, *5*, 539–543. [CrossRef]

64. Schroeder, K.B.; Schurr, T.G.; Long, J.C.; Rosenberg, N.A.; Crawford, M.H.; Tarskaia, L.A.; Osipova, L.P.; Zhadanov, S.I.; Smith, D.G. A private allele ubiquitous in the Americas. *Biol. Lett.* **2007**, *3*, 218–223. [CrossRef]

65. Bird, K.A.; An, H.; Gazave, E.; Gore, M.A.; Pires, J.C.; Robertson, L.D.; Labate, J.A. Population structure and phylogenetic relationships in a diverse panel of *Brassica rapa* L. *Front. Plant Sci.* **2017**, *8*, 321. [CrossRef]

66. Wright, S. The interpretation of population structure by F-statistics with special regard to systems of mating. *Evolution* **1965**, *19*, 395–420. [CrossRef]

67. Deng, H.H.; Li, Q.W. Utilization of CP72-1210 in sugarcane breeding program in mainland China. *Guangdong Agric. Sci.* **2007**, *11*, 18–21.

13

Molecular Assisted Selection for Pollination-Constant and Non-Astringent Type without Male Flowers in Spanish Germplasm for Persimmon Breeding

Manuel Blasco [1], Francisco Gil-Muñoz [2], María del Mar Naval [1] and María Luisa Badenes [2,*]

[1] CANSO, Mestre Serrano, 1, 46250 L'Alcúdia, Valencia, Spain; blasco_manvil@externos.gva.es (M.B.); naval_marmer@gva.es (M.d.M.N.)
[2] Instituto Valenciano de Investigaciones Agrarias, CV 315 km 10,5., 46113 Moncada, Valencia, Spain; gil_framuna@externos.gva.es
*   Correspondence: badenes_mlu@gva.es

Abstract: Persimmon (*Diospyros kaki* Thunb) species is a hexaploid genotype that has a morphologically polygamous gyonodioecious sexual system. *D. kaki* bears unisexual flowers. The presence of male flowers resulted in the presence of seeds in the varieties. The fruits of persimmon are classified according to their astringency and the pollination events that produced seeds and modify the levels of astringency in the fruit. The presence of seeds in astringent varieties as pollination variant astringent (PVA), pollination variant non-astringent (PVNA) and pollination constant astringent (PCA) resulted in fruits not marketable. Molecular markers that allow selection of the varieties according to the type of flowers at the plantlet stage would allow selection of seedless varieties. In this study, a marker developed in *D. lotus* by bulk segregant analysis (BSA) and amplified fragment length polymorphism (AFLP) markers, named DlSx-AF4, has been validated in a germplasm collection of persimmon, results obtained agree with the phenotype data. A second important trait in persimmon is the presence of astringency in ripened fruits. Fruits non-astringent at the ripen stage named pollination constant non-astringent (PCNA) are the objective of many breeding programs as they do not need removal of the astringency by a postharvest treatment. Astringency in the hexaploid persimmon is a dominant trait. The presence of at least one astringent allele confers astringency to the fruit. In this paper we checked the marker developed linked to the AST gene. Our goal has been to validate both markers in germplasm from different origins and to test the usefulness in a breeding program.

Keywords: persimmon; sex determination; fruit astringency; molecular markers

1. Introduction

Persimmon (*Diospyros kaki* Thunb) species is a hexaploid genotype that has a morphologically polygamous gyonodioecious sexual system [1]. *D. kaki* bears unisexual flowers, as do other *Diospyros* species. There are genotypes that bear only female flowers and genotypes bearing male and female flowers [2]. Furthermore, varieties bearing only male flowers were described in China [3] and occasional male flower formation was reported in varieties that usually bear only female flowers [4]. In addition to the unisexual flowers, some varieties or genotypes bear hermaphrodite flowers; however, these flowers do not function fully as female flowers [5]. Most of the commercial varieties present only female flowers [6], however the presence of the male flowers type is important in two scenarios: first when the production of seeded fruits is convenient and second in breeding activities in which crosses are requested.

The fruits of persimmon are classified according to their astringency and the pollination events that resulted in different types of fruits. The PCNA (pollination constant non-astringent) varieties are always not astringent at maturity regardless of pollination events and the presence or absence of seeds in the fruit. The presence of male flowers in these varieties is most convenient since it allows pollination and produces seeded fruits that increase the fruit size and weight. In Japan, the presence of seeds in the fruit does not affect the consumers demand [7]. However, in Europe and western countries consumers prefer seedless fruits.

Three additional variety types can be distinguished: PVNA-type (pollination variant non-astringent), which are non-astringent varieties when seeds are present; PVA-type (pollination variant astringent), which are astringent varieties in most parts of the fruit and non-astringent around the seeds if they are present, and PCA-type (pollination constant astringent), which are astringent varieties regardless the presence of seeds. The loss of astringency in these types of persimmons is associated to the ability of the seeds to produce acetaldehyde. This production resulted in browning of the flesh around the seeds (Figure 1), which interfere with the postharvest treatment for removing the astringency in the fruit, all together the PVA and PCA type fruits are unmarketable if they are pollinated and the fruits present seeds [8,9]. In PVA and PCA varieties it is crucial to avoid pollination, hence the presence of male flowers in the variety and in the vicinity of the crop should be avoided. The ability of the production of male flowers is a genetic trait that should be determined in the varieties for avoiding seeds in astringent varieties or improving the presence of them in non-astringent varieties.

Figure 1. Phenotypes of the traits selected. (**A,B**) flowers from the variety 'Cal Fuyu', female and male respectively and (**C,D**) results of pollination on astringent fruits from 'Rojo Brillante' a pollination variant astringent (PVA) variety: (**C**) parthenocarpic non-pollinated fruits and (**D**) pollinated fruits in which presence of seeds resulted in no marketable fruits.

Elucidation of the genetic and molecular basis of sex expression in *D. kaki* leading to the development of molecular markers would allow selection of the varieties according to the type of flowers, being a great contribution for persimmon production and *Diospyros* breeding. The hexaploidy of *D. kaki* made elucidation of this question more difficult than in diploid genotypes. Since the genus includes more than 700 species with different levels of polyploidy, the diploid *Diospyros lotus* was used for investigation of the sex expression into the genus [10]. These authors described the model of inheritance and developed molecular markers associated to sex expression. Later small RNA acting as a sex determinant was identified [11]. Development of markers used the bulk segregant analysis (BSA) and amplified fragment length polymorphism (AFLP). An AFLP marker identified as DlSx-AF4 was sequence-characterized and converted into the sequence characterized amplified region (SCAR) [10]. In this study the marker has been tested in a germplasm selection of varieties phenotyped for sex expression and a backcross population obtained at Instituto Valenciano de Investigaciones Agrarias (IVIA). The results provide evidence of the usefulness of molecular marker assistance in identifying the genetic potential of production of male flowers in persimmon, an important trait in breeding.

PCNA varieties are highly desired because their mature fruits are not astringent, as they stop accumulating tannins at early steps of fruit development [12]. In Japanese varieties, the PCNA trait is recessive to the non-PCNA trait [13] and is controlled by a single locus, AST [14]. Due to persimmon being a hexaploid, the PCNA type should contain six recessive ast alleles [15]. In breeding programs aimed at obtaining PCNA cultivars, the hexaploidy of persimmon along with the recessive inheritance of the non-astringency trait led to breeders to develop crosses that involved only PCNA genotypes. Consequently, several generations of crosses between PCNA genotypes along with the low genetic diversity of this group of persimmons resulted in families with a high rate of inbreeding and plenty of the problems derived from this fact. To avoid inbreeding, the programs need to use non-astringent cultivars in the crosses, but the rate of PCNA obtained could be very low depending on the number of dominant AST alleles carried by the parents selected. In a backcross BC_1, the expected proportion of PCNA offspring from a non-PCNA F_1 parent with one dominant AST, two or three is 50, 20 or 5%, respectively, under an autohexaploid model. In this context, it is of high interest to be able of selecting PCNA types and non-PCNA types in the families obtained at the plantlet stage. The alternative is to select the type of the fruits in the fields after a juvenile period of four years minimum, which is extremely costly and has a low efficiency.

Many efforts have been made to target the region linked to AST [14,16–18]. The most promising results were obtained in a study that identified a region tightly linked to the AST gene [19]. These authors developed a multiplex PCR method based on primers developed from the region identified, highly reliable that allowed detecting recessive and dominant alleles. These primers have been used to test a group of varieties [15]. The region contains microsatellites that allow distinguishing 12 different alleles from 14 non PCNA genotypes. More than 200 accessions and several crosses between PCNA and non-PCNA genotypes were analyzed [20]. Based on the number of fragments detected per individual these authors were able to determine the dominant (AST) and recessive (ast) alleles in the hexaploid persimmon germplasm.

Using this methodology, in this paper we applied molecular assisted selection for discriminate PCNA cultivars and seedlings from different segregated populations obtained in the frame of the IVIA breeding program. The markers for both traits were developed from Japanese varieties, our goal is to validate the markers in a set of germplasm from different origins and different type of astringency and applied them to the IVIA breeding program, in which the involvement of varieties from Mediterranean origin is relevant.

2. Materials and Methods

2.1. Plant Materials

2.1.1. Validation of AST and DlSx-AF4S Markers

Molecular markers developed for sex expression and type of astringency in persimmon were studied in a set of 42 accessions (Table 1) from the persimmon germplasm collection maintained at IVIA, Moncada, Spain (39.588741, −0.394848). The accessions were phenotyped regarding the presence of male flowers. The phenotype of astringency type was known from previous germplasm characterization [21,22].

Table 1. Plant material studied, origin, astringency type, genotype of AST marker, flower type and genotype of the DlSx-AF4S marker.

Variety	Origin	Type [1]	AST [2]	Flowers [1]	D1Sx-AF4S [2]
Anheca	Spain	PCA	+	♀	–
Ferrán-12	Spain	PCA	+	♀	–
Reus-6	Spain	PCA	+	♀	–
Tomatero	Spain	PCA	+	♀	–
Costata	Italy	PCA	+	♀	–
Lycopersicon	Italy	PCA	+	♀	–
Aizumishirazu-A	Japan	PCA	+	♀	–
Fuji	Japan	PCA	+	♀	–
Korea Kaki	Japan	PCA	+	♀	–
Takura	Japan	PCA	+	♀	–
Yokono	Japan	PCA	+	♀	–
Cal Fuyu	Japan	PCNA	–	♀/♂	+
Fau Fau	Japan	PCNA	–	♀/♂	+
Fukuro Gosho	Japan	PCNA	–	♀/♂	+
Fuyu	Japan	PCNA	–	♀	–
Giant Fuyu	Japan	PCNA	–	♀	–
Hana fuyu	Japan	PCNA	–	♀	–
Hana Gosho	Japan	PCNA	–	♀/♂	+
Ichikikei Jiro	Japan	PCNA	–	♀	–
Isahaya	Japan	PCNA	–	♀	–
Jiro	Japan	PCNA	–	♀	–
Kawabata	Japan	PCNA	–	♀	–
Koda Gosho	Japan	PCNA	–	♀	–
Maekawa jiro	Japan	PCNA	–	♀	–
Mukaku jiro	Japan	PCNA	–	♀	–
O'Gosho	Japan	PCNA	–	♀	–
Suruga	Japan	PCNA	–	♀	–
Yamato Gosho	Japan	PCNA	–	♀	–
Bétera-2	Spain	PVA	+	♀	–
Reus-15	Spain	PVA	+	♀	–
Rojo Brillante	Spain	PVA	+	♀	–
Xato de Bonrepós	Spain	PVA	+	♀	–
Aizumishirazu-B	Japan	PVA	+	♀	–
Atago	Japan	PVA	+	♀	–
Hiratanekaki	Japan	PVA	+	♀	–
Hiratanenashi	Japan	PVA	+	♀	–
Maru	Japan	PVA	+	♀/♂	+
Tone Wase	Japan	PVA	+	♀	–
Pakistan Seedless	Pakistan	PVA	+	♀	–
Agakaki	Japan	PVNA	+	♀/♂	+
Castellani	Italy	PVNA	+	♀	–
Edoichi	Italy	PVNA	+	♀	–

[1] Data from phenotyping. [2] Data from genotyping.

2.1.2. Marker Assisted Selection

Marker assisted selection was made on 12 segregated populations obtained from ('Rojo Brillante' × 'Cal Fuyu') × 'Cal Fuyu' in 2016. The backcross was made using 'Rojo Brillante' a high-quality variety astringent (PVA) and 'Cal Fuyu' a PCNA variety with male flowers. Both parents were selected based on agronomic characteristics and adaptability to the Mediterranean environment [21].

Segregated populations screened and individuals per population are described in Table 2. All progenies and seedlings obtained were maintained in orchards at CANSO's Experimental Station, L'Alcudia, Valencia, Spain (39.189086, −0.542067).

Table 2. Results of genotypes analyzed by molecular markers. Number of offspring with the astringent allele (AST), with the DlSx-AF4S allele, with AST + DlSx-AF4S (not selected) and number of offspring with the absence of both markers (genotypes selected).

Progeny	Total	Number of Offspring			
		AST +	D1Sx-AF4S +	AST + D1Sx-AF4S +	AST − D1Sx-AF4S −
F-1.34	99	61 (61.6)	39 (39.4)	22 (22.2)	21 (21.2)
F-1.50	65	39 (60.0)	43 (66.2)	28 (43.1)	11 (16.9)
F-1.52	47	13 (27.7)	24 (51.1)	4 (8.5)	14 (29.8)
F-2.27	5	4 (80.0)	3 (60.0)	3 (60.0)	1 (20.0)
F-4.19	11	8 (72.7)	9 (81.8)	7 (63.6)	1 (9.1)
F-4.24	61	20 (32.8)	31 (50.8)	10 (16.4)	20 (32.8)
F-4.35	38	14 (36.8)	9 (23.7)	4 (10.5)	19 (50.0)
F-4.49	52	14 (26.9)	17 (32.7)	3 (5.8)	24 (46.2)
F-5.32	12	10 (83.3)	6 (50.0)	4 (33.3)	0 (0.0)
F-5.34	27	11 (40.7)	16 (59.3)	5 (18.5)	5 (18.5)
F-5.36	17	12 (70.6)	9 (52.9)	4 (23.5)	0 (0.0)
F-5.41	7	5 (71.4)	3 (42.9)	3 (42.9)	2 (28.6)
total	441	211 (47.8)	209 (47.4)	97 (22.0)	118 (26.8)

Numbers in parentheses are the rate (%) of corresponding offspring in each progeny.

2.2. Methods

2.2.1. DNA Isolation

Young fully expanded leaves were collected from trees and kept at −20 °C until DNA isolation. DNA was isolated according to the CTAB method described in [23] with minor modifications [24].

2.2.2. Molecular Markers Analysis

The capacity of producing male flowers was checked with the sequence characterized amplified region (SCAR) marker 'DlSx-AF4S' [10], primers used were: forward (DlSx-AF4-3F; 5′-ACA TCC AAA GTT CTG GAG AAT CA-3′) and reverse (DlSx-AF4-3R; 5′-ATT GGT GCT TGG TCA AAC ATA TC-3′).

Determination of PCNA genotypes used the primers described in [19] PCNA-F (CCCCTCAGTGGCAGTGCTGC) and 5R3R (GAAACACTCATCCGGAGACTTC).

Polymerase chain reactions (PCRs) were performed in a final volume of 20 μL containing 1× of DreamTaq Buffer (Thermo Fisher Scientific, Vilnius, Lithuania), 0.1 mM of each dNTPs (Promega, Madison, WI, USA), 20 ng of genomic DNA and 1 U of DreamTaq polymerase (Thermo Fisher Scientific, Vilnius, Lithuania). The PCR program consisted of pre-denaturation at 94 °C for 2 min; 35 cycles at 98 °C for 15 s, 60 °C for 20 s and 72 °C for 1 min; followed by a final extension at 72 °C for 10 min. PCR products were separated by electrophoresis on 1.5% agarose gels in 0.5× TAE buffer and visualized with GelRED® (Sigma-Aldrich, St. Louis, MI, USA).

3. Results and Discussion

3.1. Marker Assisted Selection Validation: Production of Male Flowers

A set of accessions belonging to the persimmon germplasm bank were phenotyped for the presence of male flowers and later genotyped with the marker DlSx-AF4S [10] to test the accuracy of the marker for Molecular Assisted Selection (MAS). Results of the genotype agreed with the results of the phenotype (Table 1), non-discrepancies were observed. The capacity of developing male flowers was clearly stated by the presence of the amplified band (320 bp). Figure 2a shows the results on an agarose gel of the presence of male flowers in the genotypes 'Agakaki', 'Cal Fuyu' and the selection 'F-1.34' from the IVIA breeding program. Total correlation between the phenotype and the presence/absence of the band was obtained for all the genotypes studied (Table 1). This marker is a great advantage in breeding programs in which astringent and non-astringent genotypes are involved. The presence of male flowers in astringent varieties (PVNA, PVA and PCA) resulted in the presence of seeds in the fruit. The loss of astringency in these types of persimmons is associated to the ability of the seeds to produce acetaldehyde (Figure 3). Production of acetaldehyde is a quantitative trait in which less production by the seed resulted in higher astringency on the pulp (PCA), and high production resulted in low/none astringent flesh (PVNA) being PVA intermediate. This acetaldehyde production resulted in browning of the flesh around the seeds, which interferes with the postharvest treatment for removing the astringency in the fruit [25]. All together the PVA and PCA type fruits pollinated are unmarketable. In the case of PVNA types in which the presence of seeds browned completely the flesh (Figure 3), there are specific markets in which these varieties are accepted. However, in most of the markets, the PVNA fruits are accepted with no seeds and after removing the astringency by postharvest treatment. In all breeding programs that use astringent varieties, MAS for discriminating male flowers is very important for avoiding self-pollination and/or mix of cultivars that can cross pollinated among them and produced seeds. In breeding programs that involve non-astringent varieties the discrimination of the presence of male flowers is necessary too. Some Japanese programs look for varieties with male flowers and seeds that increase the size and setting of fruits, but in western countries, where the presence of seeded fruits is not acceptable, the presence of male flowers is discarded similarly to astringent varieties. Selection of this trait in persimmon species that have a four-year juvenile period resulted in great interest to avoid plants in the fields that will be eliminated in the future and, additionally, to avoid undesired pollination in the breeding plots.

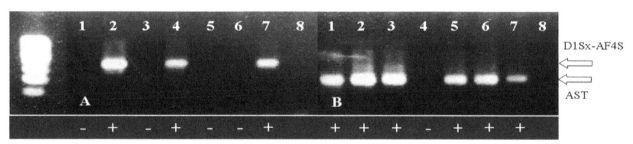

1. Rojo Brillante 2. Agakaki 3. Tone Wase 4. Cal Fuyu 5. Castellani 6. Edoichi 7. F-1.34 8. Blank

Figure 2. PCR results by electrophoresis in an agarose gel (1.5%) vs. phenotype data; (**A**) DlSx-AF4S PCR results. The (+) presence and (−) absence of male flowers from phenotype data; the DlSx-AF4S marker is present in varieties 'Agakaki', 'Cal Fuyu' and F.1-34 in agreement with the phenotype; (**B**) AST PCR results; (+) astringent fruits according to phenotype data and (−) non-astringent fruits (pollination constant non-astringent (PCNA)) according to phenotype data. The AST marker was present in all astringent varieties and absent in 'Cal Fuyu', a PCNA variety.

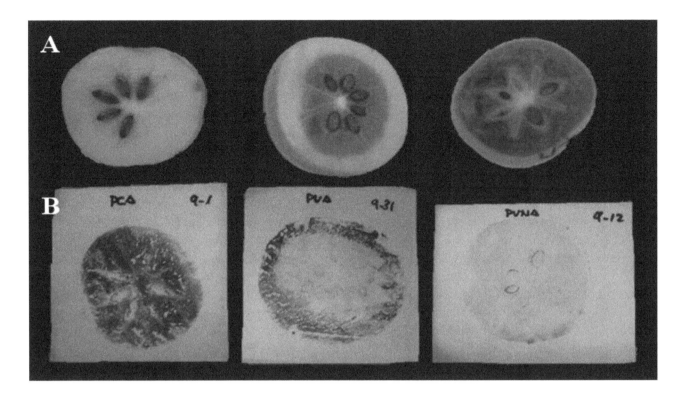

Figure 3. Three types of astringent persimmon according to the amount of acetaldehyde produced by the seed; (**A**) fruits of pollination constant astringent (PCA), PVA and pollination variant non-astringent (PVNA; from right to left) and (**B**) distribution of condensed tannins, visualized by precipitation of blue ferric chloride impregnated in a paper [25,26]. Fruits of PCA, PVA and PVNA (from right to left).

3.2. Marker Assisted Selection Validation: Selection of PCNA

In Japanese cultivars, the PCNA trait is recessive to the non-PCNA trait [13] and is controlled by a single locus, AST [14]. Due to persimmon being a hexaploid, the PCNA type should contain six recessive ast alleles [15]. Detection of at least one AST allele determines the astringency of fruit. Selection of non-astringent fruits or PCNA are the objective in most of the persimmon breeding programs currently active in the world [27–32]. In this study we validated the AST marker developed in [19] for discrimination between PCNA genotypes and the different astringent types.

Validation of AST marker was carried out in a set of cultivars from the germplasm collection with known astringency (Table 1). A total correlation between the phenotypic data of astringency and the markers obtained in the genotypes analyzed was obtained. Figure 2b shows the PCR products of a set of accessions. Two PVA cultivars 'Rojo Brillante' and 'Tone Wase', three PVNA cultivars 'Agakaki', 'Castellani', 'Edoichi' and 'F-1.34' showed a clear band for AST marker. The PCNA cultivar 'Cal Fuyu' showed no amplified product.

3.3. Marker Assisted Selection of Both Traits in the IVIA Breeding Program

After validation of the DlSx-AF4S and AST markers in a set of accessions phenotyped, we applied both markers in the breeding program for selecting the individuals of several segregated populations (Figure 4).

A total of 441 individuals belonging to 12 segregated populations obtained by a backcross that consisted in (PVA × PCNA × PCNA) were evaluated (Table 2). The cross (PVA × PCNA) was made using 'Rojo Brillante' a high-quality variety astringent and 'Cal Fuyu' a PCNA variety with male flowers.

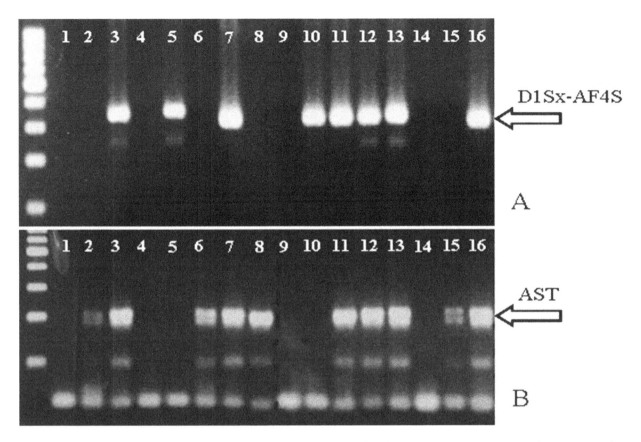

Figure 4. PCR results from agarose gel electrophoresis of 16 backcross (BC) individuals genotyped for AST and D1Sx-AF4S markers. (**A**) Presence of the marker (D1SX-AF4S) indicates ability for developing male flowers; (**B**) Absence of the marker means absence of any astringent allele (AST), which corresponds to a PCNA genotype The BC progenies are selected based on the absence of both markers, which corresponds to PCNA types without capability of the development of male flowers. Individuals 1, 4, 9 and 14 have been selected.

The astringency trait is a dominant marker and taking into account the hexaploidy of persimmon, the number of PCNA genotypes obtained in crosses that involved astringent types depends on the number of AST alleles present in the astringent parental. The validation of the AST marker was made based on different segregated families by [33]. Identification of different AST alleles was made in a set of cultivars by means of crosses with PCNA varieties, analysis of the segregation obtained and sequence of the genomic region [15]. If a non-PCNA has a single A allele (Aaaaaa) and is crossed with a PCNA individual (aaaaaaa), 50% of the offspring will be astringent. In the program the F_1 seedlings obtained and crossed with the PCNA 'Cal Fuyu' resulted in a different percentage of astringent genotypes. It has been demonstrated that the number of PCNA obtained depends on the allelic dose of the F_1 backcrossed [17]. In this study, the number of individuals per progeny was very low for studying segregation ratios and inferred the number of A alleles in the F_1 mothers. However, taking all the tested BC1 seedlings together the rate of astringent genotypes was around 50%, which indicates that the F_1 group of maternal genitors might contain one AST allele on average.

Flower gender analysis revealed that around half of the genotypes analyzed have the capacity to generate the male flower (47.4%). This proportion is as expected, since the crosses need always a parent bearing male flowers. In the IVIA breeding program the presence of male flowers is a discarded trait for all types of fruit. In astringent types as PCA, PVA and PVNA, the presence of male flowers resulted in fruits pollinated and the production of seeds that brown the flesh and difficult the postharvest treatment for removing the astringency. PCNA types are discarded as well because consumers do not accept seeds in the fruits and the presence of male flowers can pollinate astringent fruits, affecting negatively the quality and marketability of them.

Combined results of both markers, AST positive (astringency of the fruit) and DlSx-AF4S positive (presence of male flowers) resulted in a high number of discarded genotypes. In column (AST- and DlSx-AF4S) from Table 2 we indicated the genotypes that will be selected according to our breeding objectives. Only genotypes not astringent (PCNA type) and without male flowers will be selected (absence of both markers), a total of 118 from 441 (26.8%). It is important to point out that the markers segregated independently. According to the published genome of *Diospyros oleifera* [34], identified as the diploid *D. kaki* ancestor, the DNA fragments from which the markers were derived are located in different chromosomes. Therefore, the AST and D1Sx-AF4 markers must segregate independently.

The low rate of genotypes selected from the populations generated indicates the usefulness of the MAS applied in persimmon breeding. We could select at a plantlet stage the genotypes that will be planted in the fields for further agronomic selection. This MAS avoids keeping the future rejected plants during 4 years in the experimental fields. In our case near to 75% of the genotypes obtained can be discarded at the seedling stage in the greenhouse, indicating a high effectiveness of MAS in persimmon breeding.

4. Conclusions

The markers DlSx-AF4, linked to the production of male flowers and AST linked to astringency of the fruits, have been validated in a germplasm collection of persimmon. Although the markers were developed from Japanese cultivars, the correlation between the phenotype and genotype was 100% in germplasm from a different origin, which demonstrated the usefulness of the markers for selecting these important traits. Both markers have been screened in different progenies from a backcross that includes an astringent parent from non-Japanese origin. Results demonstrated that selection of both traits combined resulted in a very low rate of selection. In a context of breeding programs that involve astringent cultivars the MAS applied to discriminate PCNA genotypes is highly valuable.

Author Contributions: Conceptualization, M.B. and M.L.B.; Methodology, M.B. and F.G.-M.; Investigation, M.B. and M.d.M.N.; Writing—Original Draft, M.L.B.; Writing—Review and Editing, M.B., F.G.-M. and M.d.M.N.; Funding Acquisition, M.L.B. and M.B.; Resources, M.L.B.; Supervision, M.L.B. and M.B. All authors have read and agreed to the published version of the manuscript.

References

1. Dellaporta, S.L.; Calderon-Urrea, A. Sex determination in flowering plants. *Plant Cell* **1993**, *5*, 1241–1251. [PubMed]
2. Yonemori, K.; Sugiura, A.; Tanaka, K.; Kameda, K. Floral Ontogeny and Sex Determination in Monoecious-type Persimmons. *J. Am. Soc. Hortic. Sci.* **1993**, *118*, 293–297. [CrossRef]
3. Xu, L.Q.; Zhang, Q.L.; Luo, Z.R. Occurrence and cytological mechanism of 2n pollen formation in Chinese Diospyros spp. (Ebenaceae) staminate germplasm. *J. Hortic. Sci. Biotechnol.* **2008**, *83*, 668–672. [CrossRef]
4. Yakushiji, H.; Yamada, M.; Yonemori, K.; Sato, A.; Kimura, N. Staminate flower production on shoots of Fuyu' and "Jiro" persimmon (Diospyros kaki Thunb.). *J. Jpn. Soc. Hortic. Sci.* **1995**, *64*, 41–46. [CrossRef]
5. Giordani, E.; Picardi, E.; Radice, S. Morfologia y Fisiologia. In *El Cultivo del Caqui*; Badenes, M.L., Intrigliolo, D.S., Salvador, A., Vicent, A., Eds.; Generalitat Valenciana: Valencia, Spain, 2015; pp. 17–33; ISBN 9788448260187.
6. Kajiura, M.; Blumenfeld, A. Diospyros kaki. In *CRC Handbook of Flowering*; Halevy, A.H., Ed.; CRC Press: Boca Raton, FL, USA, 1989; Volume 6, pp. 298–306; ISBN 9781315893464.
7. Yamada, M.; Giordani, E.; Yonemori, K. Persimmon. In *Fruit Breeding*; Badanes, M.L., Byrne, D.H., Eds.; Springer: New York, NY, USA, 2012; Volume 8, pp. 663–693; ISBN 9781441907639.

8. Sugiura, A.; Yonemori, K.; Harada, H.; Tomana, T. Changes of ethanol and acetaldehyde contents in Japanese persimmon fruits and their relation to natural deastringency. *Stud. Inst. Hortic. Kyoto Univ.* **1979**, *9*, 41–47.

9. Sugiura, A.; Tomana, T. Relationships of ethanol production by seeds of different types of Japanese persimmons and their tannin content. *Hort. Sci.* **1983**, *18*, 319–321.

10. Akagi, T.; Kajita, K.; Kibe, T.; Morimura, H.; Tsujimoto, T.; Nishiyama, S.; Kawai, T.; Yamane, H.; Tao, R. Development of molecular markers associated with sexuality in Diospyros lotus L. and their application in D. kaki Thunb. *J. Jpn. Soc. Hortic. Sci.* **2014**, *83*, 214–221. [CrossRef]

11. Akagi, T.; Henry, I.M.; Tao, R.; Comai, L. A Y-chromosome-encoded small RNA acts as a sex determinant in persimmons. *Science* **2014**, *346*, 646–650. [CrossRef]

12. Yonemori, K.; Matsushima, J. Property of Development of the Tannin Cells in Non-Astringent Type Fruits of Japanese Persimmon (Diospyros kaki) and Its Relationship to Natural Deastringency. *J. Jpn. Soc. Hortic. Sci.* **1985**, *54*, 201–208. [CrossRef]

13. Ikeda, I.; Yamada, M.; Kurihara, A.; Nishida, T. Inheritance of Astringency in Japanese Persimmon. *J. Jpn. Soc. Hortic. Sci.* **1985**, *54*, 39–45. [CrossRef]

14. Kanzaki, S.; Yamada, M.; Sato, A.; Mitani, N.; Ustunomiya, N.; Yonemori, K. Conversion of RFLP markers for the selection of pollination-constant and non-astringent type persimmons (Diospyros kaki Thunb.) into PCR-based markers. *J. Jpn. Soc. Hortic. Sci.* **2009**, *78*, 68–73. [CrossRef]

15. Kono, A.; Kobayashi, S.; Onoue, N.; Sato, A. Characterization of a highly polymorphic region closely linked to the AST locus and its potential use in breeding of hexaploid persimmon (Diospyros kaki Thunb.). *Mol. Breed.* **2016**, *36*, 1–13. [CrossRef]

16. Kanzaki, S.; Yonemori, K.; Sugiura, A.; Sato, A.; Yamada, M. Identification of molecular markers linked to the trait of natural astringency loss of Japanese persimmon (Diospyros kaki) fruit. *J. Jpn. Soc. Hortic. Sci.* **2001**, *126*, 51–55. [CrossRef]

17. Ikegami, A.; Yonemori, K.; Sugiura, A.; Sato, A.; Yamada, M. Segregation of astringency in F1 progenies derived from crosses between pollination-constant, nonastringent persimmon cultivars. *HortScience* **2004**, *39*, 371–374. [CrossRef]

18. Akagi, T.; Takeda, Y.; Yonemori, K.; Ikegami, A.; Kono, A.; Yamada, M.; Kanzaki, S. Quantitative genotyping for the astringency locus in hexaploid persimmon cultivars using quantitative real-time PCR. *J. Jpn. Soc. Hortic. Sci.* **2010**, *135*, 59–66. [CrossRef]

19. Kanzaki, S.; Akagi, T.; Masuko, T.; Kimura, M.; Yamada, M.; Sato, A.; Mitani, N.; Ustunomiya, N.; Yonemori, K. SCAR markers for practical application of marker-assisted selection in persimmon (Diospyros kaki Thunb.) breeding. *J. Jpn. Soc. Hortic. Sci.* **2010**, *79*, 150–155. [CrossRef]

20. Onoue, N.; Kobayashi, S.; Kono, A.; Sato, A. SSR-based molecular profiling of 237 persimmon (Diospyros kaki Thunb.) germplasms using an ASTRINGENCY-linked marker. *Tree Genet. Genomes* **2018**, *14*. [CrossRef]

21. Martínez-Calvo, J.; Naval, M.; Zuriaga, E.; Llácer, G.; Badenes, M.L. Morphological characterization of the IVIA persimmon (Diospyros kaki Thunb.) germplasm collection by multivariate analysis. *Genet. Resour. Crop Evol.* **2013**, *60*, 233–241. [CrossRef]

22. Del Naval, M.M.; Zuriaga, E.; Pecchioli, S.; Llácer, G.; Giordani, E.; Badenes, M.L. Analysis of genetic diversity among persimmon cultivars using microsatellite markers. *Tree Genet. Genomes* **2010**, *6*, 677–687. [CrossRef]

23. Doyle, J.J.; Doyle, J.L. A rapid DNA isolation procedure for small quantities of fresh leaf tissue. *Phytolog. Bull* **1987**, *19*, 11–15. [CrossRef]

24. Soriano, J.M.; Pecchioli, S.; Romero, C.; Vilanova, S.; Llácer, G.; Giordani, E.; Badenes, M.L. Development of microsatellite markers in polyploid persimmon (Diospyros kaki Thunb) from an enriched genomic library. *Mol. Ecol. Notes* **2006**, *6*, 368–370. [CrossRef]

25. Besada, C.; Novillo, P.; Navarro, P.; Salvador, A. Causes of flesh browning in persimmon–A review. *Acta Hortic.* **2018**, *1195*, 203–210. [CrossRef]

26. Munera, S.; Besada, C.; Blasco, J.; Cubero, S.; Salvador, A.; Talens, P.; Aleixos, N. Astringency assessment of persimmon by hyperspectral imaging. *Postharvest Biol. Technol.* **2017**, *125*, 35–41. [CrossRef]

27. Yamada, M. Persimmon Breeding in Japan. *Jpn. Agric. Res. Q.* **1993**, *27*, 33–37.

28. Yamada, M.; Sato, A.; Yakushiji, H.; Yoshinaga, K.; Yamane, H.; Endo, M. Characteristics of "Luo Tian Tian Shi", a non-astringent cultivar of oriental persimmon (Diospyros kaki Thunb.) of Chinese origin in relation to non-astringent cultivars of Japanese origin. *Bull. Fruit Tree Res. Stn.* **1993**, *25*, 19–32.

29.   Bellini, E.; Giordani, E. Germplasm and breeding of persimmon in Europe. *Acta Hortic.* **2005**, *685*, 65–74. [CrossRef]

30.   Badenes, M.L.; Martinez-Calvo, J.; Naval, M.M. The persimmon breeding program at IVIA: Alternatives to conventional breeding of persimmon. *Acta Hortic.* **2013**, *996*, 71–76. [CrossRef]

31.   Luo, Z.; Zhang, Q.; Luo, C.; Xie, F. Recent advances of persimmon research and industry in China. *Acta Hortic.* **2013**, *996*, 43–48. [CrossRef]

32.   Ma, K.B.; Lee, I.B.; Kim, Y.K.; Won, K.H.; Cho, K.S.; Choi, J.J.; Lee, B.H.N.; Kim, M.S. 'Jowan', an early maturing PCNA (pollination constant non-astringent) persimmon (Diospyros kaki Thunb.). *Acta Hortic.* **2018**, *1195*, 61–64. [CrossRef]

33.   Mitani, N.; Kono, A.; Yamada, M.; Sato, A.; Kobayashi, S.; Ban, Y.; Ueno, T.; Shiraishi, M.; Kanzaki, S.; Tsujimoto, T.; et al. Application of marker-assisted selection in persimmon breeding of PCNA offspring using SCAR markers among the population from the cross between Non-PCNA 'taigetsu' and PCNA 'kanshu'. *HortScience* **2014**, *49*, 1132–1135. [CrossRef]

34.   Zhu, Q.G.; Xu, Y.; Yang, Y.; Guan, C.F.; Zhang, Q.Y.; Huang, J.W.; Grierson, D.; Chen, K.S.; Gong, B.C.; Yin, X.R. The persimmon (Diospyros oleifera Cheng) genome provides new insights into the inheritance of astringency and ancestral evolution. *Hortic. Res.* **2019**. [CrossRef] [PubMed]

14

Exploring the Genetic Architecture of Root-Related Traits in Mediterranean Bread Wheat Landraces by Genome-Wide Association Analysis

Rubén Rufo [1], Silvio Salvi [2], Conxita Royo [1] and Jose Miguel Soriano [1,*]

[1] Sustainable Field Crops Programme, IRTA (Institute for Food and Agricultural Research and Technology), 25198 Lleida, Spain; ruben.rufo@irta.cat (R.R.); conxita.royo@irta.cat (C.R.)

[2] Department of Agricultural and Food Sciences, University of Bologna, Viale Fanin 44, 40127 Bologna, Italy; silvio.salvi@unibo.it

[*] Correspondence: josemiguel.soriano@irta.cat

Abstract: Background: Roots are essential for drought adaptation because of their involvement in water and nutrient uptake. As the study of the root system architecture (RSA) is costly and time-consuming, it is not generally considered in breeding programs. Thus, the identification of molecular markers linked to RSA traits is of special interest to the breeding community. The reported correlation between the RSA of seedlings and adult plants simplifies its assessment. Methods: In this study, a panel of 170 bread wheat landraces from 24 Mediterranean countries was used to identify molecular markers associated with the seminal RSA and related traits: seminal root angle, total root number, root dry weight, seed weight and shoot length, and grain yield (GY). Results: A genome-wide association study identified 135 marker-trait associations explaining 6% to 15% of the phenotypic variances for root related traits and 112 for GY. Fifteen QTL hotspots were identified as the most important for controlling root trait variation and were shown to include 31 candidate genes related to RSA traits, seed size, root development, and abiotic stress tolerance (mainly drought). Co-location for root related traits and GY was found in 17 genome regions. In addition, only four out of the fifteen QTL hotspots were reported previously. Conclusions: The variability found in the Mediterranean wheat landraces is a valuable source of root traits to introgress into adapted phenotypes through marker-assisted breeding. The study reveals new loci affecting root development in wheat.

Keywords: drought stress; association mapping; root system architecture; QTL hotspot; seminal root

1. Introduction

Wheat is the most widely cultivated crop in the world, covering around 219 million ha (Faostat 2017, http://www.fao.org/faostat/). It is a staple food for humans, as it provides 18% of daily human intake of calories and 20% of protein (http://www.fao.org/faostat/). Global wheat demand is estimated to increase by 60% by the year 2050 [1], so wheat production will need to rise by 1.7% per year until then. Achieving this objective is a great challenge under the current climate change scenario, as the prediction models estimate a precipitation decrease of 25% to 30% and a temperature increase of 4 °C to 5 °C for the Mediterranean region [2]. It is well known that wheat production is greatly affected by environmental stresses such as drought and heat [3] that negatively affect yield and grain quality [4]. Drought is considered the greatest environmental constraint to yield and yield stability in rainfed production systems [5]. Environmental effects on yield in the Mediterranean Basin have been estimated at 60% for bread wheat [6] and 98% for durum wheat [7]. The expected effects of climate change and the declining availability of water and chemical fertilizers will require the release of cultivars with an enhanced genetic capacity to maintain acceptable yield levels and yield stability under harmful environmental conditions [8,9]. To cope with the challenges of climate change, breeders are particularly challenged to stretch the adaptability and performance stability of new cultivars, so many improvement programs are focussing on breeding for adaptation [10].

Plants respond and adapt to water deficit using various strategies that have evolved at several levels of function and are components of the conceptual framework developed by Reynolds et al. [11], which defines drought resistance in terms of dehydration escape, tolerance, and avoidance. Traits defining root system architecture (RSA) are critical for wheat adaptation to drought environments and non-optimal nutritional supply conditions [12]. Besides, water-use efficiency (WUE) can be significantly increased by optimizing the anatomy and growth features of roots [13]. Root traits are critical for drought tolerance due to its role in plant performance and the acquisition of nutrients and water from dry soils [14]. The wheat plant includes two types of roots: seminal (embryonal) and nodal (crown or adventitious or adult root system). The seminal roots are the first to penetrate the soil and remain functional during the whole plant cycle [9,15]. A correlation between seminal and adult roots in terms of size, dry-weight, or even specific architectural features have been reported [9,13]. Since the evaluation of RSA features in the field is very difficult, expensive, and time-consuming when a large number of genotypes need to be phenotyped, several studies have been carried out at early growth stages to allow an optimal screening of RSA traits [8,12,16–18]. Maccaferri et al. [9] observed that among RSA traits, those involving the root structure and related to the uptake of nutrients and water are root length, surface area and volume, and the number of roots, while root diameter is significantly associated with drought tolerance. Another RSA trait of interest in wheat is the seminal root angle (SRA), whose features suggest that narrow angles could lead to deeper root growth to obtain water from deeper soil layers and hence maintain higher yields [5,13].

Identifying quantitative trait loci (QTLs) and applying marker-assisted selection is of particular interest for RSA because the trait is important but difficult to phenotype. In the last few years, genome-wide association studies (GWAS) have become very popular because of their use of germplasm collections with wider variability than the classical bi-parental crosses. These collections allow many recombination events to be detected, making the association between genotype and phenotype more accurate. Collections of landraces are an ideal subject of GWAS [19] since they are genetically diverse repositories of unique traits that have evolved in local environments characterized by a wide range of biotic and abiotic conditions. Several studies have shown that Mediterranean wheat landraces possess a wide genetic background for root architecture, yield formation, stress tolerance, and quality traits [17–22]. In the current study, a GWAS for three RSA traits and two related traits was performed on a panel of 170 bread wheat (*Triticum aestivum* L.) landraces from 24 Mediterranean countries with the following goals: (1) to detect differences in RSA among genetic subpopulations previously distinguished in the panel, (2) to identify correlations among RSA and grain yield under rainfed conditions, and (3) to identify molecular markers and candidate genes linked to root-related traits and candidate gene models for the associations.

2. Materials and Methods

2.1. Plant Material

A germplasm collection of 170 bread wheat (*Triticum aestivum* L.) genotypes from the MED6WHEAT IRTA panel described by Rufo et al. [23] was used in this study. The panel was genotyped and characterized using the Illumina Infinium 15K Wheat SNP Chip at Trait Genetics GmbH (Gatersleben, Germany), and markers were ordered according to the SNP map developed by Wang et al. [24]. The collection was previously structured into three subpopulations (SPs) matching their geographical origin [23]: western (SP1, WM), northern (SP2, NM), and eastern Mediterranean (SP3,

EM) (Supplementary Materials, Table S1). Additionally, the cultivars 'Arthur Nick', 'Anza', 'Soissons', and 'Chinese Spring' were included as checks.

2.2. Root Morphology and Statistical Analysis

Root analysis was performed following the protocol described by Canè et al. [8], which was slightly modified in the current study (Figure 1). Ten representative seeds were randomly chosen from each genotype, weighed, sterilized in a 10% sodium hypochlorite solution for 5–10 min, washed thoroughly in distilled water and placed on hydrated filter paper in a 140 mm Petri dish at 28 °C for 24 h. Subsequently, five seedlings were selected on the basis of a normal seminal root emergence and were spaced 8 cm from each other on a filter paper sheet placed on a vertical black rectangular polycarbonate plate (42.5 × 38.5 cm). Finally, each plate was covered with another wet sheet of filter paper. Distilled water was used for the plantlets' growth. The plantlets were grown in a growth chamber for 14 days at 22 °C under a 16-h light photoperiod. In addition to the ten seed weight (SW), four other traits were scored for each genotype: total root number (TRN), shoot length (SL) from the seed to the tip of the longest leaf and SRA, obtained using a digital camera following the methodology described in Canè et al. [8]. The images were processed with ImageJ software [25]. The angle between the two external roots of each plantlet was measured at a distance of 3.5 cm from the tip of the seed. Finally, the roots were desiccated at 70 °C for 24 h to obtain the root dry weight (RDW).

The experimental design followed a randomized complete block with two replications in time. Means of five observational units for each genotype were used for TRN, RDW, and SL, while only three observational units were used for SRA because the two external ones were considered as border plantlets for root angle.

Figure 1. Experimental setup for the analysis of seminal root traits. Seeds were placed 8 cm apart on moist filter paper (**A**) and kept in a box with distilled water in a growth chamber for 14 days at 22 °C under a 16-h light photoperiod (**B**). (**C**) Example of seminal root angle measurement, using ImageJ software.

2.3. Grain Yield

Field experiments were carried out in 2016, 2017, and 2018 harvesting seasons in Gimenells, Lleida, north-east Spain (41°38′ N and 0°22′ E, 260 m a.s.l) under rainfed conditions. The experiments followed a non-replicated augmented design with two replicated checks (the cultivars 'Anza' and 'Soissons') and plots of 3.6 m². The experimental design is shown in Supplementary Materials, Figure S1. Sowing density was adjusted to 250 germinable seeds m². Weeds and diseases were controlled following standard practices at the site. The anthesis date was determined in each plot. Grain yield (GY, t ha^{-1}) was determined by mechanically harvesting the plots at maturity and expressed on a 12% moisture level.

2.4. Statistical Analysis

Phenotypic data for GY was fitted to a linear mixed model with the check cultivars as fixed effects and the row number, column number and cultivar as random effects following the SAS PROC MIXED procedure:

$$y = X\beta + Z\gamma + \varepsilon \tag{1}$$

where β is an unknown vector of fixed-effects parameters with known design matrix X, γ is an unknown vector of random-effects parameters with known design matrix Z, and ε is an unknown random error vector whose elements are no longer required to be independent and homogeneous.

Restricted maximum likelihood was used to estimate the variance components and to produce the best linear unbiased predictors (BLUPs) for the traits of each cultivar and year with the SAS-STAT statistical package (SAS Institute Inc, Cary, NC, USA).

Analyses of variance (ANOVA) were performed for the root traits, considering the genotypes and the replication as random effects in the model. Additionally, for a subset of 55 of the 141 structured landraces, selected as having an SP membership q > 0.8 (WM, 17; NM, 15; EM, 23), the sum of squares of the cultivar effect in the ANOVAs was partitioned into differences between SPs and differences within them. ANOVA for grain yield was performed for the complete collection, considering genotype, year, and the combination of genotype and year the sources of variation. Least squares means were calculated and compared using the Tukey HSD test at P < 0.05. Pearson correlation coefficients among root traits were computed. Repeatability (H) was calculated on a mean basis across two replications following the formula described by Harper [26] r = (B − W) / (B + ((n −1) W)), where n is the number of genotypes and B and W the two variances from the ANOVA table: between (B) and within (W). Frequency distributions, ANOVAs, the Tukey test, and the Pearson correlation coefficients were calculated using the JMP v13.1.0 statistical package (SAS Institute Inc, Cary, NC, USA).

2.5. Genome-Wide Association Analysis

A GWAS was performed for the mean of measured root traits and from the BLUPs for GY per year and across years with TASSEL 5.0 software [27]. A mixed linear model (MLM) was conducted using the information of the genetic structure reported in Rufo et al. [23] as the fixed effect and a kinship (K) matrix, calculated using Haploview [28], as the random effect (Q + K model) at the optimum compression level. In addition, the anthesis date was incorporated as a cofactor in the analysis. As reported in other studies [29–32], an adjusted −log10 P > 3 was established as a threshold for considering a marker-trait association (MTA) statistically significant. A moderate threshold at −log10 P > 2.5 was also established for GY. Confidence intervals (CI) for MTAs were calculated for each chromosome according to the linkage disequilibrium (LD) decay reported by Rufo et al. [23]. In order to simplify the MTA information, the associations were grouped into QTL hotspots when at least two MTAs belonging to different traits overlapped their CIs. Circular Manhattan plots were performed using the R package "CMplot".

2.6. Gene Annotation

Gene annotation within the CIs of the QTL hotspots was performed using the gene models for high-confidence genes reported for the wheat genome sequence [33], available at https://wheat-urgi.versailles.inra.fr/Seq-Repository/Assemblies. Markers flanking the CIs were used to estimate physical distances from genetic distances.

3. Results

3.1. Phenotypic Data of Root Traits

A summary of the genetic variation of the root traits is shown in Table 1. The genotypes showed a low coefficient of variation (CV) with a narrow range of variation among traits, from 10.4 for SW to 18.8 for RDW and repeatability (H) ranging from 48.5% for RDW to 75.4% for SW.

Table 1. Statistics of the seminal root traits.

Table	TRN (N)	RDW (mg)	SRA (°)	SW (g)	SL (cm)
Min	3.2	43	53.1	0.27	14.9
Max	5.4	20.5	125.9	0.63	30.2
Mean	4.4	11.7	98.6	0.48	22.0
SD	0.5	2.2	13.3	0.05	2.8
CV (%)	10.9	18.8	13.5	10.4	12.7
H (%)	52.0	48.5	70.0	75.4	50.0

SD, standard deviation; CV, coefficient of variation; H, repeatability; TRN, total root number; RDW, root dry weight; SRA, seminal root angle; SW, seed weight; SL, seed length.

The ANOVA (Table 2) for cultivars with a high membership coefficient ($q > 0.8$) showed that for all traits the total variability was mainly explained by the genotype effect, in a range from 63.5% for SL to 88.8% for SW. When the sum of squares of the genotype effect was partitioned into differences between and within SPs, the results revealed that the genetic variability was mainly explained by differences within SPs in a range from 47.8% for TRN to 71.8% for SRA (Table 2). Differences between SPs were statistically significant for SRA, TRN, SW, and SL, in a range from 6.0% of the genotype effect for SL to 25.3% for TRN (Table 2). The sum of squares within SPs was partitioned into western (WM), northern (NM), and eastern (EM) effects, being statistically significant for SRA (40.8%), TRN (28.3%), SW (38.3%), and SL (26.8%) in the western SP, SRA (34.4%) in the northern SP and RDW (53.6%) and SW (55.4%) in the eastern SP.

Table 2. Percentage of the sum of squares of the ANOVA in a set of 55 bread wheat landraces structured into three genetic subpopulations with membership coefficient $q > 0.8$.

Source of Variation	df	TRN	RDW	SRA	SW	SL
Replicate	1	0.1	0.7	0.0	0.1	0.3
Genotype	54	73.1 ***	64.8 *	82.1 ***	88.8 ***	63.5 *
Between SPs	2	25.3 ***	0.6	10.3 ***	18.9 ***	6.0 *
Within SPs	52	47.8 *	64.2 *	71.8 ***	69.9 ***	57.5 *
WM	16	28.3 ***	31.6	40.8 ***	38.3 ***	26.8 *
NM	14	16.1	14.8	34.4 ***	6.3	25.2
EM	22	55.6	53.6 *	24.8	55.4 ***	48.0
Replicate x Genotype	54	26.8	34.5	17.9	11.1	36.2
Total	108					

WM, western Mediterranean; NM, northern Mediterranean; EM, eastern Mediterranean; TRN, total root number; RDW, root dry weight; SRA, seminal root angle; SW, seed weight; SL, seed length. * $P < 0.05$, *** $P < 0.001$.

The ANOVA for grain yield revealed that the genotype effect was the most important in the phenotypic expression of traits, accounting for 59% of the total phenotypic variation, whereas the year effect accounted only for 5%. The interaction accounted for almost 36% of the phenotypic variation although it was not significant (Table 3).

Table 3. Percentage of the sum of squares for grain yield of the ANOVA in the collection of 170 bread wheat landraces.

Source of Variation	df	Grain Yield	P
Genotype	169	59.2	< 0.001
Year	2	5.1	< 0.001
Genotype x Year	338	35.7	No significant
Total	509		

The landraces from northern Mediterranean countries showed the highest number of seminal roots with a root angle not statistically different from the western Mediterranean ones. On the other hand, eastern Mediterranean landraces showed the lowest number of roots but the widest angle. These landraces reported the lowest SW and the longest shoots. No differences were reported for RDW among the three SPs (Table 4).

Table 4. Means comparison of seminal root traits measured in a set of 55 Mediterranean wheat landraces structured into three genetic subpopulations [23] with $q > 0.8$. Means within columns with different letters are significantly different at $P < 0.05$ following a Tukey test.

	TRN (N)	RDW (mg)	SRA (°)	SW (g)	SL (cm)
Northern Mediterranean	4.7 a	0.011 a	98.5 b	0.50 a	20.8 b
Western Mediterranean	4.3 b	0.011 a	96.2 b	0.49 a	21.4 ab
Eastern Mediterranean	4.0 c	0.011 a	106.5 a	0.45 b	22.5 a

TRN, total root number; RDW, root dry weight; SRA, seminal root angle; SW, seed weight; SL, seed length.

Correlation coefficients between root traits were calculated, showing highly significant correlation coefficients between RDW and SW and RDW and SL ($r = 0.47$ and 0.45 respectively; $P < 0.0001$). Moderate significant correlations were reported for TRN with RDW, SW and SRA ($r = 0.20$, 0.28 and 0.28, respectively), and for SW with SL ($r = 0.27$). Finally, a negative correlation coefficient ($r = -0.12$) was found between SRA and SW (Figure 2). GY showed a moderate significant correlation with TRN and SW ($r = 0.28$ and 0.29, respectively; $P < 0.0005$).

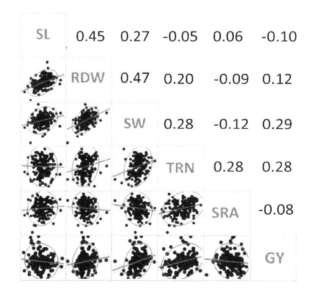

Figure 2. Correlations between seminal root traits and grain yield. On the right side are shown the values of the correlation coefficients (r). SL, seed length; RDW, root dry weight; SW, seed weight; TRN, total root number; SRA, seminal root angle; GY, grain yield.

3.2. Marker-Trait Associations

After filtering for duplicated patterns, missing values, and minor frequency alleles, a total of 10,458 SNPs were used to genotype the panel of 170 wheat landraces [23].

The results of the GWAS for root related traits are reported in Figure 3 and Supplementary Materials, Table S2. Using a common threshold of $-\log10 P > 3$, as reported by other authors [29–32], a total of 135 MTAs were identified for the analyzed traits. Of these, 50 MTAs corresponded to SW, 39 to RDW, 18 to SL, 17 to SRA, and 11 to TRN. The A and B genomes harbored 46% and 48% of MTAs, respectively, whereas the D genome harbored only 6% of MTAs. The number of MTAs per chromosome ranged from 1 in chromosomes 4D, 5D, and 6D to 14 in chromosome 1B, with a mean of 7 MTAs per chromosome. Most of the MTAs (88%) showed a phenotypic variance explained (PVE) by each MTA in a range of 5% to 10%, and only 2% showed a PVE higher than 15%. Among traits, the PVE mean was stable in a range of 7% (SL) to 9% (RDW).

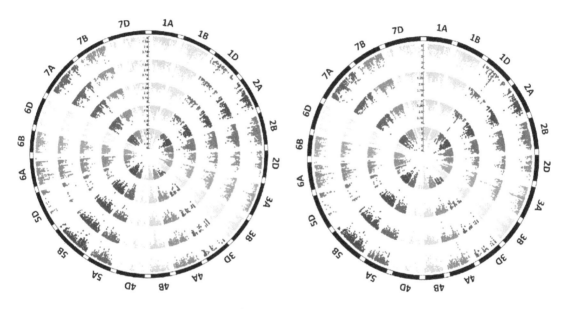

Figure 3. GWAS for root related traits (left circle) and grain yield for 3 years and across years (right circle). From the inside out, root traits correspond to RDW, SW, TRN, SRA, and SL, whereas for GY corresponds to 2016, 2017, 2018 harvesting seasons and the mean across years.

In order to identify and summarize the genomic regions most involved in trait variation, QTL hotspots were defined when two or more MTAs from different traits were grouped together within the same LD block. LD was previously estimated for locus pairs in each chromosome, and its decay was set to 1 to 10 cM depending on the chromosome [23]. Using this approach, 15 QTL hotspots grouping 43 MTAs were identified (Table 5), while 92 MTAs remained as singletons.

The results of the GWAS for GY are reported in Figure 3 and Supplementary Materials, Table S3. A common threshold of $-\log10 P > 3$, detected a total of 40 MTAs, thus a moderate threshold at $-\log10 P > 2.5$ was applied, increasing the number of significant associations to 112. Of these, 32 MTAs corresponded to the year 2016, 30 to 2017, 18 to 2018, and 32 across years. The A and B genomes harbored 43% and 38% of MTAs, respectively, whereas the D genome harbored only 18% of MTAs. The number of MTAs per chromosome ranged from 1 in chromosomes 3D and 6B to 16 in chromosome 1D. Chromosomes 1A, 4D, 5D, and 7D did not show any association. All of MTAs showed a phenotypic variance explained (PVE) by each MTA in a range from 5% to 11%. Most of the MTAs with a PVE > 8% were located on chromosome 1D (76%, 13 out of 17), whereas the percentage increased to 80% among MTAs with a PVE > 10% (4 out of 5).

In order to identify and summarize the genomic regions with a pleiotropic effect for root traits and grain yield, QTL hotspots were defined as previously but including the MTAs for GY. Using this approach, 17 QTL hotspots grouping 81 MTAs were identified (Table 6). From them, five were in common with those reported only with root traits (rootQTL1B.3, rootQTL2A.2, rootQTL3B.2, rootQTL6A.1, and rootQTL6A.2). GY shared 8 genomic regions with SW and 9 with RDW, 4 with SL, and 3 with SRA, whereas no regions were in common with TRN. In 59% of these genomic regions, GY co-localize with only one root trait, whereas the other 41% co-localize with two different root traits.

Table 5. Root QTL hotspots. Positions are indicated in cm.

QTL Hotspot	MTAs	Trait	Chromosome	Peak	CI Left	CI Right
rootQTL1B.1	2	RDW, SW	1B	70.6	69.6	71.6
rootQTL1B.2	2	RDW, TRN	1B	77.5	75.9	79.1
rootQTL1B.3	2	RDW, SL	1B	83.0	81.4	84.6
rootQTL2A.1	2	RDW, SW	2A	47.8	46.7	48.9
rootQTL2A.2	2	RDW, SW	2A	104.1	103.6	104.6
rootQTL2A.3	2	SW, SL	2A	177.5	176.9	178.2
rootQTL2B.1	2	SW, SL	2B	109.5	109.0	110.0
rootQTL3B.1	3	RDW, SW, TRN	3B	62.3	61.4	63.2
rootQTL3B.2	2	SRA, SW	3B	80.6	79.6	81.5
rootQTL5A.1	2	RDW, SL	5A	56.5	56.0	57.0
rootQTL5B.1	2	RDW, SL	5B	95.7	94.5	96.9
rootQTL6A.1	2	RDW, SL	6A	45.8	40.0	51.6
rootQTL6A.2	2	RDW, TRN	6A	76.7	70.7	82.6
rootQTL6A.3	2	RDW, SW	6A	138.4	132.3	144.6
rootQTL7A.1	3	RDW, SRA, TRN	7A	216.6	215.3	218.0

TRN, total root number; RDW, root dry weight; SRA, seminal root angle; SW, seed weight; SL, seed length.

Table 6. QTL hotspots including grain yield. Positions are indicated in cm.

QTL Hotspot	MTAs	Trait	Chromosome	Peak	CI Left	CI Right
QTL yield/root_1B.1	3	GY, SRA	1B	8.4	7.4	9.4
QTL yield/root_1B.2	3	GY, SW	1B	43.9	42.9	44.9
QTL yield/root_1B.3	3	GY, SW	1B	63.5	61.5	65.5
QTL yield/root_1B.4	8	GY, RDW, SL	1B	83.3	82.3	84.3
QTL yield/root_2A.1	8	GY, RDW, SW	2A	104.1	103.6	104.6
QTL yield/root_2A2	3	GY, SRA	2A	151.3	150.8	151.9
QTL yield/root_3A.1	8	GY, RDW, SL	3A	84.3	81.9	86.7
QTL yield/root_3B.1	5	GY, SW	3B	72.8	70.8	74.8
QTL yield/root_3B.2	3	GY, SRA, SW	3B	80.5	79.6	81.5
QTL yield/root_4B.1	4	GY, SW	4B	76.6	74.1	79.2
QTL yield/root_5B.1	3	GY, SL	5B	57.8	56.8	58.9
QTL yield/root_5B.2	7	GY, RDW	5B	77.3	75.9	78.8
QTL yield/root_5B.3	2	GY, RDW	5B	176.2	175.2	177.2
QTL yield/root_6A.1	5	GY, RDW, SL	6A	45.6	39.6	51.6
QTL yield/root_6A.2	10	GY, RDW, SW	6A	76.6	70.7	82.6
QTL yield/root_7A.1	3	GY, RDW, SW	7A	135	133.2	136.8
QTL yield/root_7B.1	3	GY, RDW	7B	70.0	67.8	72.3

TRN, total root number; RDW, root dry weight; SRA, seminal root angle; SW, seed weight; SL, seed length; GY, grain yield.

In order to identify the most useful markers for selecting for the root traits, extreme phenotypes were identified in the upper and lower 10th percentile of genotypes within the collection for each trait (Figure 4). Among the most significant MTAs for each trait, markers with different alleles between extreme genotypes were identified (Table 7, Figure 5). The frequency of the most common allele among genotypes from the upper 10th percentile ranged from 78% for RDW to 88% for SW, while for the lower 10th percentile it ranged from 65% for TRN and SRA to 92% for RDW (Figure 2).

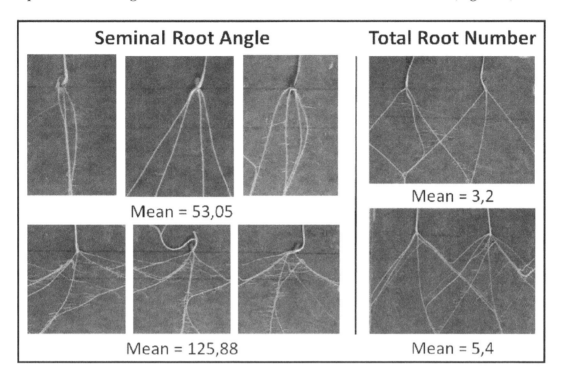

Figure 4. Extreme phenotypes for SRA and TRN. The means correspond for 3 observational units of the genotype for SRA and 5 observational units of the genotype for TRN.

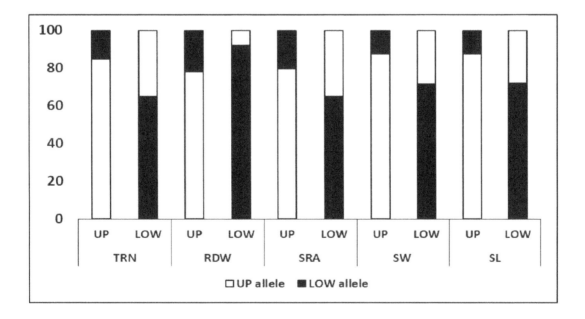

Figure 5. Marker allele frequency means from landraces within the upper and lower 10th percentile for the analyzed traits. All significant markers shown in Table 5 are included. TRN, total root number; RDW, root dry weight; SRA, seminal root angle; SW, seed weight; SL, seed length.

Table 7. Selected significant markers from the GWAS with different allele composition for the upper (UP) and lower (LOW) 10th percentile of genotypes. Different letters on the UP and LOW 10th phenotype indicate that means are significantly different at $P < 0.01$ following a Tukey test.

Trait	Phenotype			Marker	Chr	Position	R^2	Most Frequent Allele			
	Mean	UP 10th	LOW 10th					UP	Freq	LOW	Freq
TRN (N)	4.4	5.2[a]	3.4[b]	Excalibur_c8613_1266	1B	78.1	0.09	T	0.70	C	0.80
				Tdurum_contig12995_792	5B	98.3	0.12	T	1.00	Y	0.50
RDW (g)	0.012	0.016[a]	0.007[b]	RAC875_c8482_160	1A	95.2	0.14	A	0.60	A	0.90
				GENE-0392_97	1A	95.5	0.11	T	0.60	C	0.70
				RAC875_c39401_502	1A	104.1	0.15	T	1.00	C	1.00
				BS00023071_51	1B	95.5	0.15	G	1.00	A	1.00
				Kukri_c14877_303	6A	79.1	0.14	C	0.70	T	1.00
SRA (°)	98.65	119.98[a]	69.88[b]	Kukri_c34453_332	4B	70.9	0.11	A	0.60	C	0.70
				BS00028183_51	5B	51.0	0.11	T	1.00	C	0.60
SW (g)	0.487	0.59[a]	0.37[b]	RAC875_c37540_583	2B	139.5	0.14	T	1.00	C	0.77
				Ku_c101046_1063	4B	77.7	0.11	Y	0.50	C	0.80
				BS00094770_51	4D	80.4	0.12	C	1.00	A	0.60
				GENE-2220_165	7D	161.1	0.12	T	1.00	T	0.70
SL (cm)	22.04	27.72[a]	16.80[b]	Tdurum_contig8158_269	1B	82.4	0.09	A	0.80	G	0.90
				RFL_Contig5277_888	2A	177.6	0.09	C	1.00	C	0.60
				BobWhite_c6966_236	5A	130.9	0.09	T	0.80	C	0.90
				BS00023161_51	5B	48.3	0.08	T	0.90	Y	0.50
				RAC875_rep_c106439_1159	6A	48.1	0.09	A	0.88	C	0.70

Chr, chromosome. TRN, total root number. RDW, root dry weight. SRA, seminal root angle. SW, seed weight. SL, shoot length.

3.3. Gene Annotation

As reported in Supplementary Materials, Table S4, a total of 1489 gene models were identified within the 15 QTL hotspots using the high-confidence gene annotation from the wheat genome sequence [33]. Genetic distances were converted into physical distances using the position of common flanking markers on the genetic map [24] and the genome sequence. The number of gene models ranged from 224 in rootQTL_2A.2 to 9 in rootQTL_5B.1. Based on the high number of gene models, a selection was made according to gene families involved in root traits, growth and development, and abiotic stress resistance (Table 8). Thus, 31 gene families with a total of 96 gene models remained for subsequent analysis. Among them, F-box and zinc finger family proteins were identified in 12 of the 15 QTL hotspots, whereas 10 gene families were present in only one QTL hotspot. Among chromosomes with QTL hotspots, chromosome 2A had the highest number of gene models (22), whereas chromosomes 5A and 5B had the lowest number (4).

Table 8. Selected gene model families.

Description	N QTL Hotspots	Function
F-box family protein	12	Salt and drought stress responses
RING/FYVE/PHD zinc finger protein	12	Salt and drought stress responses
MYB-related transcription factor	8	Salt and drought stress responses
NAC domain-containing proteins	8	Induced by biotic and abiotic stresses
Cytochrome P450 family protein	5	Involved in seed size
BZIP transcription factor	5	Regulated by abiotic stress
Ethylene-responsive transcription factor	4	Induced by biotic and abiotic stresses
Calmodulin	4	Heat shock transduction pathway
Peroxidase	4	Root growth
ABC transporter	4	Control root development
Nucleoside triphosphate hydrolase	3	Associated with drought stress
E3 ubiquitin-protein ligase	3	Associated with drought stress
Glycine-rich protein	2	Enhance drought stress tolerance
Xyloglucan endotransglucosylase/hydrolase	2	Response dehydration, salinity, cold
Aquaporin	2	Drought stress tolerance
Expansin protein	2	Drought tolerance in wheat
Trihelix transcription factor	2	Stomatal development, drought
VQ motif family protein	2	Involved in seed size
Heat shock family protein	2	Induced by abiotic stress
Protein root UVB sensitive 6	2	Early seedling morphogenesis
SAUR-like auxin-responsive family protein	2	Maintain growth during abiotic stress
Bax inhibitor-1 family protein	1	Tolerance to abiotic stresses
Formin-like protein	1	Structure organization in drought-stressed plants
Late embryogenesis abundant protein	1	Participate in drought response
Cell wall invertase	1	Downregulated by drought
Senescence regulator	1	Related to drought stress
Plastid-lipid associated protein PAP/fibrillin	1	Induced by drought
Protein STAY-GREEN LIKE, chloroplastic	1	Improves drought resistance
PI-PLC X domain-containing protein	1	Induced by abiotic stresses
Histidine-containing phosphotransfer protein	1	Enhance tolerance to drought stress
Phospholipase D	1	Enhance drought stress tolerance

4. Discussion

Breeding for drought adaptation is one of the main challenges to be addressed in the coming years in order to increase wheat production and ensure sufficient food supply in the current scenario of climate change. Roots are crucial in this adaptation, as they are responsible for water and nutrient uptake. The wide morphological plasticity of the root system to different soil conditions and the role of root traits in drought environments are well known [34,35]. Wheat roots reduce their growth in water-limited conditions but increase the water uptake rate, extracting the water from deep soil layers [36]. The shape and spatial arrangement of the RSA can provide a growth advantage and increasing yield performance during periods of water scarcity [37]. Thus, it is necessary to increase the knowledge of the genetics of root architecture in order to improve wheat yield stability under stress conditions by introgressing favorable alleles through breeding programs.

The current study evaluated root-related traits in a collection of Mediterranean bread wheat landraces representative of the variability existing for the species in the Mediterranean Basin [23] with the aim of providing QTL information for these traits regarding seminal roots. Seminal roots are important for early vigor and crop establishment in dryland areas because they explore the soil for nutrients and water [38]. Moreover, it has been reported that under drought stress, seminal roots activity is more important than that of nodal roots [39]. Additionally, field phenotyping of hundreds of genotypes is a complex and expensive task. As the root geometry of adult plants is strongly related to the SRA [5], it may be assumed that genotypes that differ in root architecture at an early developmental stage would also differ in the field at later growth stages, when nutrient and/or water capture become critical for yield performance [8].

The range of variation for the traits analyzed in the present study (from 10.9% for TRN to 18.8% for RDW) is in agreement with those reported for elite durum wheat cultivars by Canè et al. [8], who explained this variability as an adaptive value for the environmental conditions of the region of origin of the cultivars. Moreover, the high repeatability found for the traits supports the approach followed to analyze the seminal roots under controlled conditions.

Landraces from the eastern Mediterranean Basin showed the widest SRA, the lowest SW, the longest SL, and the lowest number of roots. According to previous studies in durum wheat [18,40], landraces from southeastern Mediterranean countries corresponding to the warmest and driest areas of the Mediterranean Basin, reported more grains per unit area and lighter grains than those developed in cooler and wetter zones of the region. Although it has been reported that in water-limited environments a vigorous root system could have benefits at the beginning of the growing season because it offers a more efficient water capture [41], no significant differences were observed for RDW among the SPs in the current study. Moreover, our results for SRA are in agreement with those reported by Roselló et al. [18], who found that durum wheat landraces from the eastern Mediterranean have the widest root angle, which probably allows them to cover a larger soil area and be more efficient in water uptake than landraces that originated in wetter areas.

Although not significant, probably due to the very early stage when the root traits were measured, the correlation between SRA and SW was negative. The same result was also reported by Canè et al. [8], who suggested that it could be due to the influence of the root angle on the distribution of the roots on soil layers and, therefore, the water uptake from deeper layers. On the other hand, the correlation between RDW and SW was positive, in agreement with the findings of Fang et al. [42], thus indicating the effectiveness of greater root mass for obtaining more soil water for plant growth and grain filling in drought. Seedling growth has also been related to SW in wheat [43]. The vertical distribution of the root system can have a strong effect on yield [44], so mass root concentrated in upper layers can be more effective for resource capture, while roots in deeper layers have more access to deep water.

The complexity of the genetic control of root traits was confirmed with 135 marker-trait associations identified in the current study. Their distribution across genomes was similar in the A and B genomes (46% and 48%, respectively), leaving only 6% of MTAs in the D genome. These results agree with the lower genetic diversity and higher LD found in the D genome, as reported previously [23]. According to Chao et al. [45], the different levels of diversity in wheat genomes could be due to different rates of gene flow from the ancestors of wheat, since polyploidy bottleneck resulting from speciation reduced diversity and increased the levels of LD in the D genome in comparison with the A and B genomes.

In order to simplify and to integrate closely linked MTAs in a consensus region, QTL hotspots were identified based on the results of LD decay reported in [23]. LD decay was used to define the CIs for the QTL hotspots. Following this approach, 43 MTAs were grouped in 15 QTL hotspots. The genomic position of QTL hotspots was compared with previous studies reporting meta-QTLs for root traits [46] and MTAs from GWAS studies in order to detect previously identified regions controlling root traits. Among the 15 QTL hotspots, only rootQTL6A.3 was located in the same region of a previously mapped meta-QTL, RootMQTL74 [46]. When compared with MTA-QTLs reported by [18] in durum wheat Mediterranean landraces, the QTL hotspot rootQTL6A.3 corresponded to the MTA-QTLs mtaq-6A.3 and mtaq-6A.6. This hotspot was also in the same region of a major SRA QTL identified by Alahmad et al. [47] and by a QTL controlling root growth angle identified by Maccaferri et al. [9], who also found a QTL for grain weight that is located in a common region with the hotspot rootQTL2A.2, which includes an MTA for SW. rootQTL3B.1 shared a common position with an MTA reported by Ayalew et al. [48] on chromosome 3B under stress conditions. rootQTL7A.1, including an MTA for RDW, was located in a similar position as MLM-RDWB-10 reported by Li et al. [49] and associated with RDW at the booting stage. Finally, no genomic regions were shared with the study carried out by Beyer et al. [50]. Only four of the 15 QTL hotspots identified in this work had been detected previously, suggesting the importance of wheat Mediterranean landraces for the identification of new loci controlling root-related traits.

As reported in previous studies, at early developmental stages [8,18] the co-location of MTAs for grain yield and root related traits within the same QTL hotspot suggests their pleiotropic effect, however, deeper analyses should be necessary to confirm it. In durum wheat elite cultivars, Canè et al. [8] found that 30% of the QTLs affecting root system architecture were included within QTLs for agronomic traits. More recently, Roselló et al. [18] using a collection of Mediterranean durum wheat landraces found that 45% of QTL hotspots for root related traits were mapped in similar regions to yield-related traits reported for the same collection of landraces.

From a breeding standpoint, exploiting genetic diversity from local landraces is a valuable approach for recovering and broadening allelic variation for traits of interest [19]. Therefore, identifying the genotypes showing the extreme phenotypes within the pool of Mediterranean landraces and the associated markers provide the opportunity for introgressing suitable traits in elite cultivars by marker-assisted breeding using the most recent technologies to speed the process.

The availability of a high-quality reference wheat genome sequence [33] enabled us to quickly identify gene models corresponding to QTLs. Thus, the genetic position of the CIs of the QTL hotspots was projected into physical distances on the reference sequence to search for putative candidate gene models. To narrow the number of candidates, only gene models involved in the development and abiotic stress according to the literature were taken into consideration. Therefore, of 1489 gene models identified within the 15 QTL hotspots, only 31 gene families were selected.

F-box and zinc finger family proteins were the most represented, each one appearing in 12 hotspots. F-box proteins play important roles in plant development and abiotic stress responses via the ubiquitin pathway [51] and the ABA signaling pathway [52]. In wheat, the F-box protein *TaFBA1* is involved in plant hormone signaling and response to abiotic stresses and is expressed in all plant organs, including roots [53]. The overexpression of *TaFBA1* in transgenic tobacco reported by Li et al. [54] to improve heat tolerance resulted in increased root length in the transgenic plants. Zinc finger proteins are involved in several processes, such as regulation of plant growth and development, and response to abiotic stresses [46]. In *Arabidopsis* and rice, they play a role in tolerance to drought and salt stresses [55], while in wheat the overexpression of *TaZFP34* enhances root-to-shoot ratio during plant adaptation to drying soil [56].

Other kinds of gene models found in a high number of QTL hotspots were *MYB* transcription factors and *NAC* domain-containing proteins, each of them presents in 8 hotspots. *MYB* domain-containing transcription factors are involved in salt and drought stress adaptation in wheat. Some examples in wheat are the genes *TaMyb1*, *TaMYBsdu1*, and *TaMYB33*. The expression of *TaMyb1* in roots is strongly related to responses to abiotic stresses [57]. The gene *TaMYBsdu1* was found to be upregulated in leaves and roots of wheat under long-term drought stress [58]. Finally, the overexpression of *TaMYB33* in *Arabidopsis* enhances tolerance to drought and salt stresses [59]. *NAC* domain-containing proteins have been described to play many important roles in abiotic stress adaptation [46]. Xie et al. [60]

reported that *NAC1* promoted the development of lateral roots. Similarly, He et al. [61] found that the expression of *AtNAC2* in response to salt stress led to an increase in the development of lateral roots. Xia et al. [62] demonstrated that the gene *TaNAC4* is a transcriptional activator involved in wheat's response to biotic and abiotic stresses.

Proteins belonging to the cytochrome *P450* family and *bZIP* transcription factors were present in five QTL hotspots. The first class of proteins belongs to one of the largest families of plant proteins, with genes affecting important traits for crop improvement such as *TaCYP78A3*, which is involved in the control of seed size [63]. *bZIP* transcription factors are involved in abiotic stress response [64]. In *Arabidopsis*, it has been observed that the overexpression of *TabZIP14-B*, involved in salt and freezing tolerance, hindered root growth in transgenic plants in comparison with the control plants [65].

Other proteins involved in root growth and development are the peroxidases and *ABC* transporters that were identified in four QTL hotspots. Extracellular peroxidases are involved in plant defense reactions against biotic and abiotic stresses through the generation of reactive oxygen species in wounded root cells [66]. In *Arabidopsis*, the *ABC* transporter *AtPGP4* is expressed mainly during early root development, and its loss of function enhances lateral root initiation and root hair development [67]. Gaedeke et al. [68] reported a new member of the *ABC* transporter superfamily of *Arabidopsis thaliana*, *AtMRP5*. Using reverse genetics, these authors found that the recessive allele *mrp5* exhibited decreased root growth and increased lateral root formation. In addition to peroxidases and *ABC* transporters, other proteins identified in four QTLs were the ethylene-responsive transcription factors (ERFs), found to be involved in the response to abiotic stresses. In wheat, the *ERF TaERFL1a* is induced in wheat seedlings in response to salt, cold, and water deficiency [69].

Other family proteins involved in drought stress, seed size, or early development were represented in a lower number of QTL hotspots. Among them, aquaporins are known to affect drought tolerance influencing the capacity of roots to take up the soil water [70]. The expansins were suggested to be involved in root development, as the overexpression of the wheat expansin *TaEXPB23* improved drought tolerance by stimulating the growth of the root system in tobacco [71].

5. Conclusions

The exploitation of unexplored genetic variation present in local landraces can potentially contribute to breeding programs aimed at enhancing drought tolerance in wheat. Roots are crucial for adaptation to drought stress because they are the plant organ responsible for water and nutrient uptake and interaction with soil microbes. Thus, designing and developing novel root system ideotypes could be one of the targets of wheat breeding for the coming years. The variability found in the Mediterranean wheat landraces together with the newly identified QTL hotspots shows landraces as a valuable source of favorable root traits to introgress into adapted phenotypes through marker-assisted breeding. Among the different marker trait associations, those reported in extreme genotypes could result as a starting point to develop new mapping populations to fine map the corresponding traits.

Author Contributions: Conceptualization, S.S. and J.M.S.; Data curation, R.R.; Formal analysis, R.R.; Funding acquisition, C.R. and J.M.S.; Investigation, R.R., S.S. and J.M.S.; Methodology, R.R., S.S. and J.M.S.; Project administration, C.R. and J.M.S.; Supervision, S.S. and J.M.S.; Writing—original draft, R.R.; Writing—review & editing, S.S., C.R. and J.M.S. All authors have read and agreed to the published version of the manuscript.

Acknowledgments: The authors acknowledge the contribution of the CERCA programme (Generalitat de Catalunya). Thanks are given to the group of Agricultural Genetics of University of Bologna for technical support.

Abbreviations

EM	Eastern Mediterranean
GWAS	Genome wide association study
GY	Grain yield
MTA	Marker-trait association
NM	Northern Mediterranean
QTL	Quantitative trait locus
RDW	Root dry weight
RSA	Root system architecture
SL	Shoot length
SP	Sub-population
SRA	Seminal root angle
SW	Seed weight
TRN	Total root number
WM	Western Mediterranean

References

1. Leegood, R.C.; Evans, J.R.; Furbank, R.T. Food security requires genetic advances to increase farm yields. *Nature* **2010**, *464*, 831. [CrossRef]

2. Giorgi, F.; Lionello, P. Climate change projections for the Mediterranean region. *Glob. Planet. Chang.* **2008**, *63*, 90–104. [CrossRef]

3. Uga, Y.; Kitomi, Y.; Ishikawa, S.; Yano, M. Genetic improvement for root growth angle to enhance crop production. *Breed. Sci.* **2015**, *65*, 111–119. [CrossRef] [PubMed]

4. Kulkarni, M.; Soolanayakanahally, R.; Ogawa, S.; Uga, Y. Drought Response in Wheat: Key Genes and Regulatory Mechanisms Controlling Root System Architecture and Transpiration Efficiency. *Front. Chem.* **2017**, *5*, 1–13. [CrossRef] [PubMed]

5. Manschadi, A.M.; Hammer, G.L.; Christopher, J.T.; de Voil, P. Genotypic variation in seedling root architectural traits and implications for drought adaptation in wheat (*Triticum aestivum* L.). *Plant Soil* **2008**, *303*, 115–129. [CrossRef]

6. Sanchez-Garcia, M.; Álvaro, F.; Martín-Sánchez, J.A.; Sillero, J.C.; Escribano, J.; Royo, C. Breeding effects on the genotype×environment interaction for yield of bread wheat grown in Spain during the 20th century. *Field Crop. Res.* **2012**, *126*, 79–86. [CrossRef]

7. Royo, C.; Maccaferri, M.; Álvaro, F.; Moragues, M.; Sanguineti, M.C.; Tuberosa, R.; Maalouf, F.; del Moral, L.F.G.; Demontis, A.; Rhouma, S.; et al. Understanding the relationships between genetic and phenotypic structures of a collection of elite durum wheat accessions. *Field Crop. Res.* **2010**, *119*, 91–105. [CrossRef]

8. Canè, M.A.; Maccaferri, M.; Nazemi, G.; Salvi, S.; Francia, R.; Colalongo, C.; Tuberosa, R. Association mapping for root architectural traits in durum wheat seedlings as related to agronomic performance. *Mol. Breed.* **2014**, *34*, 1629–1645. [CrossRef]

9. Maccaferri, M.; El-Feki, W.; Nazemi, G.; Salvi, S.; Canè, M.A.; Colalongo, M.C.; Stefanelli, S.; Tuberosa, R. Prioritizing quantitative trait loci for root system architecture in tetraploid wheat. *J. Exp. Bot.* **2016**, *67*, 1161–1178. [CrossRef]

10. Bhatta, M.; Morgounov, A.; Belamkar, V.; Baenziger, P. Genome-Wide Association Study Reveals Novel Genomic Regions for Grain Yield and Yield-Related Traits in Drought-Stressed Synthetic Hexaploid Wheat. *Int. J. Mol. Sci.* **2018**, *19*, 3011. [CrossRef]

11. Reynolds, M.P.; Mujeeb-Kazi, A.; Sawkins, M. Prospects for utilising plant-adaptive mechanisms to improve wheat and other crops in drought- and salinity-prone environments. *Ann. Appl. Biol.* **2005**, *146*, 239–259. [CrossRef]

12. Sanguineti, M.C.; Li, S.; Maccaferri, M.; Corneti, S.; Rotondo, F.; Chiari, T.; Tuberosa, R. Genetic dissection of seminal root architecture in elite durum wheat germplasm. *Ann. Appl. Biol.* **2007**, *151*, 291–305. [CrossRef]

13. Wasson, A.P.; Richards, R.A.; Chatrath, R.; Misra, S.C.; Prasad, S.V.S.; Rebetzke, G.J.; Kirkegaard, J.A.; Christopher, J.; Watt, M. Traits and selection strategies to improve root systems and water uptake in water-limited wheat crops. *J. Exp. Bot.* **2012**, *63*, 3485–3498. [CrossRef] [PubMed]

14. Liu, P.; Jin, Y.; Liu, J.; Liu, C.; Yao, H.; Luo, F.; Guo, Z.; Xia, X.; He, Z. Genome-wide association mapping of root system architecture traits in common wheat (*Triticum aestivum* L.). *Euphytica* **2019**, *215*, 1–12. [CrossRef]

15. Chochois, V.; Voge, J.P.; Rebetzke, G.J.; Watt, M. Variation in adult plant phenotypes and partitioning among seed and stem-borne roots across Brachypodium distachyon accessions to exploit in breeding cereals for well-watered and drought environments. *Plant Physiol.* **2015**, *168*, 953–967. [CrossRef]

16. Mace, E.S.; Singh, V.; van Oosterom, E.J.; Hammer, G.L.; Hunt, C.H.; Jordan, D.R. QTL for nodal root angle in sorghum (Sorghum bicolor L. Moench) co-locate with QTL for traits associated with drought adaptation. *Theor. Appl. Genet.* **2012**, *124*, 97–109. [CrossRef]

17. Ruiz, M.; Giraldo, P.; González, J.M. Phenotypic variation in root architecture traits and their relationship with eco-geographical and agronomic features in a core collection of tetraploid wheat landraces (Triticum turgidum L.). *Euphytica* **2018**, *214*, 54. [CrossRef]

18. Roselló, M.; Royo, C.; Sanchez-Garcia, M.; Soriano, J.M. Genetic Dissection of the Seminal Root System Architecture in Mediterranean Durum Wheat Landraces by Genome-Wide Association Study. *Agronomy* **2019**, *9*, 364. [CrossRef]

19. Lopes, M.S.; El-Basyoni, I.; Baenziger, P.S.; Singh, S.; Royo, C.; Ozbek, K.; Aktas, H.; Ozer, E.; Ozdemir, F.; Manickavelu, A.; et al. Exploiting genetic diversity from landraces in wheat breeding for adaptation to climate change. *J. Exp. Bot.* **2015**, *66*, 3477–3486. [CrossRef]

20. Nazco, R.; Peña, R.J.; Ammar, K.; Villegas, D.; Crossa, J.; Royo, C. Durum wheat (Triticum durum Desf.) Mediterranean landraces as sources of variability for allelic combinations at Glu-1/Glu-3 loci affecting gluten strength and pasta cooking quality. *Genet. Resour. Crop Evol.* **2014**, *61*, 1219–1236. [CrossRef]

21. Moragues, M.; Del Moral, L.F.G.; Moralejo, M.; Royo, C. Yield formation strategies of durum wheat landraces with distinct pattern of dispersal within the Mediterranean basin I: Yield components. *Field Crop. Res.* **2006**, *95*, 194–205. [CrossRef]

22. Roselló, M.; Villegas, D.; Álvaro, F.; Soriano, J.M.; Lopes, M.S.; Nazco, R.; Royo, C. Unravelling the relationship between adaptation pattern and yield formation strategies in Mediterranean durum wheat landraces. *Eur. J. Agron.* **2019**, *107*, 43–52. [CrossRef]

23. Rufo, R.; Alvaro, F.; Royo, C.; Soriano, J.M. From landraces to improved cultivars: Assessment of genetic diversity and population structure of Mediterranean wheat using SNP markers. *PLoS ONE* **2019**, *14*, e0219867. [CrossRef] [PubMed]

24. Wang, S.; Wong, D.; Forrest, K.; Allen, A.; Chao, S.; Huang, B.E.; Maccaferri, M.; Salvi, S.; Milner, S.G.; Cattivelli, L.; et al. Characterization of polyploid wheat genomic diversity using a high-density 90 000 single nucleotide polymorphism array. *Plant Biotechnol. J.* **2014**, *12*, 787–796. [CrossRef] [PubMed]

25. Abràmoff, M.D.; Magalhães, P.J.; Ram, S.J. *Image Processing with ImageJ*, 2nd ed.; Packt Publishing: Birmingham, UK, 2004; Volume 11, ISBN 9781785889837.

26. Harper, D.G.C. Some comments on the repeatability of measurements. *Ringing Migr.* **1994**, *15*, 84–90. [CrossRef]

27. Bradbury, P.J.; Zhang, Z.; Kroon, D.E.; Casstevens, T.M.; Ramdoss, Y.; Buckler, E.S. TASSEL: Software for association mapping of complex traits in diverse samples. *Bioinformatics* **2007**, *23*, 2633–2635. [CrossRef]

28. Barrett, J.C.; Fry, B.; Maller, J.; Daly, M.J. Haploview: Analysis and visualization of LD and haplotype maps. *Bioinformatics* **2005**, *21*, 263–265. [CrossRef]

29. Wang, S.-X.; Zhu, Y.-L.; Zhang, D.-X.; Shao, H.; Liu, P.; Hu, J.-B.; Zhang, H.; Zhang, H.-P.; Chang, C.; Lu, J.; et al. Genome-wide association study for grain yield and related traits in elite wheat varieties and advanced lines using SNP markers. *PLoS ONE* **2017**, *12*, e0188662. [CrossRef]

30. Mangini, G.; Gadaleta, A.; Colasuonno, P.; Marcotuli, I.; Signorile, A.M.; Simeone, R.; De Vita, P.; Mastrangelo, A.M.; Laidò, G.; Pecchioni, N.; et al. Genetic dissection of the relationships between grain yield components by genome-wide association mapping in a collection of tetraploid wheats. *PLoS ONE* **2018**, *13*, e0190162. [CrossRef]

31. Condorelli, G.E.; Maccaferri, M.; Newcomb, M.; Andrade-Sanchez, P.; White, J.W.; French, A.N.; Sciara, G.; Ward, R.; Tuberosa, R. Comparative Aerial and Ground Based High Throughput Phenotyping for the Genetic Dissection of NDVI as a Proxy for Drought Adaptive Traits in Durum Wheat. *Front. Plant Sci.* **2018**, *9*. [CrossRef] [PubMed]

32. Sukumaran, S.; Reynolds, M.P.; Sansaloni, C. Genome-Wide Association Analyses Identify QTL Hotspots for Yield and Component Traits in Durum Wheat Grown under Yield Potential, Drought, and Heat Stress Environments. *Front. Plant Sci.* **2018**, *9*, 81. [CrossRef] [PubMed]

33. The International Wheat Genome Sequencing Consortium (IWGSC); Appels, R.; Eversole, K.; Stein, N.; Feuillet, C.; Keller, B.; Rogers, J.; Pozniak, C.J.; Choulet, F.; Distelfeld, A.; et al. Shifting the limits in wheat research and breeding using a fully annotated reference genome. *Science* **2018**, *361*, eaar7191. [CrossRef] [PubMed]

34. Christopher, J.; Christopher, M.; Jennings, R.; Jones, S.; Fletcher, S.; Borrell, A.; Manschadi, A.M.; Jordan, D.; Mace, E.; Hammer, G. QTL for root angle and number in a population developed from bread wheats (Triticum aestivum) with contrasting adaptation to water-limited environments. *Theor. Appl. Genet.* **2013**, *126*, 1563–1574. [CrossRef] [PubMed]

35. Paez-Garcia, A.; Motes, C.; Scheible, W.-R.; Chen, R.; Blancaflor, E.; Monteros, M. Root Traits and Phenotyping Strategies for Plant Improvement. *Plants* **2015**, *4*, 334–355. [CrossRef] [PubMed]

36. Asseng, S.; Ritchie, J.T.; Smucker, A.J.M.; Robertson, M.J. Root growth and water uptake during water deficit and recovering in wheat. *Plant Soil* **1998**, *201*, 265–273. [CrossRef]

37. Rogers, E.D.; Benfey, P.N. Regulation of plant root system architecture: Implications for crop advancement. *Curr. Opin. Biotechnol.* **2015**, *32*, 93–98. [CrossRef]

38. Reynolds, M.; Tuberosa, R. Translational research impacting on crop productivity in drought-prone environments. *Curr. Opin. Plant Biol.* **2008**, *11*, 171–179. [CrossRef]

39. Manschadi, A.M.; Christopher, J.T.; Hammer, G.L.; Devoil, P. Experimental and modelling studies of drought-adaptive root architectural traits in wheat (*Triticum aestivum* L.). *Plant Biosyst. Int. J. Deal. Asp. Plant Biol.* **2010**, *144*, 458–462.

40. Royo, C.; Nazco, R.; Villegas, D. The climate of the zone of origin of Mediterranean durum wheat (Triticum durum Desf.) landraces affects their agronomic performance. *Genet. Resour. Crop Evol.* **2014**, *61*, 1345–1358. [CrossRef]

41. Liao, M.; Palta, J.A.; Fillery, I.R.P. Root characteristics of vigorous wheat improve early nitrogen uptake. *Aust. J. Agric. Res.* **2006**, *57*, 1097. [CrossRef]

42. Fang, Y.; Du, Y.; Wang, J.; Wu, A.; Qiao, S.; Xu, B.; Zhang, S.; Siddique, K.H.M.; Chen, Y. Moderate Drought Stress Affected Root Growth and Grain Yield in Old, Modern and Newly Released Cultivars of Winter Wheat. *Front. Plant Sci.* **2017**, *8*, 672. [CrossRef] [PubMed]

43. Aparicio, N.; Villegas, D.; Araus, J.L.; Blanco, R.; Royo, C. Seedling development and biomass as affected by seed size and morphology in durum wheat. *J. Agric. Sci.* **2002**, *139*, 143–150. [CrossRef]

44. King, J.; Gay, A.; Sylvester-Bradley, R.; Bingham, I.; Foulkes, J.; Gregory, P.; Robinson, D. Modelling Cereal Root Systems for Water and Nitrogen Capture: Towards an Economic Optimum. *Ann. Bot.* **2003**, *91*, 383–390. [CrossRef] [PubMed]

45. Chao, S.; Dubcovsky, J.; Dvorak, J.; Luo, M.-C.; Baenziger, S.P.; Matnyazov, R.; Clark, D.R.; Talbert, L.E.; Anderson, J.A.; Dreisigacker, S.; et al. Population- and genome-specific patterns of linkage disequilibrium and SNP variation in spring and winter wheat (*Triticum aestivum* L.). *BMC Genomics* **2010**, *11*, 727. [CrossRef] [PubMed]

46. Soriano, J.M.; Alvaro, F. Discovering consensus genomic regions in wheat for root-related traits by QTL meta-analysis. *Sci. Rep.* **2019**, *9*, 10537. [CrossRef]

47. Alahmad, S.; El Hassouni, K.; Bassi, F.M.; Dinglasan, E.; Youssef, C.; Quarry, G.; Aksoy, A.; Mazzucotelli, E.; Juhász, A.; Able, J.A.; et al. A major root architecture QTL responding to water limitation in durum wheat. *Front. Plant Sci.* **2019**, *10*, 436. [CrossRef]

48. Ayalew, H.; Liu, H.; Börner, A.; Kobiljski, B.; Liu, C.; Yan, G. Genome-Wide Association Mapping of Major Root Length QTLs Under PEG Induced Water Stress in Wheat. *Front. Plant Sci.* **2018**, *9*, 1759. [CrossRef]

49. Li, L.; Peng, Z.; Mao, X.; Wang, J.; Chang, X.; Reynolds, M.; Jing, R. Genome-wide association study reveals genomic regions controlling root and shoot traits at late growth stages in wheat. *Ann. Bot.* **2019**, *124*, 993–1006. [CrossRef]

50. Beyer, S.; Daba, S.; Tyagi, P.; Bockelman, H.; Brown-Guedira, G.; Mohammadi, M. Loci and candidate genes controlling root traits in wheat seedlings—A wheat root GWAS. *Funct. Integr. Genomics* **2019**, *19*, 91–107. [CrossRef]

51. Jia, Y.; Gu, H.; Wang, X.; Chen, Q.; Shi, S.; Zhang, J.; Ma, L.; Zhang, H.; Ma, H. Molecular cloning and characterization of an F-box family gene CarF-box1 from chickpea (*Cicer arietinum* L.). *Mol. Biol. Rep.* **2012**, *39*, 2337–2345. [CrossRef] [PubMed]

52. Koops, P.; Pelser, S.; Ignatz, M.; Klose, C.; Marrocco-Selden, K.; Kretsch, T. EDL3 is an F-box protein involved in the regulation of abscisic acid signalling in Arabidopsis thaliana. *J. Exp. Bot.* **2011**, *62*, 5547–5560. [CrossRef] [PubMed]

53. Zhou, S.; Sun, X.; Yin, S.; Kong, X.; Zhou, S.; Xu, Y.; Luo, Y.; Wang, W. The role of the F-box gene *TaFBA1* from wheat (*Triticum aestivum* L.) in drought tolerance. *Plant Physiol. Biochem.* **2014**, *84*, 213–223. [CrossRef] [PubMed]

54. Li, Q.; Wang, W.; Wang, W.; Zhang, G.; Liu, Y.; Wang, Y.; Wang, W. Wheat F-Box Protein Gene TaFBA1 Is Involved in Plant Tolerance to Heat Stress. *Front. Plant Sci.* **2018**, *9*. [CrossRef] [PubMed]

55. Xu, D.-Q.; Huang, J.; Guo, S.-Q.; Yang, X.; Bao, Y.-M.; Tang, H.-J.; Zhang, H.-S. Overexpression of a TFIIIA-type zinc finger protein gene *ZFP252* enhances drought and salt tolerance in rice (*Oryza sativa* L.). *FEBS Lett.* **2008**, *582*, 1037–1043. [CrossRef]

56. Chang, H.; Chen, D.; Kam, J.; Richardson, T.; Drenth, J.; Guo, X.; McIntyre, C.L.; Chai, S.; Rae, A.L.; Xue, G.-P. Abiotic stress upregulated TaZFP34 represses the expression of type-B response regulator and SHY2 genes and enhances root to shoot ratio in wheat. *Plant Sci.* **2016**, *252*, 88–102. [CrossRef]

57. Lee, T.G.; Jang, C.S.; Kim, J.Y.; Kim, D.S.; Park, J.H.; Kim, D.Y.; Seo, Y.W. A Myb transcription factor (TaMyb1) from wheat roots is expressed during hypoxia: Roles in response to the oxygen concentration in root environment and abiotic stresses. *Physiol. Plant.* **2007**, *129*, 375–385. [CrossRef]

58. Rahaie, M.; Xue, G.-P.; Naghavi, M.R.; Alizadeh, H.; Schenk, P.M. A MYB gene from wheat (*Triticum aestivum* L.) is up-regulated during salt and drought stresses and differentially regulated between salt-tolerant and sensitive genotypes. *Plant Cell Rep.* **2010**, *29*, 835–844. [CrossRef]

59. Qin, Y.; Wang, M.; Tian, Y.; He, W.; Han, L.; Xia, G. Over-expression of TaMYB33 encoding a novel wheat MYB transcription factor increases salt and drought tolerance in Arabidopsis. *Mol. Biol. Rep.* **2012**, *39*, 7183–7192. [CrossRef]

60. Xie, Q.; Frugis, G.; Colgan, D.; Chua, N.H. Arabidopsis NAC1 transduces auxin signal downstream of TIR1 to promote lateral root development. *Genes Dev.* **2000**, *14*, 3024–3036. [CrossRef]

61. He, X.-J.; Mu, R.-L.; Cao, W.-H.; Zhang, Z.-G.; Zhang, J.-S.; Chen, S.-Y. AtNAC2, a transcription factor downstream of ethylene and auxin signaling pathways, is involved in salt stress response and lateral root development. *Plant J.* **2005**, *44*, 903–916. [CrossRef] [PubMed]

62. Xia, N.; Zhang, G.; Liu, X.-Y.; Deng, L.; Cai, G.-L.; Zhang, Y.; Wang, X.-J.; Zhao, J.; Huang, L.-L.; Kang, Z.-S. Characterization of a novel wheat NAC transcription factor gene involved in defense response against stripe rust pathogen infection and abiotic stresses. *Mol. Biol. Rep.* **2010**, *37*, 3703–3712. [CrossRef] [PubMed]

63. Ma, M.; Wang, Q.; Li, Z.; Cheng, H.; Li, Z.; Liu, X.; Song, W.; Appels, R.; Zhao, H. Expression of *TaCYP78A3*, a gene encoding cytochrome P450 CYP78A3 protein in wheat (*Triticum aestivum* L.), affects seed size. *Plant J.* **2015**, *83*, 312–325. [CrossRef] [PubMed]

64. Sornaraj, P.; Luang, S.; Lopato, S.; Hrmova, M. Basic leucine zipper (bZIP) transcription factors involved in abiotic stresses: A molecular model of a wheat bZIP factor and implications of its structure in function. *Biochim. Biophys. Acta Gen. Subj.* **2016**, *1860*, 46–56. [CrossRef] [PubMed]

65. Zhang, L.; Zhang, L.; Xia, C.; Gao, L.; Hao, C.; Zhao, G.; Jia, J.; Kong, X. A Novel Wheat C-bZIP Gene, TabZIP14-B, Participates in Salt and Freezing Tolerance in Transgenic Plants. *Front. Plant Sci.* **2017**, *8*. [CrossRef] [PubMed]

66. Minibayeva, F.V.; Gordon, L.K.; Kolesnikov, O.P.; Chasov, A.V. Role of extracellular peroxidase in the superoxide production by wheat root cells. *Protoplasma* **2001**, *217*, 125–128. [CrossRef] [PubMed]

67. Santelia, D.; Vincenzetti, V.; Azzarello, E.; Bovet, L.; Fukao, Y.; Düchtig, P.; Mancuso, S.; Martinoia, E.; Geisler, M. MDR-like ABC transporter AtPGP4 is involved in auxin-mediated lateral root and root hair development. *FEBS Lett.* **2005**, *579*, 5399–5406. [CrossRef]

68. Gaedeke, N.; Klein, M.; Kolukisaoglu, U.; Forestier, C.; Müller, A.; Ansorge, M.; Becker, D.; Mamnun, Y.; Kuchler, K.; Schulz, B.; et al. The Arabidopsis thaliana ABC transporter AtMRP5 controls root development and stomata movement. *EMBO J.* **2001**, *20*, 1875–1887. [CrossRef]

69. Gao, T.; Li, G.-Z.; Wang, C.-R.; Dong, J.; Yuan, S.-S.; Wang, Y.-H.; Kang, G.-Z. Function of the ERFL1a Transcription Factor in Wheat Responses to Water Deficiency. *Int. J. Mol. Sci.* **2018**, *19*, 1465. [CrossRef]

70. Javot, H. The Role of Aquaporins in Root Water Uptake. *Ann. Bot.* **2002**, *90*, 301–313. [CrossRef]

71. Li, A.X.; Han, Y.Y.; Wang, X.; Chen, Y.H.; Zhao, M.R.; Zhou, S.M.; Wang, W. Root-specific expression of wheat expansin gene TaEXPB23 enhances root growth and water stress tolerance in tobacco. *Environ. Exp. Bot.* **2015**, *110*, 73–84. [CrossRef]

15

Assessment of Genetic Diversity in Differently Colored Raspberry Cultivars using SSR Markers Located in Flavonoid Biosynthesis Genes

Vadim G. Lebedev [1,2], **Natalya M. Subbotina** [1,2], **Oleg P. Maluchenko** [3],
Konstantin V. Krutovsky [4,5,6,7,8,*] **and Konstantin A. Shestibratov** [2]

[1] Pushchino State Institute of Natural Sciences, Prospekt Nauki 3, 142290 Pushchino, Moscow Region, Russia

[2] Branch of the Shemyakin-Ovchinnikov Institute of Bioorganic Chemistry of the Russian Academy of Sciences, Prospekt Nauki 6, 142290 Pushchino, Moscow Region, Russia

[3] All-Russian Research Institute of Agricultural Biotechnology, Timiriazevskaya Str. 42, 127550 Moscow, Russia

[4] Department of Forest Genetics and Forest Tree Breeding, Faculty of Forest Sciences and Forest Ecology, Georg-August University of Göttingen, Büsgenweg 2, D-37077 Göttingen, Germany

[5] Center for Integrated Breeding Research (CiBreed), Georg-August University of Göttingen, Albrecht-Thaer-Weg 3, D-37075 Göttingen, Germany

[6] Laboratory of Population Genetics, N. I. Vavilov Institute of General Genetics, Russian Academy of Sciences, Gubkin Str. 3, 119333 Moscow, Russia

[7] Laboratory of Forest Genomics, Genome Research and Education Center, Siberian Federal University, 660036 Krasnoyarsk, Russia

[8] Department of Ecosystem Science and Management, Texas A&M University, 2138 TAMU, College Station, TX 77843-2138, USA

[*] Correspondence: konstantin.krutovsky@forst.uni-goettingen.de

Abstract: Raspberry is a valuable berry crop containing a large amount of antioxidants that correlates with the color of the berries. We evaluated the genetic diversity of differently colored raspberry cultivars by the microsatellite markers developed using the flavonoid biosynthesis structural and regulatory genes. Among nine tested markers, seven were polymorphic. In total, 26 alleles were found at seven loci in 19 red (*Rubus idaeus* L.) and two black (*R. occidentalis* L.) raspberry cultivars. The most polymorphic marker was *RiMY01* located in the MYB10 transcription factor intron region. Its polymorphic information content (PIC) equalled 0.82. The *RiG001* marker that previously failed to amplify in blackberry also failed in black raspberry. The raspberry cultivar clustering in the UPGMA dendrogram was unrelated to geographical and genetic origin, but significantly correlated with the color of berries. The black raspberry cultivars had a higher homozygosity and clustered separately from other cultivars, while at the same time they differed from each other. In addition, some of the raspberry cultivars with a yellow-orange color of berries formed a separate cluster. This suggests that there may be not a single genetic mechanism for the formation of yellow-orange berries. The data obtained can be used prospectively in future breeding programs to improve the nutritional qualities of raspberry fruits.

Keywords: flavonoid biosynthesis; fruit coloration; marker-assisted selection; microsatellites; *Rubus*

1. Introduction

The genus *Rubus* L. (Rosaceae, Rosoideae) is one of the most diverse in the plant kingdom and contains between 600 and 800 species grouped in 12 subgenera, which are widely distributed throughout the world from the lowland tropics to subarctic regions [1]. Among these species, red raspberry (*Rubus idaeus* L.) and blackberry (several species in the genus *Rubus*) grown world-wide, and black raspberry (*R. occidentalis* L.) grown mainly in the United States, are of the greatest economic importance. Their berries are in great demand due to their flavor, color, and taste. In addition, they are very healthy providing a good source of antioxidants, including phenolic acids, flavonoids, anthocyanins, and carotenoids [2]. Berries contain four times more antioxidants than non-berry fruits, 10 times more than vegetables, and 40 times than cereals [3]. For this reason, berries and their products (i.e., berry juice and jam) are very often recognized as "superfoods" [4]. The popularity of this crop can be indicated by the fact that their harvest increased 1.5 times from 2010 to 2017 worldwide and exceeded 800,000 tons [5]. Russia consistently ranks first in the world for the raspberry production. The growing interest in raspberry has led not only to an increase in its production, but also to the expansion of breeding programs for the development of new cultivars. However, classical selection takes a lot of time: in red raspberry, it can take up to 15 years for development and release of a new cultivar [6]. Moreover, a specific feature in the *Rubus* spp. breeding system is that multiple species are often utilized in breeding programs [7]. Scientific achievements in molecular biology, and use of molecular markers, in particular, can accelerate the selection process, as they will allow for the assessment of the seedlings with valuable traits at a much earlier stage. Molecular genetic markers provide more reliable cultivar identification of *Rubus* species than morphological markers [8].

In order to speed up the breeding process, it is useful to have genetic linkage maps containing information about the markers associated with the most important traits, including disease and pest resistance, plant habitus, nutritional and sensory fruit quality, and plant architecture. The first genetic linkage map of *Rubus* was constructed from a cross between two *Rubus* subspecies, *R. idaeus* (cv. Glen Moy) × *R strigosus* (cv. Latham), in 2004 [9]. After that, other molecular maps for red raspberry [10–12], black raspberry [13] and tetraploid blackberry [14] appeared. Quantitative trait loci (QTL) have been identified for important traits including resistance to diseases [11,15] and pests [10], fruit anthocyanin content [16], growth characteristics [10,17], fruit color and quality traits [18]. Currently, molecular markers are routinely used in breeding raspberries for resistance to the Phytophthora root rot at the James Hutton Institute (UK), and two promising genotypes are under commercial trials, as well as markers for the quality of berries are in the process of validation [19]. If in the first reports a combination of various types of molecular markers such as AFLP and simple sequence repeat (SSR) [9,10], RAPD, and RGAP [11] were used, then the most recent molecular maps were produced using only molecular markers designed from sequenced DNA such as microsatellites or SSR markers [13,20]. SSRs are DNA tandem repeats of the 1–6 nucleotide long motifs that are very frequent in genomes. They are very polymorphic with high information content, co-dominant inheritance, locus specificity, extensive genome coverage and simple detection using labelled primers that flank the microsatellite [9,21], and their ability to distinguish even closely related individuals is particularly important for many crop species [21]. Raspberry researchers have noted the benefits of the SSR markers, but very few molecular markers still exist for *Rubus* [7,22]. It should be also acknowledged that the breeding process can be accelerated using genomic selection (e.g., [23]), an approach under rapid adoption in many species, which is based on multiple marker–trait associations and does not require linkage maps.

The color of the berries not only affects their attractiveness but also serves as an indicator of the content of biologically active compounds. For example, the content of anthocyanins in raspberry berries varies widely from 2 to 325 mg/100 g depending on the color of the berries [24]. Flavonols and anthocyanins are synthesized in the flavonoid pathway, and its enzymes are well characterized. Kassim et al. [16] mapped QTLs for individual anthocyanin pigments in raspberry. The genes of various enzymes of flavonoid biosynthesis were also identified in red [18] and black [25] raspberry and blackberry [26]. Besides the structural genes, regulatory genes are important in the biosynthesis of flavonoids. The late flavonoid biosynthetic genes are activated by the ternary transcriptional MYB-bHLH-WD40 (MBW) complex comprising three classes of regulatory proteins including R2R3-MYBs, bHLHs, and TTG1 (WD40) [27]. Transcription factor genes, such as *MYB10*, *bHLH* and *bZIP*, have also been identified in the *Rubus* species [18,26].

There are several studies that used random genomic SSR markers to assess genetic diversity in cultivars within [8,28] and between [29] different species. However, we are unaware of studies in which genetic diversity would be assessed using markers located in genes of any metabolic pathway and the biosynthesis of flavonoids, in particular. In this study, we developed SSR markers using nucleotide sequences of structural and regulatory genes of flavonoid biosynthesis in *Rubus* and *Fragaria* (strawberry) available at the National Center for Biotechnology Information (NCBI) GenBank database to test whether genetic variation associated with these genes correlate with a variation of berry colors. These markers were genotyped in 19 raspberry cultivars from different geographic regions (Russia, Poland, Italy, Switzerland, UK, and USA) and two cultivars of black raspberry. If alleles at these loci correlate with the content of biologically active substances, they could subsequently be used to optimize selection for valuable traits associated with color and, indirectly, with the content of flavonoids, by accelerating selection via screening genotypes at early stages.

2. Materials and Methods

2.1. Plant Materials

Nineteen cultivars of red raspberry (Amira, Anne, Babye Leto II, Beglyanka, Brilliantovaya, Bryanskoe Divo, Gerakl, Glen Ample, Marosejka, Meteor, Oranzhevoe Chudo, Pingvin, Polka, Poranna Rosa, Solnyshko, Sugana, Tarusa, Zheltyj Gigant, and Zolotaya Osen) and two cultivars of black raspberry (Cumberland and Jewel) were chosen to genotype SSR loci located in the flavonoid biosynthesis genes. These cultivars have a wide range of fruit color from yellow to black with various geographic and genetic origins, but cultivars of Russian origin from two raspberry breeding centers (Bryansk and Moscow) dominated in the list (Table 1). Raspberry plants used in this study were kindly provided by Dr. I. A. Pozdniakov (OOO Microklon, Pushchino, Russia). Each cultivar represented a microclonally vegetatively propagated line containing practically genetically identical plants. Therefore, a single specimen per culture was used for further DNA isolation and genotyping.

Table 1. Parentage and fruit color of the *Rubus* cultivars used in the study.

Cultivar	Abbr.	Genetic Origin and Background	Fruit Color	Origin
		R. idaeus (red raspberry)		
Amira	Ami	Polka × Tulameen	red	Italy
Anne	Ann	Amity × Glenn Garry	yellow	USA
Babye Leto II	BL2	Autumn Bliss × Babye Leto	red	Russia (Bryansk)
Beglyanka	Beg	Kostinbrodskaya × Novost Kuzmina	orange	Russia (Bryansk)
Brilliantovaya	Bri	open pollination of interspecific hybrids	red	Russia (Bryansk)
Bryanskoe Divo	BrD	47-18-4 (open pollination)	light-red	Russia (Bryansk)
Gerakl	Ger	Autumn Bliss × 14-205-4	red	Russia (Bryansk)
Glen Ample	GAm	SCRI7326EI × SCRI7412H16	dark red	UK
Marosejka	Mar	7324/50 × 7331/3	light-red	Russia (Moscow)
Meteor	Met	Kostinbrodskaya × Novost Kuzmina	red	Russia (Bryansk)
Oranzhevoe Chudo	OrC	Shapka Monomaha (open pollination)	orange	Russia (Bryansk)
Pingvin	Pin	interspecific hybrid	dark red	Russia (Bryansk)
Polka	Pol	P89141(open pollination)	red	Poland
Poranna Rosa	PoR	83291 × ORUS 1098-1	yellow	Poland
Solnyshko	Sol	Kostinbrodskaya × Novost Kuzmina	red	Russia (Bryansk)
Sugana	Sug	Autumn Bliss × Tulameen	light-red	Switzerland
Tarusa	Tar	Stolichnaya × Shtambovyj-1	red	Russia (Moscow)
Zheltyj Gigant	ZhG	Marosejka × Ivanovskaya	yellow	Russia (Moscow)
Zolotaya Osen	ZOs	13-39-11 (open pollination)	yellow	Russia (Bryansk)
		R. occidentalis (black raspberry)		
Cumberland	Cum	Gregg selfed	blue-black	USA
Jewel	Jew	(Bristol × Dundee) × Dundee	black	USA

2.2. Simple Sequence Repeat (SSR) Marker and Polymerase Chain Reaction (PCR) Primer Development

The WebSat software [30] was used to detect SSR loci in the nucleotide sequences of *Rubus* and *Fragaria* × *ananassa* (the garden strawberry or simply strawberry, a widely grown hybrid species of the genus *Fragaria*) flavonoid biosynthesis genes available at the NCBI GenBank database (http://www.ncbi.nlm.nih.gov) (Table 2). The Primer 3 software (http://primer3.org) was used to design appropriate polymerase chain reaction (PCR) primers based on the sequences flanking the SSR loci. The minimum number of motifs used to select the SSR locus was nine for mono-nucleotide repeats, five for di-nucleotide motifs, three for tri-, and tetra-, and two for penta-, and hexa-nucleotide repeats. Primers were designed using the following criteria: primer length of 18–27 bp (optimally 22 bp), GC content of 40%–80%, annealing temperature of 57–68 °C (optimally 60 °C), and expected amplified product size of 100–400 bp. Primers for the *RiG001* locus were as in [8]. Primers were synthesized by Syntol Company (Moscow, Russia) and are summarized in Table 2.

2.3. DNA Isolation, PCR Amplification and Fragment Analysis

A single DNA sample per each cultivar was produced from young expanding leaves representing a single plant per each cultivar. Total genomic DNA was extracted using the STAB method [31]. The quality and quantity of extracted DNA were determined by the NanoDrop 2000 spectrophotometer (ThermoFisher). The final concentration of each DNA sample was adjusted to 50 ng/µL in TE buffer before the PCR amplification.

For genotyping, PCR was performed separately for each primer pair using a forward primer labeled with the fluorescent dye 6-FAM and an unlabeled reverse primer (Syntol, Russia). The PCR amplification was performed in a total volume of 20 µL consisted of 50 ng of genomic DNA, 10 pmol of the labeled forward primer, 10 pmol of an unlabeled reverse primer, and PCR Mixture Screenmix (Eurogen, Russia). After an initial denaturation at 95 °C for 3 min, DNA was amplified during 33 cycles in a gradient thermal cycler (Bio-Rad, Hercules, CA, USA) programmed for a 30 s denaturation step at 95 °C, a 20 s annealing step at the optimal annealing temperature of the primer pair and a 35 s extension step at 72 °C. A final extension step was done at 72 °C for 5 min.

The PCR generating clear, stable, and specific DNA fragments within an expected length (200–400 bp) were considered as successful PCR amplifications. If a primer pair failed three times to amplify template DNA that was amplified with other primers, then it was scored as a null genotype.

Separation of amplified DNA fragments was performed in an ABI 3130xl Genetic Analyzer using S450 LIZ size standard (Syntol Company, Moscow, Russia). Peak identification and fragment sizing were done using the Gene Mapper v4.0 software (Applied Biosystems, Foster, CA, USA).

2.4. Genetic Data Analysis

Genetic parameters were calculated for 21 raspberry cultivars based on seven SSR polymorphic loci. The allele frequencies, number of alleles, observed (H_o) and expected (H_e) heterozygosities, and polymorphic information content (PIC) were calculated using the PowerMarker v.3.25 software [32]. This software was also used to estimate pairwise Nei's standard genetic distances between each pair of cultivars and to generate a UPGMA dendrogram, which was visualized using the Statistica software (TIBCO Software Inc., Palo Alto, CA, USA).

Table 2. Data on nine simple sequence repeat (SSR) loci located in the flavonoid biosynthesis genes and their polymerase chain reaction (PCR) primer pairs used to study genetic diversity in *Rubus* cultivars.

Locus	Gene	Species	NCBI GenBank Accession Number	Motif and Number of Repeats	Location in the Gene	PCR Primer Nucleotide Sequence Forward	PCR Primer Nucleotide Sequence Reverse	$T*$, °C	Allele Size, bp Expected	Observed Red Raspberry	Observed Black Raspberry
RiG001	aromatic polyketide synthase (PKS3)	*R. idaeus*	AF292369	(AT)$_6$	intron	TGTCCGATCCTTTTCTTTGG	CGCTTCTTGATCCTTGACTTGT	55	345	349, 350, 351	0
RcFH01	flavanone-3-hydroxylase (F3H)	*R. coreanus*	EU255776	(TATG)$_3$	intron	GGTCCAAGTGCATTCCATATTAC	GTTCTTGAATCTCCCGTTGCT	60	262	255, 265, 271	255, 271
FaFS01	flavonol synthase (FLS)	*Fragaria × ananassa*	DQ834905	(CT)$_{12}$	intron	CATCCTAATGCCCTAGTCATC	TGTACTTCGGTGGATTCTCCTT	60	304	323, 328	323
FaFS02	flavonol synthase (FLS)	*F. ananassa*	DQ834905	(GGAAG)$_2$	exon	AAGCTCCTCAAACAAATCTTCG	GTAGTTAATGGCAGAAGGTGGC	60	273	255, 271	271
RiAS01	anthocyanidin synthase (ANS)	*R. idaeus*	KX950789	(ATCTC)$_2$	exon	TCAACAAGGAGAAGGTGAGGAT	CCGTTAGGAGGAGATGAAAGCAG	60	334	309, 333, 358	309
FaAR01 **	anthocyanidin reductase (ANR)	*F. ananassa*	DQ664193	(TGCTG)$_2$ (CATTT)$_2$	exon intron	AATCTGCTTCTGGTCGGTACAT	AGAGAGTATGGTCTTCGCCTTG	60	244	250	250
RhUF01 **	UDP-glucose flavonoid 3-O-glycosyltransferase-like protein (UFGT)	*R. hybrid*	JF764808	(GAG)$_7$ (ACAAGC)$_2$	exon	AGGAGCTGAAGAAAAGACTCCA	AAAGTCCTCTAGGTTTCCCCTG	60	275	267, 270	269, 270
RiMY01 **	transcription factor MYB10	*R. idaeus*	EU155165	(TAATA)$_2$ (CT)$_7$ (AT)$_{15}$	introns	GTTCCTCTCCAAGCAGGTTATT	TGCAAAGTCTCTCTCTTGATG	59	330	323, 325, 327, 329, 331, 333, 341, 342	358
RiTT01	transparent testa glabra 1 (TTG1) protein	*R. idaeus*	HM579852	(CAC)$_5$	exon	ACTCCACACAAGAATCCCATCT	CTGTTGTTCAAGACCGAAATTG	60	379	379	379

* Optimal annealing temperature. ** Two or three SSRs in these loci were amplified simultaneously by a single pair of primers.

3. Results

3.1. Polymorphism and Genetic Diversity Analysis

Nine SSR markers (six based on *Rubus* and three on *Fragaria* nucleotide sequences of the flavonoid biosynthesis genes) were used to estimated genetic diversity in 19 raspberry (*R. idaeus*) and two black raspberry (*R. occidentalis*) cultivars. All PCR primer pairs amplified one or two alleles. In raspberries, two loci (*RiTT01* and *FaAR01*) were monomorphic, and other seven were polymorphic. In black raspberry cultivars, the *RiG001* was not amplified at all, six loci were monomorphic and only two polymorphic (Table 2). In total, 26 alleles were found in seven polymorphic microsatellite loci. The number of alleles per locus varied from two per locus (*FaFS02* and *FaFL01*) to nine per locus (*RiMY01*) with an average number of 3.7 alleles per locus (Table 3). The *RiMY01* locus was the most polymorphic. In general, the SSR loci located in introns were more polymorphic than loci in exons.

Table 3. Parameters of genetic variation for seven polymorphic SSR loci in 21 *Rubus* cultivars.

Locus	Location in the Gene	Major Allele Frequency	Number of Alleles	Heterozygosity		Polymorphism Information Content (PIC)
				Expected (H_e)	Observed (H_o)	
RiG001	intron	0.81	4	0.33	0.19	0.31
RcFH01	intron	0.74	3	0.41	0.52	0.35
FaFS01	intron	0.76	2	0.36	0.48	0.30
FaFS02	exon	0.98	2	0.05	0.05	0.05
RiAS01	exon	0.79	3	0.36	0.19	0.33
RhUF01	exon	0.90	3	0.18	0.00	0.17
RiMY01	introns	0.29	9	0.84	0.57	0.82
Mean		0.75	3.71	0.36	0.29	0.33

There were cultivar-specific alleles, such as a unique allele 358 at the *RiMY01* locus found only in black raspberry, and alleles 267 and 269 at the *RhUF01* locus found only in the red raspberry Meteor and Jewel cultivars, respectively. Meteor contained also a unique allele 333 at the *RiMY01* locus.

Parameters of genetic variation for seven polymorphic SSR loci in 21 *Rubus* cultivars are presented in Table 3. Expected heterozygosity (H_e) ranged from 0.05 in the *RiMY01* locus up to 0.84 in the *RiMY01* locus with an average value of 0.36. Observed heterozygosity was zero in the *RhUF01* locus and ranged from 0.05 in the *FaFS02* locus to 0.57 in the *RiMY01* locus with an average value of 0.29. The observed heterozygosity was lower than expected in four microsatellite loci and on average (Table 3). On average, the expected and observed heterozygosities were higher for the SSRs in introns (0.49 and 0.44, respectively) compared to the SSRs in exons (0.20 and 0.08, respectively). The average PIC was 0.332 and varied from 0.05 in the *FaFS02* locus to 0.82 in the *RiMY01* locus (Table 3).

3.2. Cluster Analysis

A UPGMA dendrogram was constructed for 21 raspberry cultivars based on seven SSR markers located in the genes of the flavonoid biosynthesis (Figure 1). The dendrogram clearly separates red and black raspberries. Among the red raspberry cultivars, there is a group of cultivars with yellow-orange colored berries (Anne, Poranna Rosa, Orangevoe Chudo, and Zolotaya Osen), which forms a separate cluster. The same group includes also the Bryanskoe Divo cultivar with light red berries. At the same time, the Zheltyj Gigant (yellow berries) and Beglyanka (orange berries) were not included in this group. Separation of cultivars did not follow their genetic origin. The cultivars Beglyanka, Solnyshko, and Meteor having the same genetic origin from the Kostinbrodskaya × Novost Kuzmina cross were completely separated from each other. In addition, the Babye Leto 2 also having an ancestral hybrid (Autumn Bliss × (September × (Kostinbrodskaya × Novost Kuzmina))) turned out to differ mostly from other raspberry cultivars. Gerakl and Sugana both also having Autumn Bliss as their parent species were significantly separated. At the same time, close similarities have been observed for cultivars from different geographic regions. No genetic differences were found between the Orangevoe Chudo (Russia) and Poranna Rosa (Poland) cultivars, and between the Amira (Italy) and Tarusa (Russia) cultivars, although they have different genetic origins. The Brilliantovaya and Pingvin cultivars were also identical and were obtained with the use of interspecific hybrids.

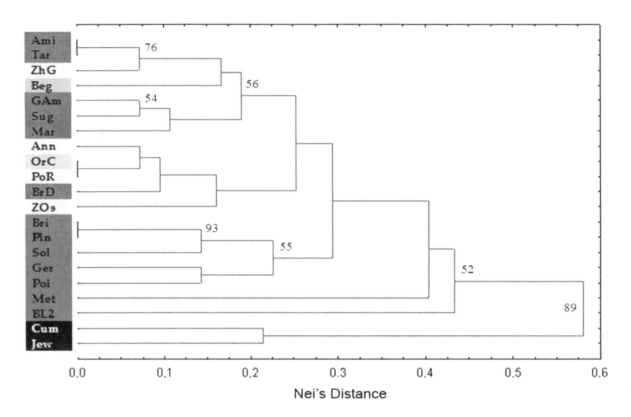

Figure 1. The UPGMA dendrogram of the 21 *Rubus* cultivars based on pairwise Nei's standard genetic distances calculated using seven SSR markers located in the flavonoid biosynthesis genes. Left column shows the colors of the cultivar berries. Only bootstrap values larger than 50% are presented. See Table 1 for the full cultivar names.

4. Discussion

SSR markers (microsatellites) are widely used in genetic diversity studies, QTL and genetic mapping, molecular-assisted selection (MAS), and cultivar identification, because they are multi-allelic, co-dominant, highly informative, relatively accurate and easily detected [33]. SSR markers have been often used to map different types of *Rubus* [9,13], fingerprinting germplasm [34], and in studies of the genetic diversity and population structure within [28] and among [29] *Rubus* species. However, genetic diversity has not previously been studied in terms of any specific metabolic pathway genes that determine valuable breeding traits.

In this study, we report on the evaluation of a number of red and black raspberry cultivars using SSR loci representing known sequences of the flavonoid biosynthesis pathway genes, which synthesize biologically active substances with high antioxidant activity—flavonols and anthocyanins. Among these microsatellite loci, six (*RcFH01, FaFS01, FaFS02, RiAS01, FaAR01,* and *RhUF01*) were located in the structural genes of the flavonoid biosynthesis (*F3H, FLS, ANS, ANR,* and *UFGT*) and two (*RiMY01* and *RiTT01*) in the regulatory genes (*MYB10* and *TTG1*). Flavanone-3-hydroxylase (F3H) is a key enzyme in the flavonoid biosynthesis in plants, as it catalyzes formation of 3-hydroxy flavonol, a common precursor of anthocyanins, flavanols, and proanthocyanidins [35]. Particular attention was paid to the flavonol synthase gene, for which two loci were used. Flavonol synthase (FLS) is an important enzyme of flavonoid pathway that catalyzes the formation of flavonols from dihydroflavonols, and thus may influence anthocyanin levels, as dihydroflavonols are intermediates in the production of both colored anthocyanins and colorless flavonols [36]. The anthocyanidin synthase (ANS) leads to the synthesis of the anthocyanidin, the first colored compound in the anthocyanin biosynthetic pathway, from which anthocyanidin reductase catalyzes the formation of proanthocyanidins (condensed tannins) [37]. The last common step for the production of stable anthocyanins is the glycosylation by the enzyme UDP-glucose/flavonoid 3-O-glucosyl transferase (UFGT) [38].

In addition, loci were used on the sequence of two transcription factors (MYB 10 and TTG1) that belong to the MBW complex, which regulates the production of the late biosynthetic genes [27]. For comparison, we also used a pair of primers designed for the *RiG001* locus using the sequence of the *R. idaeus* aromatic polyketide synthase (*PiPKS3*) gene, which was not amplified in blackberry cultivars [8]. The *RiPKS3* gene differed from the *RiPKS1* gene, encoding a typical chalcone synthase (CHS) catalyzing the first step of flavonoid biosynthesis, in four amino acid positions and produced in vitro predominantly p-coumaryltriacetic acid lactone and low levels of chalcone [39]. Within the PCR fragment amplified by the primers for the *RiG001* locus the sequence of the *RiPKS3* gene (NCBI GenBank AF292369) differed from the *RiPKS1* gene sequence (AF292367) by a two nucleotide long deletion (2 bp) and a single nucleotide insertion. Three alleles (349, 350, and 351 bp) were obtained for this locus (Table 2).

In addition to the sequences of the genes of the *Rubus* plants (*R. idaeus*, *R. coreanus*, and *R. hybrid*), we used the sequences of the genes from *Fragaria* × *ananassa*, which is a close relative of *Rubus* from the same sub-family, Rosoideae. The *Rubus* and *Fragaria* both have the same base chromosome number $1n = 7$, similar morphology and chloroplast and nuclear DNA phylogenies [13].

Among three most economically important types of raspberry, 19 cultivars of red raspberry with a wide range of berry color from various world breeding centers and two cultivars of black raspberry are mostly used. Both species, red (*R. idaeus*) and black (*R. occidentalis*) raspberry belong to the same subgenus *Idaeobatus* (raspberries) and are diploids ($2n = 2x = 14$), while blackberry species vary greatly in ploidy [34].

In our study, the average number of alleles for seven polymorphic SSR loci in the flavonoid biosynthesis genes was 3.71, the mean H_o and H_e were 0.286 and 0.360, respectively, and the mean PIC was 0.332. These values were generally lower than previously reported for *R. idaeus* [8] and *R. coreanus* [29], but quite comparable with the data for black raspberry cultivars [28]. Perhaps, this is due to the fact that red raspberry cultivars are, for the most part, complex hybrids with a limited genetic pool [34], and the selection for berries quality has further reduced their diversity. The level of expected heterozygosity (H_e) was higher than observed (H_o) both on average and in most individual loci. These data are different from other studies of the *Rubus* species, where these parameters were approximately equal [8,29], or even higher [28]. However, unlike those studies, where population samples were used, a collection of different cultures was used in this study, which is not a population sample, but a mixture of genotypes with different genetic background and origin. Therefore, it is expected to observe excess of expected heterozygosity in comparison to observed heterozygosity due to Wahlund effect.

Only the *RiMY01* locus was highly polymorphic (PIC = 0.82). This locus had three SSR regions, two of which represent dinucleotide repeats. These data coincide with the results of Castillo et al. [8], in which all three highly informative markers (PIC = 0.78–0.82) represented dinucleotide repeats. In *R. coreanus*, among five highly polymorphic markers (PIC > 0.7), four represented dinucleotide repeats, and one trinucleotide repeats [29]. The high variation of the *RiMY01* locus can be explained by its location in the first intron of the transcription factor MYB10. SSR markers located in introns were more variable in comparison to those located in exons (expected and observed heterozygosities averaged 0.49 and 0.44 vs. 0.20 and 0.08, respectively). Our results are in agreement with those of Garcia-Gomez et al. [40], which showed that SSRs in introns had a higher level of heterozygosity compared to SSRs in exons in Prunus species—0.65 vs. 0.17, respectively. Similar results were also obtained in maize [41]. Significantly higher variation was observed also for SNPs in noncoding regions compared to coding ones [42]. In general, introns are more variable than exons, as they are under less selection pressure during the evolutionary process [43].

The length of most alleles at the *RiMY01* locus differ from each other by two nucleotide-long steps, which is consistent with dinucleotide repeats of the SSR motifs in this locus. However, imperfect repeats also often occur in the raspberry SSR loci. For instance, Fernandez et al. [34] has previously reported the alleles with length different by consecutive one nucleotide-long steps in the *Rubus57a* and *Rub5a* markers. This single nucleotide stepwise variation is expected for *Rub5a*, which is a SSR marker with a mononucleotide motif, but *Rubus57a* is a SSR marker with a dinucleotide motif. We also observed a few alleles with imperfect repeats, such as the unique allele 267 of the *RhUF01* locus in the Meteor cultivar, for which the perfect allele size is 270 following the trinucleotide motif GAG stepwise allelic variation.

The black raspberry cultivars were highly homozygous: six out of eight loci were monomorphic (Table 2). High homozygous in black raspberry has been also found earlier by Lewers and Weber [44]. They noticed that the level of homozygosity for the black raspberry was 80%, but only 40% for the red raspberry. The 21 SSR loci were unable to distinguish between six of the black raspberry cultivars [28]. However, the black raspberry cultivars Cumberland and Jewel were well discriminated in this study. Despite the small number of loci used in our study, these two cultivars were also separated by two loci: *RcFH01* and *RhUF01*. In our study the red raspberry cultivars were easily discriminated from the black raspberry cultivars by a unique black raspberry specific allele 358 at the *RiMY01* locus and the allele 309 at the *RiAS01* locus, which occurred almost exclusively in the black raspberry cultivars, except the red raspberry cultivar Babye Leto 2. In addition, the *RiG001* locus was not amplified in black raspberry. The same was observed also in 48 earlier tested blackberry cultivars [8]. Thus, in respect to this locus, the black raspberry is closer to the wild blackberry than to the red raspberry, although it belongs to different subgenera. No amplification of RiG001 and the unique allele 358 at the *RiMY01* locus can be used to separate the red raspberry cultivars from the black ones.

Cluster analysis of the SSR markers located in the genes of the biosynthesis of flavonoids showed a clear separation of the black raspberry (*R. occidentalis*) cultivars with black colored berries from the red raspberry (*R. idaeus*) cultivars with berries colored from yellow to dark red (Figure 1). It is important to note also that five cultivars with berries of similar shades of light red color (three with yellow berries, one with orange, and another with light red color) having completely different origin still clustered together into one sub-group. Perhaps, gene-targeted markers [45] such as SSR loci in the genes of the biosynthesis of flavonoids reflect better their genetic similarity for traits, such as color of their berries, likely controlled or affected by these genes, than random genomic SSR markers.

Castillo et al. [8] found that the primocane fruiting (fall fruiting) raspberry cultivars were grouped into a separate cluster. In Fernandez et al. [34] studies, it was shown that the majority of primocane-fruiting material from various breeding programs, as well as some very early ripening floricane-fruiting genotypes are grouped into one cluster. This shows that cultivars can be grouped according to a particular trait regardless of their origin. At the same time, two cultivars with yellow and orange-colored fruits (Zheltyj Gigant and Beglyanka) fell into another group of red-colored fruits. Perhaps, for a clearer separation, it is necessary to use additionally more polymorphic markers, including other genes of the biosynthesis of flavonoids not represented in this study.

Moreover, it is possible that the yellow color of the raspberry fruits can be obtained by two or more mechanisms. For example, primocane fruiting cultivars were also distributed in two different groups [34]. The genetic mechanisms for the formation of yellow color in raspberry fruit have not yet been fully studied. Although assumptions on this topic were made back in the 1930s, it was not until 2016 when an inactive anthocyanidin synthase (ANS) allele was identified in yellow raspberry [46]. A 5 bp insertion in the coding region of gene creates a premature stop codon resulting in a truncated amino acid sequence of the defective ANS protein. However, other mechanisms are also possible, such as the combinations of recessive and dominant alleles, or the transcription factors that may lead to a huge variety of berry colors in raspberry.

The clustering along the flavonoid pathway also showed that there is a lack of connections between cultivars of the related origin. This is exactly the opposite data compared to the analyses carried out on randomly selected SSR markers evenly distributed across the genome. For example, Fernandez et al. [34] demonstrated that one cluster is almost entirely composed of cultivars from the Scottish raspberry breeding program or cultivars based on their germplasm. From the point of view of MAS the use of gene-targeted markers to assess genotypes for particular breeding traits is preferable to the use of random SSR markers. Graham et al. [9] suggested in 2004 that *Rubus idaeus* due to the diploid set of chromosomes ($2n = 2x = 14$) and a very small genome (275 Mb) may be used as a model species for the Rosaceae. For many years, this was impeded by the lack of the full-genome *Rubus* sequence, although the genomes of other Rosaceae species have been already sequenced, such as apple in 2010, strawberry in 2011, pear and peach in 2013 [47]. However, the situation is changing with genomes of *R. occidentalis* [48] and *R. idaeus* [49] having been recently published. This will facilitate developing gene-targeted markers that can advance breeding *Rubus* for important traits including those related to the nutritional value of their berries.

5. Conclusions

In this study, we demonstrated that a set of gene-targeted SSR markers representing structural and regulatory genes of flavonoid biosynthesis could potentially allow more informative and meaningful evaluation of the genetic relationship between different cultivars of red and black raspberries that reflect the color of their berries and possibly also their nutritional value. However, the study did not compare this set of gene-targeted markers with an analysis of the same germplasm set using neutral markers. A comparative analysis using a set of neutral SSR markers would seem to be important to support this particular conclusion. The developed primer set can be potentially used for MAS in the *Rubus* breeding programs for improving the nutritional quality of fruits. This first requires confirmation that the SSR alleles identified correlate with differences in the content of flavonoids. Additional studies and further development of these gene-targeted markers are needed to validate this approach.

Author Contributions: Conceptualization, V.G.L. and K.A.S.; Data curation, V.G.L., K.V.K. and K.A.S.; Formal Analysis, V.G.L. and O.P.M.; Funding Acquisition, V.G.L., K.V.K. and K.A.S.; Investigation, V.G.L., N.M.S., O.P.M. and K.A.S.; Methodology, V.G.L. and K.A.S.; Project Administration, V.G.L. and K.A.S.; Resources, V.G.L. and K.A.S.; Supervision, V.G.L. and K.A.S.; Writing, V.G.L., K.V.K. and K.A.S.

Acknowledgments: We thank I. A. Pozdniakov (OOO Microklon, Pushchino, Russia) for providing us with raspberry plants used in this study.

References

1. Thompson, M.M. Chromosome numbers of *Rubus* species at the National Clonal Germplasm Repository. *HortScience* **1995**, *30*, 1447–1452. [CrossRef]
2. Skrovankova, S.; Sumczynski, D.; Mlcek, J.; Jurikova, T.; Sochor, J. Bioactive compounds and antioxidant activity in different types of berries. *Int. J. Mol. Sci.* **2015**, *16*, 24673–24706. [CrossRef]
3. Halvorsen, B.L.; Myhrstad, M.C.W.; Wold, A.B.; Jacobs, D.R.; Haffner, K.; Holte, K.; Andersen, L.F.; Baugerod, H.; Barikmo, I.; Hvattum, E.; et al. A systematic screening of total antioxidants in dietary plants. *J. Nutr.* **2002**, *132*, 461–471. [CrossRef]
4. Olas, B. Berry phenolic antioxidants—Implications for human health? *Front. Pharmacol.* **2018**, *9*, 78. [CrossRef]
5. FAOSTAT. 2019. Available online: http://www.fao.org/faostat/ (accessed on 29 July 2019).

6.   Graham, J.; Jennings, S.N. Raspberry breeding. In *Breeding Tree Crops*; Jain, S.M., Priyadarshan, M., Eds.; IBH & Science Publication: Oxford, UK, 2009; pp. 233–248.

7.   Bushakra, J.M.; Lewers, K.S.; Staton, M.E.; Zhebentyayeva, T.; Saski, C.A. Developing expressed sequence tag libraries and the discovery of simple sequence repeat markers for two species of raspberry (*Rubus* L.). *BMC Plant Biol.* **2015**, *15*, 258. [CrossRef]

8.   Castillo, N.R.F.; Reed, B.M.; Graham, J.; Fernandez-Fernandez, F.; Bassil, N.V. Microsatellite markers for raspberry and blackberry. *J. Am. Soc. Hortic. Sci.* **2010**, *135*, 271–278. [CrossRef]

9.   Graham, J.; Smith, K.; MacKenzie, K.; Jorgenson, L.; Hackett, C.; Powell, W. The construction of a genetic linkage map of red raspberry (*Rubus idaeus* subsp. *idaeus*) based on AFLPs, genomic-SSR and EST-SSR markers. *Theor. Appl. Genet.* **2004**, *109*, 740–749. [CrossRef]

10.   Sargent, D.J.; Fernández-Fernández, F.; Rys, A.; Knight, V.H.; Simpson, D.W.; Tobutt, K.R. Mapping of A1 conferring resistance to the aphid *Amphorophora idaei* and *dw* (dwarfing habit) in red raspberry (*Rubus idaeus* L.) using AFLP and microsatellite markers. *BMC Plant Biol.* **2007**, *7*, 15. [CrossRef]

11.   Pattison, J.A.; Samuelian, S.K.; Weber, C.A. Inheritance of *Phytophthora* root rot resistance in red raspberry determined by generation means and molecular linkage analysis. *Theor. Appl. Genet.* **2007**, *115*, 225–236. [CrossRef]

12.   Woodhead, M.; McCallum, S.; Smith, K.; Cardle, L.; Mazzitelli, L.; Graham, J. Identification, characterisation and mapping of simple sequence repeat (SSR) markers from raspberry root and bud ESTs. *Mol. Breed.* **2008**, *22*, 555–563. [CrossRef]

13.   Bushakra, J.M.; Stephens, M.J.; Atmadjaja, A.N.; Lewers, K.S.; Symonds, V.V.; Udall, J.A.; Chagne, D.; Buck, E.J.; Gardiner, S.E. Construction of black (*Rubus occidentalis*) and red (*R. idaeus*) raspberry linkage maps and their comparison to the genomes of strawberry, apple, and peach. *Theor. Appl. Genet.* **2012**, *125*, 311–327. [CrossRef]

14.   Castro, P.; Stafne, E.T.; Clark, J.R.; Lewers, K.S. Genetic map of the primocane-fruiting and thornless traits of tetraploid blackberry. *Theor. Appl. Genet.* **2013**, *126*, 2521–2532. [CrossRef]

15.   Graham, J.; Smith, K.; Tierney, I.; MacKenzie, K.; Hackett, C.A. Mapping gene *H* controlling cane pubescence in raspberry and its association with resistance to cane botrytis and spur blight, rust and cane spot. *Theor. Appl. Genet.* **2006**, *112*, 818–831. [CrossRef]

16.   Kassim, A.; Poette, J.; Paterson, A.; Zait, D.; McCallum, S.; Woodhead, M.; Smith, K.; Hackett, C.; Graham, J. Environmental and seasonal influences on red raspberry anthocyanin antioxidant contents and identification of quantitative traits loci (QTL). *Mol. Nutr. Food Res.* **2009**, *53*, 625–634. [CrossRef]

17.   Graham, J.; Hackett, C.A.; Smith, K.; Woodhead, M.; Hein, I.; McCallum, S. Mapping QTL for developmental traits in raspberry from bud break to ripe fruit. *Theor. Appl. Genet.* **2009**, *118*, 1143–1155. [CrossRef]

18.   McCallum, S.; Smith, K.; Woodhead, M.; Hackett, C.; Paterson, A.; Graham, J. Developing molecular markers for quality traits in red raspberry. *Theor. Appl. Genet.* **2010**, *121*, 611–627. [CrossRef]

19.   Jennings, S.N.; Graham, J.; Ferguson, L.; Young, V. New developments in raspberry breeding in Scotland. *Acta Hortic.* **2016**, *1133*, 23–28. [CrossRef]

20.   Bushakra, J.M.; Bryant, D.B.; Dossett, M.; Vining, K.J.; VanBuren, R.; Gilmore, B.S.; Lee, J.; Mockler, T.C.; Finn, C.E.; Bassil, N.V. A genetic linkage map of black raspberry (*Rubus occidentalis*) and the mapping of Ag_4 conferring resistance to the aphid *Amphorophora agathonica*. *Theor. Appl. Genet.* **2015**, *128*, 1631–1646. [CrossRef]

21.   Powell, W.; Machray, G.C.; Provan, J. Polymorphism revealed by simple sequence repeats. *Trends Plant Sci.* **1996**, *1*, 215–222. [CrossRef]

22.   Graham, J.; Smith, K.; Woodhead, M.; Russell, J. Development and use of simple sequence repeat SSR markers in Rubus species. *Mol. Ecol. Notes* **2002**, *2*, 250–252. [CrossRef]

23.   Gezan, S.; Osorio, L.; Verma, S.; Whitaker, V. An experimental validation of genomic selection in octoploid strawberry. *Hortic. Res.* **2017**, *4*, 16070. [CrossRef]

24.   Bobinaite, R.; Viskelis, P.; Venskutonis, P.R. Variation of total phenolics, anthocyanins, ellagic acid and radical scavenging capacity in various raspberry (*Rubus* spp.) cultivars. *Food Chem.* **2012**, *132*, 1495–1501. [CrossRef]

25.   Lee, S.S.; Lee, E.M.; An, B.C.; Barampuram, S.; Kim, J.-S.; Cho, J.Y.; Lee, I.-C.; Chung, B.Y. Molecular cloning and characterization of a flavanone-3-hydroxylase gene from *Rubus occidentalis* L. *J. Radiat. Ind.* **2008**, *2*, 121–128.

26. Chen, Q.; Yu, H.W.; Wang, X.R.; Xie, X.L.; Yue, X.Y.; Tang, H.R. An alternative cetyltrimethylammonium bromide-based protocol for RNA isolation from blackberry (*Rubus* L.). *Genet. Mol. Res.* **2012**, *11*, 1773–1782. [CrossRef]

27. Li, S. Transcriptional control of flavonoid biosynthesis: Fine-tuning of the MYB-bHLH-WD40 (MBW) complex. *Plant Signal. Behav.* **2014**, *9*, e27522. [CrossRef]

28. Dossett, M.; Bassil, N.V.; Lewers, K.S.; Finn, C.E. Genetic diversity in wild and cultivated black raspberry (*Rubus occidentalis* L.) evaluated by simple sequence repeat markers. *Genet. Resour. Crop Evol.* **2012**, *59*, 1849–1865. [CrossRef]

29. Lee, G.-A.; Song, J.Y.; Choi, H.-R.; Chung, J.-W.; Jeon, Y.-A.; Lee, J.-R.; Ma, K.-H.; Lee, M.-C. Novel microsatellite markers acquired from *Rubus coreanus* Miq. and cross-amplification in other *Rubus* species. *Molecules* **2015**, *20*, 6432–6442. [CrossRef]

30. Martins, W.S.; Lucas, D.C.S.; Neves, K.F.S.; Bertioli, D.J. WebSat—A web software for microsatellite marker development. *Bioinformation* **2009**, *3*, 282–283. [CrossRef]

31. Nunes, C.F.; Ferreira, J.L.; Nunes-Fernandes, M.C.; de Souza Breves, S.; Generoso, A.L.; Fontes-Soares, B.D.; Carvalho-Dias, M.S.; Pasqual, M.; Borem, A.; de Almeida Cancado, G.M. An improved method for genomic DNA extraction from strawberry leaves. *Ciência Rural* **2011**, *41*, 1383–1389. [CrossRef]

32. Liu, K.; Muse, S.V. PowerMarker: An integrated analysis environment for genetic marker analysis. *Bioinformatics* **2005**, *21*, 2128–2129. [CrossRef]

33. Ahmad, A.; Wang, J.-D.; Pan, Y.-B.; Rahat Sharif, R.; Gao, S.-J. Development and use of simple sequence repeats (SSRs) markers for sugarcane breeding and genetic studies. *Agronomy* **2018**, *8*, 260. [CrossRef]

34. Fernandez-Fernandez, F.; Antanaviciute, L.; Govan, C.L.; Sargent, D.J. Development of a multiplexed microsatellite set for fingerprinting red raspberry (*Rubus idaeus*) germplasm and its transferability to other Rubus species. *J. Berry Res.* **2011**, *1*, 177–187. [CrossRef]

35. Han, Y.; Huang, K.; Liu, Y.; Jiao, T.; Ma, G.; Qian, Y.; Wang, P.; Dai, X.; Gao, L.; Xia, T. Functional analysis of two flavanone-3-hydroxylase genes from *Camellia sinensis*: A critical role in flavonoid accumulation. *Genes* **2017**, *8*, 300. [CrossRef]

36. Tian, J.; Han, Z.; Zhang, J.; Hu, Y.; Song, T.; Yao, Y. The balance of expression of dihydroflavonol 4-reductase and flavonol synthase regulates flavonoid biosynthesis and red foliage coloration in crabapples. *Sci. Rep.* **2015**, *5*, 12228. [CrossRef]

37. Saito, K.; Yonekura-Sakakibara, K.; Nakabayashi, R.; Higashi, Y.; Yamazaki, M.; Tohge, T.; Fernie, A.R. The flavonoid biosynthetic pathway in *Arabidopsis*: Structural and genetic diversity. *Plant Physiol. Biochem.* **2013**, *72*, 21–34. [CrossRef]

38. Petrussa, E.; Braidot, E.; Zancani, M.; Peresson, C.; Bertolini, A.; Patui, S.; Vianello, A. Plant Flavonoids—Biosynthesis, transport and involvement in stress responses. *Int. J. Mol. Sci.* **2013**, *14*, 14950–14973. [CrossRef]

39. Zheng, D.; Schröder, G.; Schröder, J.; Hrazdina, G. Molecular and biochemical characterization of three aromatic polyketide synthase genes from *Rubus idaeus*. *Plant Mol. Biol.* **2001**, *46*, 1–15. [CrossRef]

40. García-Gómez, B.; Razi, M.; Salazar, J.A.; Prudencio, A.S.; Ruiz, D.; Dondini, L.; Martínez-Gómez, P. Comparative analysis of SSR markers developed in exon, intron, and intergenic regions and distributed in regions controlling fruit quality traits in Prunus species: Genetic diversity and association studies. *Plant Mol. Biol. Rep.* **2018**, *36*, 23–35. [CrossRef]

41. Holland, J.B.; Helland, S.J.; Sharopova, N.; Rhyne, D.C. Polymorphism of PCR-based markers targeting exons, introns, promoter regions, and SSRs in maize and introns and repeat sequences in oat. *Genome* **2001**, *44*, 1065–1076. [CrossRef]

42. Krutovsky, K.V.; Neale, D.B. Nucleotide diversity and linkage disequilibrium in cold-hardiness and wood quality-related candidate genes in Douglas-fir. *Genetics* **2005**, *171*, 2029–2041. [CrossRef]

43. Cai, C.; Wu, S.; Niu, E.; Cheng, C.; Guo, W. Identification of genes related to salt stress tolerance using intron-length polymorphic markers, association mapping and virus-induced gene silencing in cotton. *Sci. Rep.* **2017**, *7*, 528. [CrossRef]

44. Lewers, K.S.; Weber, C.A. The trouble with genetic mapping of raspberry. *HortScience* **2005**, *40*, 1108. [CrossRef]

45. Anderson, J.R.; Lübberstedt, T. Functional markers in plants. *Trends Plant Sci.* **2003**, *8*, 554–560. [CrossRef]

46. Rafique, M.Z.; Carvalho, E.; Stracke, R.; Palmieri, L.; Herrera, L.; Feller, A.; Malnoy, M.; Martens, S. Nonsense mutation inside anthocyanidin synthase gene controls pigmentation in yellow raspberry (*Rubus idaeus* L.). *Front. Plant Sci.* **2016**, *7*, 1892. [CrossRef]

47. Michael, T.P.; VanBuren, R. Progress, challenges and the future of crop genomes. *Curr. Opin. Plant Biol.* **2015**, *24*, 71–81. [CrossRef]

48. VanBuren, R.; Bryant, D.; Bushakra, J.M.; Vining, K.J.; Edger, P.P.; Rowley, E.R.; Priest, H.D.; Michael, T.P.; Lyons, E.; Filichkin, S.A.; et al. The genome of black raspberry (*Rubus occidentalis*). *Plant J.* **2016**, *87*, 535–547. [CrossRef]

49. Wight, H.; Zhou, J.; Li, M.; Hannenhalli, S.; Mount, S.M.; Liu, Z. Draft genome assembly and annotation of red raspberry *Rubus idaeus*. *BioRxiv* **2019**, *546135*, 1–22. [CrossRef]

Permissions

All chapters in this book were first published in MDPI; hereby published with permission under the Creative Commons Attribution License or equivalent. Every chapter published in this book has been scrutinized by our experts. Their significance has been extensively debated. The topics covered herein carry significant findings which will fuel the growth of the discipline. They may even be implemented as practical applications or may be referred to as a beginning point for another development.

The contributors of this book come from diverse backgrounds, making this book a truly international effort. This book will bring forth new frontiers with its revolutionizing research information and detailed analysis of the nascent developments around the world.

We would like to thank all the contributing authors for lending their expertise to make the book truly unique. They have played a crucial role in the development of this book. Without their invaluable contributions this book wouldn't have been possible. They have made vital efforts to compile up to date information on the varied aspects of this subject to make this book a valuable addition to the collection of many professionals and students.

This book was conceptualized with the vision of imparting up-to-date information and advanced data in this field. To ensure the same, a matchless editorial board was set up. Every individual on the board went through rigorous rounds of assessment to prove their worth. After which they invested a large part of their time researching and compiling the most relevant data for our readers.

The editorial board has been involved in producing this book since its inception. They have spent rigorous hours researching and exploring the diverse topics which have resulted in the successful publishing of this book. They have passed on their knowledge of decades through this book. To expedite this challenging task, the publisher supported the team at every step. A small team of assistant editors was also appointed to further simplify the editing procedure and attain best results for the readers.

Apart from the editorial board, the designing team has also invested a significant amount of their time in understanding the subject and creating the most relevant covers. They scrutinized every image to scout for the most suitable representation of the subject and create an appropriate cover for the book.

The publishing team has been an ardent support to the editorial, designing and production team. Their endless efforts to recruit the best for this project, has resulted in the accomplishment of this book. They are a veteran in the field of academics and their pool of knowledge is as vast as their experience in printing. Their expertise and guidance has proved useful at every step. Their uncompromising quality standards have made this book an exceptional effort. Their encouragement from time to time has been an inspiration for everyone.

The publisher and the editorial board hope that this book will prove to be a valuable piece of knowledge for researchers, students, practitioners and scholars across the globe.

List of Contributors

David Ross Appleton and Harikrishna Kulaveerasingam
Biotechnology & Breeding Department, Sime Darby Plantation R&D Centre, Serdang 43400, Selangor Darul Ehsan, Malaysia

Sean Mayes
School of Biosciences, University of Nottingham, Sutton Bonington Campus, Leicestershire LE12 5RD, UK

Ai-Ling Ong and Chee-Keng Teh
Biotechnology & Breeding Department, Sime Darby Plantation R&D Centre, Serdang 43400, Selangor Darul Ehsan, Malaysia
School of Biosciences, University of Nottingham, Sutton Bonington Campus, Leicestershire LE12 5RD, UK

Festo Massawe
School of Biosciences, University of Nottingham Malaysia, Semenyih 43500, Selangor Darul Ehsan, Malaysia

Elena Dubina, Margarita Ruban and Sergey Lesnyak
Federal Scientific Rice Centre, Belozerny, 3, 350921 Krasnodar, Russia

Pavel Kostylev and Elena Krasnova
Agrarian Research Center "Donskoy", Nauchny Gorodok, 3, 347740 Zernograd, Russia

Kirill Azarin
Department of Genetics, Southern Federal University, 344006 Rostov-on-Don, Russia

Carlos Maldonado and Freddy Mora
Institute of Biological Sciences, University of Talca, 2 Norte 685, Talca 3460000, Chile

Filipe Augusto Bengosi Bertagna and Maurício Carlos Kuki
Genetic and Plant Breeding Post-Graduate Program, Universidade Estadual de Maringá, Maringá PR 87020-900, Brazil

Carlos Alberto Scapim
Departamento de Agronomia, Universidade Estadual de Maringá, Maringá PR 87020-900, Brazil

Korachan Thanasilungura, Tidarat Monkham, Sompong Chankaew and Jirawat Sanitchon
Department of Agronomy, Faculty of Agriculture, Khon Kaen University, Khon Kaen 40002, Thailand

Sukanya Kranto
Ratchaburi Rice Research Center, Muang, Ratchaburi 70000, Thailand

Yuliya Genievskaya
Laboratory of Molecular Genetics, Institute of Plant Biology and Biotechnology, Almaty 050040, Kazakhstan

Saule Abugalieva
Laboratory of Molecular Genetics, Institute of Plant Biology and Biotechnology, Almaty 050040, Kazakhstan
Department of Biodiversity and Bioresources, Faculty of Biology and Biotechnology, al-Farabi Kazakh National University, Almaty 050040, Kazakhstan

Aralbek Rsaliyev and Gulbahar Yskakova
Laboratory of Phytosanitary Safety, Research Institute of Biological Safety Problems, Gvardeisky 080409, Zhambyl Region, Kazakhstan

Yerlan Turuspekov
Laboratory of Molecular Genetics, Institute of Plant Biology and Biotechnology, Almaty 050040, Kazakhstan
Agrobiology Faculty, Kazakh National Agrarian University, Almaty 050010, Kazakhstan

Martina Rosello, Conxita Royo and Jose Miguel Soriano
Sustainable Field Crops Programme, Institute for Food and Agricultural Research and Technology (IRTA), 25198 Lleida, Spain

Miguel Sanchez-Garcia
International Centre for Agricultural Research in Dry Areas (ICARDA), Rabat 10112, Morocco

Xiaoxia Yu, Mingfei Zhang, Zhuo Yu, Dongsheng Yang, Jingwei Li, Guofang Wu and Jiaqi Li
Agricultural College, Inner Mongolia Agricultural University, Hohhot 010000, China

Sirjan Sapkota, J. Lucas Boatwright and Stephe Kresovich
Advanced Plant Technology Program, Clemson University, Clemson, SC 29634, USA
Department of Plant and Environmental Sciences, Clemson University, Clemson, SC 29634, USA

Kathleen Jordan
Advanced Plant Technology Program, Clemson University, Clemson, SC 29634, USA

Richard Boyles
Department of Plant and Environmental Sciences, Clemson University, Clemson, SC 29634, USA
Pee Dee Research and Education Center, Clemson University, Florence, SC 29506, USA

Yu Ge, Xiaoping Zang, Lin Tan, Jiashui Wang, Yuanzheng Liu, Yanxia Li, Nan Wang, Di Chen, Rulin Zhan and Weihong Ma
Haikou Experimental Station, Chinese Academy of Tropical Agricultural Sciences, Haikou 570102, China

Ángela Polo-Oltra, Inmaculada López and María Luisa Badenes
Citriculture and Plant Production Center, Instituto Valenciano de Investigaciones Agrarias (IVIA), CV-315, km 10.7, 46113 Moncada, Valencia, Spain

Carlos Romero
Instituto de Biología Molecular y Celular de Plantas (IBMCP), Consejo Superior de Investigaciones Científicas (CSIC) — Universidad Politécnica de Valencia (UPV), Ingeniero Fausto Elio s/n, 46022 Valencia, Spain

Hui Zhang
College of Crop Science, Fujian Agriculture and Forestry University, Fuzhou 350002, China
State Key Laboratory of Rice Biology and Chinese National Center for Rice Improvement, China National Rice Research Institute, Hangzhou 310006, China
Rice Research Institute and Fuzhou Branch of the National Center for Rice Improvement, Fujian Academy of Agricultural Sciences, Fuzhou 350018, China

Yu-Jun Zhu, An-Dong Zhu, Ye-Yang Fan and Jie-Yun Zhuang
State Key Laboratory of Rice Biology and Chinese National Center for Rice Improvement, China National Rice Research Institute, Hangzhou 310006, China

Ting-Xu Huang and Jian-Fu Zhang
Rice Research Institute and Fuzhou Branch of the National Center for Rice Improvement, Fujian Academy of Agricultural Sciences, Fuzhou 350018, China

Hua-An Xie
College of Crop Science, Fujian Agriculture and Forestry University, Fuzhou 350002, China
Rice Research Institute and Fuzhou Branch of the National Center for Rice Improvement, Fujian Academy of Agricultural Sciences, Fuzhou 350018, China

Jiantao Wu, Qinnan Wang, Jing Xie, Feng Zhou, Yuqiang Guo, Hailong Chang, Huanying Xu, Wei Zhang, Chuiming Zhang and Yongsheng Qiu
Guangdong Provincial Bioengineering Institute (Guangzhou Sugarcane Industry Research Institute), Guangzhou 510316, China

Yong-Bao Pan
Sugarcane Research Unit, USDA-ARS, Houma, LA 70360, USA

Manuel Blasco and María del Mar Naval
CANSO, Mestre Serrano, 1, 46250 L'Alcúdia, Valencia, Spain

Francisco Gil-Muñoz
Instituto Valenciano de Investigaciones Agrarias, CV 315 km 10,5., 46113 Moncada, Valencia, Spain

Rubén Rufo
Sustainable Field Crops Programme, IRTA (Institute for Food and Agricultural Research and Technology), 25198 Lleida, Spain

Silvio Salvi
Department of Agricultural and Food Sciences, University of Bologna, Viale Fanin 44, 40127 Bologna, Italy

Vadim G. Lebedev and Natalya M. Subbotina
Pushchino State Institute of Natural Sciences, Prospekt Nauki 3, 142290 Pushchino, Moscow Region, Russia
Branch of the Shemyakin-Ovchinnikov Institute of Bioorganic Chemistry of the Russian Academy of Sciences, Prospekt Nauki 6, 142290 Pushchino, Moscow Region, Russia

Konstantin A. Shestibratov
Branch of the Shemyakin-Ovchinnikov Institute of Bioorganic Chemistry of the Russian Academy of Sciences, Prospekt Nauki 6, 142290 Pushchino, Moscow Region, Russia

Oleg P. Maluchenko
All-Russian Research Institute of Agricultural Biotechnology, Timiriazevskaya Str. 42, 127550 Moscow, Russia

Konstantin V. Krutovsky
Department of Forest Genetics and Forest Tree Breeding, Faculty of Forest Sciences and Forest Ecology, Georg-August University of Göttingen, Büsgenweg 2, D-37077 Göttingen, Germany
Center for Integrated Breeding Research (CiBreed), Georg-August University of Göttingen, Albrecht-Thaer-Weg 3, D-37075 Göttingen, Germany

Laboratory of Population Genetics, N. I. Vavilov Institute of General Genetics, Russian Academy of Sciences, Gubkin Str. 3, 119333 Moscow, Russia
Laboratory of Forest Genomics, Genome Research and Education Center, Siberian Federal University, 660036 Krasnoyarsk, Russia
Department of Ecosystem Science and Management, Texas A&M University, 2138 TAMU, College Station, TX 77843-2138, USA

Index

Printed in the USA
CPSIA information can be obtained
at www.ICGtesting.com
JSHW062237071123
51533JS00031B/97